TRANSLATION SERIES
IN
MATHEMATICS AND ENGINEERING

TRANSLATION SERIES IN MATHEMATICS AND ENGINEERING

M.I. Yadrenko
SPECTRAL THEORY OF RANDOM FIELDS
1983, viii + 259 pp.
ISBN 0-911575-00-6 Optimization Software, Inc.
ISBN 0-387-90823-4 Springer-Verlag New York Berlin Heidelberg Tokyo
ISBN 3-540-90823-4 Springer-Verlag Berlin Heidelberg New York Tokyo

G.I. Marchuk
MATHEMATICAL MODELS IN IMMUNOLOGY
1983, xxv + 353 pp.
ISBN 0-911575-01-4 Optimization Software, Inc.
ISBN 0-387-90901-X Springer-Verlag New York Berlin Heidelberg Tokyo
ISBN 3-540-90901-X Springer-Verlag Berlin Heidelberg New York Tokyo

A.A. Borovkov, Ed.
ADVANCES IN PROBABILITY THEORY: LIMIT THEOREMS AND RELATED PROBLEMS
1984, xiv + 378 pp.
ISBN 0-911575-03-0 Optimization Software, Inc.
ISBN 0-387-90945-1 Springer-Verlag New York Berlin Heidelberg Tokyo
ISBN 3-540-90945-1 Springer-Verlag Berlin Heidelberg New York Tokyo

R.F. Gabasov, and F.M. Kirillova
OPTIMIZATION METHODS
1984, approx. 350 pp.
ISBN 0-911575-02-2 Optimization Software, Inc.
ISBN 0-387-90902-8 Springer-Verlag New York Berlin Heidelberg Tokyo
ISBN 3-540-90902-8 Springer-Verlag Berlin Heidelberg New York Tokyo

Yu.G. Evtushenko
METHODS FOR SOLVING EXTREMAL PROBLEMS AND THEIR APPLICATION IN OPTIMIZATION SYSTEMS
1984, approx. 400 pp.
ISBN 0-911575-07-3 Optimization Software, Inc.
ISBN 0-387-90949-4 Springer-Verlag New York Berlin Heidelberg Tokyo
ISBN 3-540-90949-4 Springer-Verlag Berlin Heidelberg New York Tokyo

A.N. Tikhonov, Ed.
CURRENT PROBLEMS IN MATHEMATICAL PHYSICS AND COMPUTATIONAL MATHEMATICS
1984, approx. 500 pp.
ISBN 0-911575-10-3 Optimization Software, Inc.
ISBN 0-387-90952-4 Springer-Verlag New York Berlin Heidelberg Tokyo
ISBN 3-540-90952-4 Springer-Verlag Berlin Heidelberg New York Tokyo

ADVANCES IN
PROBABILITY THEORY:

LIMIT THEOREMS & RELATED PROBLEMS

Edited by A.A. Borovkov

OPTIMIZATION SOFTWARE, INC.
PUBLICATIONS DIVISION, NEW YORK

Editor
A.A. Borovkov
Institute of Mathematics
Siberian Branch
USSR Academy of Sciences
Novosibirsk - 90
USSR

Editor
A.V. Balakrishnan
School of Engineering
University of California
Los Angeles
California 90024
USA

Library of Congress Cataloging in Publication Data

Predel'nye teoremy teorii veroiatnostei i smezhnye voprosy. English.
Advances in probability theory.

(Translation series in mathematics and engineering)
Translation of: Predel'nye teoremy teorii veroiatnostei i smezhnye voprosy.

1. Limit theorems (Probability theory)--Addresses, essays, lectures. I. Title. II. Series.
QA273.67.P74413 1984 519.2 84-2366
ISBN 0-911575-03-0
ISBN 0-387-90945-1 (Springer-Verlag)

Exclusively authorized English translation of the original Russian edition of *Predel'nye teoremy teorii veroiatnostei i smezhnye voprosy*, Volume I of The Proceedings of the Institute of Mathematics, the Siberian Branch of the USSR Academy of Sciences, edited by Academician *Sergej L'vovich SOBOLEV*, published in 1982 by the Nauka Publishing House, Novosibirsk, USSR.

Worldwide Distribution Rights by Springer-Verlag New York, Inc., 175 Fifth Avenue, New York, New York 10010, USA and Springer-Verlag Berlin Heidelberg New York Tokyo, Heidelberg Platz 3, Berlin-Wilmersdorf-33, West Germany.

ISBN 0-911575-03-0 Optimization Software, Inc.
ISBN 0-387-90945-1 Springer-Verlag New York Berlin Heidelberg Tokyo
ISBN 3-540-90945-1 Springer-Verlag Berlin Heidelberg New York Tokyo

ABOUT THE AUTHORS

ARNDT, Klaus - Associate Professor, Humboldt University of Berlin, graduate of Berlin University (1975), Candidate of Sciences (1981). Main scientific interests: *Queueing Theory and Reliability Theory, Markov Chains.*

BORISOV, Igor S. - Researcher at the Institute of Mathematics, Novosibirsk, Siberian Branch of the USSR Academy of Sciences, Associate Professor, Novosibirsk State University, graduate of Novosibirsk State University (1974), Candidate of Sciences (1978). Main scientific interests: *Limit Theorems for Sums of Random Variables in Linear Spaces.*

BOROVKOV, Alexander A. - Deputy Director of the Institute of Mathematics, Novosibirsk, Siberian Branch of the USSR Academy of Sciences, Professor, Novosibirsk State University, graduate of Moscow State University (1954), Doctor of Sciences (1964), Corresponding Member of the USSR Academy of Sciences (1966). Main scientific interests: *Mathematical Statistics, Limit Theorems of Probability Theory, Queueing Theory.*

ABOUT THE AUTHORS

BORODIKHIN, Vladimir M. - Associate Professor, Novosibirsk Electrotechnical Institute, graduate of Novosibirsk State University (1968), Candidate of Sciences (1971). Main scientific interests: *Problem of Martingales with Multidimensional Time.*

CHEBOTAREV, Vladimir I. - Associate Professor, Khabarovsk Polytechnical Institute, graduate of Novosibirsk State University (1970), Candidate of Sciences (1979). Main scientific interests: *Central Limit Theorem in Banach Spaces.*

FOSS, Sergej G. - Instructor, Novosibirsk State University, graduate of Novosibirsk State University (1975), Candidate of Sciences (1982). Main scientific interests: *Queueing Theory.*

KAREV, Georgij P. - Researcher at the Institute of Mathematics, Novosibirsk, Siberian Branch of the USSR Academy of Sciences, graduate of Novosibirsk State University (1969), Candidate of Sciences (1973). Main scientific interests: *Mathematical Simulation in Biology.*

LOTOV, Vladimir I. - Associate Professor, Novosibirsk State University, graduate of Novosibirsk State University (1971), Candidate of Sciences (1977). Main scientific interests: *Boundary Problems for Random Walks.*

MOGUL'SKIJ, Anatolij A. - Researcher at the Institute of Mathematics, Novosibirsk, Siberian Branch of the USSR Academy of Sciences, graduate of Novosibirsk State University (1969), Doctor of Sciences (1983). Main scientific interests: *Limit Theorems for Random Processes.*

NAGAEV, Sergej V. – Researcher at the Institute of Mathematics, Novosibirsk, Siberian Branch of the USSR Academy of Sciences, graduate of Tashkent State University (1955), Doctor of Sciences (1963). Main scientific interests: *Limit Theorems of Probability Theory.*

NEDOGIBCHENKO, Galina V. – Instructor, Novosibirsk Electrotechnical Institute, graduate of Novosibirsk State University (1973). Main scientific interests: *Measure Theory.*

PINELIS, Iosif F. – Instructor, Novosibirsk Institute of Railroad Engineers, graduate of Novosibirsk State University (1974), Candidate of Sciences (1982). Main scientific interests: *Limit Theorems for Sums of Independent Random Variables.*

SAKHANENKO, Alexander I. – Researcher at the Institute of Mathematics, Novosibirsk, Siberian Branch of the USSR Academy of Sciences, Associate Professor, Novosibirsk State University, graduate of Novosibirsk State University (1972), Candidate of Sciences (1975). Main scientific interests: *Limit Theorems for Sums of Independent Random Variables.*

SAVEL'EV, Lev Ya. – Researcher at the Institute of Mathematics, Novosibirsk, Siberian Branch of the USSR Academy of Sciences, Associate Professor, Novosibirsk State University, graduate of Moscow State University (1952), Candidate of Sciences (1958). Main scientific interests: *Measure Theory.*

ABOUT THE AUTHORS

TOPCHIJ, Valentin A. - Researcher at Computer Center, Omsk, Siberian Branch of the USSR Academy of Sciences, Associate Professor, Omsk State University, graduate of Novosibirsk State University (1972), Candidate of Sciences (1978). Main scientific interests: *Limit Theorems for Branching Random Processes.*

YURINSKIJ, Vadim V. - Researcher at the Institute of Mathematics, Novosibirsk, Siberian Branch of the USSR Academy of Sciences, Professor, Novosibirsk State University, graduate of Moscow Physico-Technical Institute (1968), Doctor of Sciences (1974). Main scientific interests: *Sums of Independent Random Variables, Differential Equations with Random Coefficients.*

CONTENTS

Part 2

LIMIT THEOREMS FOR RANDOM PROCESSES AND THEIR APPLICATIONS

Part 3

PROPERTIES OF DISTRIBUTIONS AND APPLIED PROBLEMS

PREFACE

The articles in this volume focus on recent advances in one of the leading branches of probability theory: Limit Theorems. They are grouped into three parts for convenience.

Part 1 deals with the well-known "Invariance Principle" of Donsker and Prokhorov. V.I. Lotov, I.F. Pinelis, and A.I. Sakhanenko study the rate of convergence in the Invariance Principle. I.S. Borisov considers the problem of constructing Markov processes on a probability space, generated by the Invariance Principle. A.A. Mogul'skij discusses large deviations in boundary value problems for a Wiener process.

Part 2 is grouped around limit theorems for random processes and their applications. A.A. Borovkov and A.I. Sakhanenko develop a general "Invariance Principle" for problems of mathematical statistics, related to the problems of testing close hypotheses, which makes it possible to construct asymptotically optimal criteria. V.I. Chebotarev estimates probabilities with which the sums of infinite-dimensional vectors enter a ball. A.A. Mogul'skij deals with estimation of the probability of large deviations from an invariant measure for mean sojourns in sets of Markov processes of a rather general kind. V.A. Topchij generates local limit theorems for critical branching processes. Random processes described by differential equations are the subject of V.V. Yurinskij.

Part 3 is somewhat less homogeneous and is devoted to the following problems: martingales on a plane (V.M. Borodikhin); quadratic variation of random sequences (G.P. Karev); probability inequalities for sums of Banach-space-valued random variables (S.V. Nagaev); colimits of directed extended measures (G.V. Nedogibchenko and L.Ya. Savel'ev); properties of the density of functionals of the integral type (A.I. Sakhanenko); and the theory of queues (K. Arndt, S.G. Foss).

Alexander A. BOROVKOV
Editor

Part 1

LIMIT THEOREMS RELATED TO THE INVARIANCE PRINCIPLE

METHODS OF A COMMON PROBABILITY SPACE FOR MARKOV PROCESSES

I.S. Borisov

1. INTRODUCTION AND STATEMENT OF BASIC RESULTS

Recently, the problems related to the accuracy of approximation in limit theorems of probability theory and, in particular, in limit theorems for random processes, have been in focus of interest. For the solution of problems of this type, the so-called methods of a common probability space are quite effective. They emerged in late 1950's; usually they are associated with the names of Yu.V. Prokhorov and A.V. Skorokhod (see [1, 2]).

The methods of this type consist in constructing prelimit and limit processes on a common probability space with, as "close" as possible, trajectories. These methods easily imply the "proximity" of distributions proper.

The methods of a common probability space were further developed by A.A. Borovkov [3], J. Komlos, P. Major, and G. Tusnady [4-7], who, essentially, solved completely many problems related to the investigation of the rate of convergence in the Donsker-Prokhorov invariance principle and of that of distributions of empirical processes. We note that all these methods are of "specifically individual" nature in the sense that their application is limited by the bounds of the specific problem to be solved. At the same time, some problems arising in applications of the probability theory (for example, in queueing theory) go beyond the available framework and require new, more universal methods.

This article is devoted to the methods of a common probability space for arbitrary Markov processes on the real line. The results that we have obtained generalize, to a certain degree, the results obtained in [1, 4]. At the same time, our results enable one to study the rate of convergence of distributions of Markov processes differing essentially from the random "step broken line" construction from sums of independent random variables as well as from the empirical process.

Thus, let ξt and $\eta(t)$, $t \in B \leqslant R$, be arbitrary separable Markov processes with values in R. To be more precise, let $B=[0,1]$. We shall assume in the sequel that $\xi(\cdot) \equiv \eta(\cdot) \equiv 0$ on $R \setminus [0,1]$.

By distributions of the processes $\xi(t)$ and $\eta(t)$, we mean distributions in the space $F[0,1]$ of all numerical functions on the interval $[0,1]$ with σ-algebra generated by cylinder sets.

For the separable random processes $\xi(t)$ and $\eta(t)$ given on a common probability space we put

$$\rho(\xi, \eta) = \sup_{0 \leqslant t \leqslant 1} |\xi(t) - \eta(t)|;$$

$$\varkappa_\rho(\xi, \eta) = \inf\{\varepsilon > 0 : \mathbf{P}(\rho(\xi, \eta) > \varepsilon) < \varepsilon\},$$

Note that the functional $\rho(\xi,\eta)$ is a random variable due to the separability of $\xi(t)$ and $\eta(t)$.

The quantity $\kappa_\rho(\xi,\eta)$ is called the Ky-Fan distance between $\xi(t)$ and $\eta(t)$.† It is in terms $\kappa_\rho(\xi,\eta)$ that we shall estimate the proximity of distributions of the random processes $\xi(t)$ and $\eta(t)$.

Before we formulate Theorem 1, we put:

$$F^{\Delta}_{l(t)}(x \mid l(t-\Delta),\ l(t+\Delta)) = \mathbf{P}(l(t) < x \mid l(t-\Delta),\ l(t+\Delta)),$$

where $t\in(0,1]$; $\Delta>0$, and $\ell(\cdot)$ is either $\ell(\cdot)$ or $\eta(\cdot)$. To avoid new notation, we agree that for any $\Delta>0$

$$F^{\Delta}_{l(0)}(x \mid l(-\Delta),\ l(\Delta)) \equiv \mathbf{P}(l(0) < x).$$

We assume without loss of generality that $F^{\Delta}_{\ell(t)}(x|u,v)$ as a function of x for any fixed u,v,Δ,t is a distribution function.

Next, by t^k_d we denote the binary-rational point $k2^{-d}$, where $d=0,1,\dots,$ $k=0,1,\dots,2^d$. To make the notation convenient, we introduce the points t^{-1}_d, t^k_{-1} and put them equal to zero by definition.

For any point $t^k_d\in(0,1]$ we define the set $\Gamma(t^k_d)$ of binary-rational points of $[0,1]$, this set being the combination of the left and right end-points of intervals imbedded in each

---

† Under the appropriate constraints on $\xi(t)$, $\eta(t)$, it is possible to consider $\kappa_\rho(\xi,\eta)$ for other "metrics" $\rho(\cdot,\cdot)$ as well. The choice of the "uniform metric" is mainly determined by the simplicity of formulation of the conditions of the theorems.

other according to the following rule: each new interval results by dividing into two halves the interval constructed on the previous step of the procedure ([0,1] is an initial interval), the point t_d^k must belong in this case to all the intervals at the same time; if the point t_d^k coincides with either end-point of the interval constructed, the procedure of division terminates.

More formally, if $[t_m^\ell, t_m^{\ell+1}]$ is the interval obtained on the mth step of the procedure, and t_d^k does not coincide with either end-point, the next interval will be either $[t_m^\ell, t_{m+1}^{2\ell+1}]$ or $[t_{m+1}^{2\ell+1}, t_m^{\ell+1}]$, depending to which interval the point t_d^k belongs. If $t_d^k = t_m^\ell$ or $t_d^k = t_m^{\ell+1}$, there is no more imbedding. For example, $\Gamma(1)=\{0,1\}$, $\Gamma(1/2)=\{0,1/2,1\}$, $\Gamma(1/4)=\{0,1/4,1/2,1\}$, $\Gamma(3/4)=\{0,1/2,3/4,1\}$, etc. Furthermore, by definition, let $\Gamma(0)=\{0\}$. We notice the obvious properties of the sets introduced:

$$\Gamma\left(t_d^k\right) = \Gamma\left(t_{d+m}^{2^m k}\right); \quad \Gamma\left(t_d^{2k-1}\right) = \Gamma\left(t_{d-1}^{k-1}\right) \cup \Gamma\left(t_{d-1}^{k}\right) \cup \{t_d^{2k-1}\}. \qquad (1)$$

We agree that throughout the entire article the symbol C (with a subscript or without) denotes positive constants. The main result is the following:

<u>THEOREM 1</u>. Let $\xi(t)$ and $\eta(t)$ be arbitrary separable Markov processes on [0,1]. Assume that there exist $\delta>0$ and an integer $N=N(\delta)$, for which the following conditions are satisfied:

$$1) \sum_{j=1}^{2^N} \mathbf{P}\left(\sup_{t_N^{j-1} \leqslant t \leqslant t_N^j} \left| l(t) - l\left(t_N^{j-1}\right)\right| > \delta\right) \leqslant C_1\delta, \quad \text{where } \ell=\xi,\eta;$$

2) the function $F^{\Delta}_{\eta(t)}(x|u,v)$ is continuously differentiable over x, u, v and the following inequality holds over all x, u, v and t:

$$\left[\frac{\partial}{\partial x} F^{\Delta}_{\eta(t)}(\cdot)\right]^{-1}\left\{\left|\frac{\partial}{\partial u} F^{\Delta}_{\eta(t)}(\cdot)\right| + \left|\frac{\partial}{\partial v} F^{\Delta}_{\eta(t)}(\cdot)\right|\right\} \leqslant 1 + h(\Delta),$$

with $h(\cdot)\geqslant 0$ and $H=\sum_{m=0}^{\infty} h(2^{-m})<\infty$;

3) for any $x, u, v\in R$, t, $\Delta\in[0,1]$ we can find $\gamma=\gamma(\Delta,t,x,u,v)$ such that

$$F^{\Delta}_{\eta(t)}(x-\gamma\,|\,u, v) \leqslant F^{\Delta}_{\xi(t)}(x\,|\,u, v) \leqslant F^{\Delta}_{\eta(t)}(x+\gamma\,|\,u, v);$$

$$\sum_{h=0}^{2^N} \mathbf{P}\left(\sum_{t_j^{2i-1}\in\Gamma(t_N^h)} \gamma\left(2^{-j}, t_j^{2i-1}, \xi\left(t_j^{2i-1}\right), \xi\left(t_{j-1}^{i-1}\right), \xi\left(t_{j-1}^{i}\right)\right) > \delta\right) \leqslant C_2\delta.$$

Then there exists a representation of the processes $\xi(t)$ and $\eta(t)$ on a common probability space, such that

$$\varkappa_\rho(\xi, \eta) \leqslant C\delta,$$

where $C=\max\{2C_1+C_2,\ 2+e^H\}$.

Before formulating Theorem 2, we introduce the following:

$$\mu\left(t_d^{2k-1}\right) = \int\int \mathbf{P}\left(\xi\left(t_{d-1}^{k-1}\right)\in du;\ \xi\left(t_{d-1}^{k}\right)\in dv\right)\int dx \times$$
$$\times\left|F^{2^{-d}}_{\xi\left(t_d^{2k-1}\right)}(x\,|\,u, v) - F^{2^{-d}}_{\eta\left(t_d^{2k-1}\right)}(x\,|\,u, v)\right|.$$

<u>THEOREM 2</u>. Let conditions 1 and 2 of Theorem 1 be satisfied. In addition we assume that

3) $$\sum_{k=0}^{2^N} \sum_{t_j^{2i-1}\in\Gamma(t_N^h)} \mu\left(t_j^{2i-1}\right) \leqslant C_2\delta^2.$$

Then there exists a representation of the processes $\xi(t)$ and $\eta(t)$ on a common probability space, such that

$$\varkappa_\rho(\xi, \eta) \leqslant C\delta,$$

where the constant C is defined in Theorem 1.

Next, for any integer N let $t_k=k/N$, $k=-1,0,1,2,\dots,N$;

$$F_l^{k,N}(x \mid u) = \mathbf{P}(l(t_k) - l(t_{k-1}) < x \mid l(t_{k-1}) = u),$$

where, as before, $\ell(\cdot)$ is either $\xi(\cdot)$ or $\eta(\cdot)$;

$$\nu_{k,N}(u) = \int | F_\xi^{k,N}(x \mid u) - F_\eta^{k,N}(x \mid u) | \, dx, \quad k \geqslant 0.$$

THEOREM 3. Let $\delta>0$ and an integer N be such that

1) $\sum\limits_{k=1}^{N} \mathbf{P}\Big(\sup\limits_{t_{k-1} \leqslant t \leqslant t_k} | l(t) - l(t_{k-1}) | > \delta \Big) \leqslant C_0 \delta,$ where $\ell=\xi,\eta$;

2) the function $F_\eta^{k,N}(x|u)$ is continuously differentiable over u, x and the following inequality holds uniformly over u, x and k:

$$\left[\frac{\partial}{\partial x} F_\eta^{k,N}(\cdot) \right]^{-1} \left| \frac{\partial}{\partial u} F_\eta^{k,N}(\cdot) \right| \leqslant C_1/N;$$

3) $\max\limits_{0 \leqslant k \leqslant N} \mathbf{E}\nu_{k,N}(\xi(t_{k-1})) \leqslant C_2 \delta^2/N.$

Then there exists a representation of the processes $\xi(t)$ and $\eta(t)$ on a common probability space, such that

$$\varkappa_\rho(\xi, \eta) \leqslant C\delta,$$

where

$$C = 1 + C_0 + \sqrt{(1+C_0)^2 + 4C_2(e^{2C_1} - 1)/C_1}.$$

We note that Theorem 1 allows us to obtain estimates of the order $O(n^{-\frac{1}{2}}\log n)$ for $\kappa_\rho(\xi,\eta)$ when $\xi(t)\equiv\xi_n(t)$ is an empirical process and $\eta(t)$ is a "Brownian bridge," or when $\xi(t)\equiv\xi_n(t)$ is the random broken line (if the conditions specified in [4] are satisfied), and $\eta(t)$ is the standard Wiener process.

Theorems 2 and 3 yield satisfactory results, for example, when the process $\xi(t)$ and $\eta(t)$ have finite absolute moments of a low order.

In Sections 3 and 4 we discuss specifically the conditions 2, 3 of the Theorems formulated.

The author is deeply grateful to K.A. Borovkov who has read the article in manuscript form and made useful comments.

2. PROOF OF THEOREMS

In proving Theorems 1 and 2, we shall use a more general, compared with that in [4], "diadic scheme" which consists in specifying the sequences $\{\xi(t_N^k);\ 0\leq k\leq 2^N\}$ and $\{\eta(t_N^k);\ 0\leq k\leq 2^N\}$ on a common probability space using the so-called conditional quantile transformations.

Proof of Theorem 1. 1. The first stage of proving this Theorem is the construction of $\xi(\cdot)$ and $\eta(\cdot)$ on a common probability space.

We introduce the following:

$$Q^{\Delta}_{l(t)}(x\mid u,v)=\sup\{z:\ F^{\Delta}_{l(t)}(z\mid u,v)<x\},$$

where $t,\ \Delta\in[0,1]$, $x,\ u,\ v\in R$. By definition, let

$Q^{\Delta}_{\ell(t)}(0|\cdot,\cdot)=-\infty$.

Let $\{z_{i,j};\ i\geq 0,\ j\geq 0\}$ be the infinite two-dimensional array of independent R.V. each of which is uniformly distributed on $[0,1]$. Next we define the values $\ell(0)$ and $\ell(1)$ (recall that $\ell(t)$ is any process among $\xi(t)$ or $\eta(t)$), using the formulas

$$l^*(0)=Q_{l(0)}(z_{0,0}\,|\,\cdot);\quad l^*(1)=Q^1_{l(1)}(z_{1,0}\,|\,l^*(0),0).$$

The further construction involves the following recurrence procedure. For each binary-rational point t_d^{2k-1}, $k=1,2,\dots,2^{d-1}$, $d=1,2,\dots$, let

$$l^*\left(t_d^{2k-1}\right)=Q^{2^{-d}}_{l\left(t_d^{2k-1}\right)}\left(z_{k,d}\,|\,l^*\left(t_{d-1}^{k-1}\right),\ l^*\left(t_{d-1}^{k}\right)\right).$$

<u>LEMMA 1</u>. For any integer $d\geq 0$ the joint distributions of the families $\{\ell^*(t_d^k);\ k=0,1,\dots,2^d\}$ and $\{\ell(t_d^k;\ k=0,1,\dots,2^d\}$ coincide.

<u>Proof</u>. We prove the Lemma by induction over d. For $d=0$ we have

$$\mathbf{P}(l^*(0)<x_0;\ l^*(1)<x_1)=\int_{-\infty}^{x_0}\mathbf{P}\left(Q^1_{l(1)}(z_{1,0}\,|\,y,0)<x_1\right)d\mathbf{P}(l^*(0)<y). \tag{2}$$

It follows from the definition of the functions $Q^{\Delta}_{\ell(\cdot)}(\cdot)$ that

$$\mathbf{P}(l^*(0)<y)=\mathbf{P}\left(Q_{l(0)}(z_{0,0}\,|\,\cdot)<y\right)=\mathbf{P}(l(0)<y);$$

$$\mathbf{P}\left(Q^1_{l(1)}(z_{1,0}\,|\,y,0)<x_1\right)=F^1_{l(1)}(x_1\,|\,y,0).$$

Substituting these expressions into (2), we have what was to be proved.

Assume now that our assertion is true for $d=k$. Then for $d=k+1$ we have

$$J \equiv \mathbf{P}\left(\bigcap_{i=0}^{2^{k+1}} \{l^*(t_{k+1}^i) < x_i\}\right) =$$

$$= \mathbf{P}\left(\bigcap_{i=1}^{2^k} \left\{Q^{2^{-(k+1)}}_{l\left(t_{k+1}^{2i-1}\right)}(z_{i,k+1} \mid l^*(t_k^{i-1}),\ l^*(t_k^i)) < \right.\right.$$

$$\left.\left. < x_{2i-1};\ \bigcap_{i=0}^{2^k} \{l^*(t_k^i) < x_{2i}\}\right\}\right).$$

By the construction, R.V. $\{\ell^*(t_k^i);\ i=0,1,\ldots,2^k\}$ do not depend on the R.V. $\{z_{i,k+1};\ i=1,2,\ldots,2^k\}$. Hence, taking into account the independence of $\{z_{i,j}\}$, we have, by the induction hypothesis:

$$J = \int\limits_{\bigcap_{i=0}^{2^k}\{y_i < x_{2i}\}} \cdots \int \mathbf{P}\left(\bigcap_{i=0}^{2^k} \{l^*(t_k^i) \in dy_i\}\right) \times$$

$$\times \mathbf{P}\left(\bigcap_{i=1}^{2^k} \left\{Q^{2^{-(k+1)}}_{l\left(t_{k+1}^{2i-1}\right)}(z_{i,k+1} \mid y_{i-1},\ y_i) < x_{2i-1}\right\}\right) =$$

$$= \int\limits_{\bigcap_{i=0}^{2^k}\{y_i < x_{2i}\}} \cdots \int \mathbf{P}\left(\bigcap_{i=0}^{2^k} \{l(t_k^i) \in dy_i\}\right) \prod_{i=1}^{2^k} F^{2^{-(k+1)}}_{l\left(t_{k+1}^{2i-1}\right)}(x_{2i-1} \mid y_{i-1},\ y_i). \tag{3}$$

Next, by the Markovianness of the process $\ell(t)$ we have

$$\prod_{i=1}^{2^k} F^{2^{-(k+1)}}_{l\left(t_{k+1}^{2i-1}\right)}(x_{2i-1} \mid y_{i-1},\ y_i) = \mathbf{P}\left(\bigcap_{i=1}^{2^k} \{l(t_{k+1}^{2i-1}) < x_{2i-1}\} \,\middle|\, \bigcap_{i=0}^{2^k} \{l(t_k^i) = y_i\}\right).$$

Substituting this expression into (3), we have what was to be proved. //

If the family of binary-rational points $\mathcal{D}=\{t_d^k,\ k\geq 0,\ d\geq 0\}$ is the separability set of the processes $\xi(t)$ and $\eta(t)$ (for this to occur, it suffices to require the stochastic continuity of $\xi(t)$ and $\eta(t)$ on $[0,1]$), then Lemma 1 allows us to construct the random processes $\xi(t)$ and $\eta(t)$ on the probability space R^∞ (on which $\{z_{i,j}\}$ are specified).

If the separability set of $\xi(t)$ and $\eta(t)$ differs from $\mathcal{D}$ (more precisely, from any subset of $\mathcal{D}$), then, as can easily be seen, the Lemma proved ensures the construction of the random processes $\xi(t)$ and $\eta(t)$ on the probability space $\mathcal{F}[0,1]\times\mathcal{F}[0,1]\times R^{2d+1}$ for any $d\geq 0$. We note that on the subspace R^{2d+1}, we have a family of R.V. $\{z_{i,d};\ 0\leq i\leq 2^d\}$, using which we determine the values $\{\xi^*(t_d^k;\ 0\leq k\leq 2^d\}$ and $\{\eta^*(t_d^k);\ 0\leq k\leq 2^d\}$. On each of the "coordinate axes" $\mathcal{F}[0,1]$ we construct trajectories of either the process $\xi(t)$ or of the process $\eta(t)$, using the appropriate family of conditional distributions. In this case we assume that the processes $\xi(t)$ and $\eta(t)$ are conditionally (with respect to the σ-algebra generated by the R.V. $\{z_{i,d};\ 0\leq i\leq 2^d\}$) independent.

Next we assume in our proof that $\xi(t)$ and $\eta(t)$ are given on the probability space in the manner mentioned above.

By $\xi_N(t)$ and $\eta_N(t)$ we denote the piecewise-constant random processes defined by the equalities

$$\xi_N(t)=\xi([t2^N]2^{-N});\ \eta_N(t)=\eta([t2^N]2^{-N});\ t\in[0,\ 1],$$

where $[\alpha]$ is the largest integer $\leq\alpha$. We have

$$\begin{aligned}\mathbf{P}(\rho(\xi,\ \eta)>(2+e^H)\delta)&\leqslant\mathbf{P}(\rho(\xi,\ \xi_N)>\delta)+\mathbf{P}(\rho(\eta,\ \eta_N)>\delta)+\\&+\mathbf{P}(\rho(\xi_N,\ \eta_N)>e^H\delta)\leqslant 2C_1\delta+\mathbf{P}(\rho(\xi_N,\ \eta_N)>e^H\delta).\end{aligned}\tag{4}$$

Here we have used condition 1 of the Theorem.

Now we show that

$$\mathbf{P}(\rho(\xi_N, \eta_N) > e^H \delta) \leqslant C_2 \delta.$$

It is seen that the proof of the Theorem follows from (4) and (5).

First we introduce the following:

$$\Delta_1(t_d^{2k-1}) = Q^{2^{-d}}_{\xi(t_d^{2k-1})}(z_{k,d} \mid \xi(t_{d-1}^{k-1}), \xi(t_{d-1}^k)) - \\ - Q^{2^{-d}}_{\eta(t_d^{2k-1})}(z_{k,d} \mid \xi(t_{d-1}^{k-1}), \xi(t_{d-1}^k))$$

$$\Delta_2(t_d^{2k-1}) = Q^{2^{-d}}_{\eta(t_d^{2k-1})}(z_{k,d} \mid \xi(t_{d-1}^{k-1}), \xi(t_{d-1}^k)) - \\ - Q^{2^{-d}}_{\eta(t_d^{2k-1})}(z_{k,d} \mid \eta(t_{d-1}^{k-1}), \eta(t_{d-1}^k)); \quad k, d \geqslant 0.$$

Note that $\Delta_1(0)=\xi(0)-\eta(0)$; $\Delta_2(0)=0$.

Next, we have

$$\rho(\xi_N, \eta_N) = \max_{0 \leqslant k \leqslant 2^N} |\xi(t_N^k) - \eta(t_N^k)|. \tag{6}$$

It is easy to see that for $d \geq 0$, $k \geq 0$

$$\xi(t_d^{2k-1}) - \eta(t_d^{2k-1}) = \Delta_1(t_d^{2k-1}) + \Delta_2(t_d^{2k-1}). \tag{7}$$

LEMMA 2. For any t_d^{2k-1}, $d \geq 0$, $k \geq 1$,

$$|\Delta_2(t_d^{2k-1})| \leqslant B_1(t_d^{2k-1}) |\xi(t_{d-1}^{k-1}) - \eta(t_{d-1}^{k-1})| + B_2(t_d^{2k-1}) |\xi(t_{d-1}^k) - \\ - \eta(t_{d-1}^k)|,$$

where

$$B_1(t_d^{2k-1}) = \left[\frac{\partial}{\partial x} F^{2^{-d}}_{\eta(t_d^{2k-1})}(\theta_1 \mid \theta_2, \theta_3)\right]^{-1} \left|\frac{\partial}{\partial u} F^{2^{-d}}_{\eta(t_d^{2k-1})}(\theta_1 \mid \theta_2, \theta_3)\right|;$$

$$B_2(t_d^{2k-1}) = \left[\frac{\partial}{\partial x} F^{2^{-d}}_{\eta(t_d^{2k-1})}(\theta_1 \mid \theta_2, \theta_3)\right]^{-1} \left|\frac{\partial}{\partial v} F^{2^{-d}}_{\eta(t_d^{2k-1})}(\theta_1 \mid \theta_2, \theta_3)\right|,$$

θ_1, θ_2, θ_3 are R.V. depending on the R.V. $\{z_{i,j};\ i \leq k,\ j \leq d\}$, and also on the distributions of $\xi(\cdot)$ and $\eta(\cdot)$.

Proof. Let

$$x_1 = Q^{2^{-d}}_{\eta\left(t_d^{2k-1}\right)}\left(z_{k,d} \mid \xi\left(t_{d-1}^{k-1}\right),\ \xi\left(t_{d-1}^{k}\right)\right);$$

$$x_2 = Q^{2^{-d}}_{\eta\left(t_d^{2k-1}\right)}\left(z_{k,d} \mid \eta\left(t_{d-1}^{k-1}\right),\ \eta\left(t_{d-1}^{k}\right)\right).$$

By the definition of the functions $Q^{\Delta}_{\eta(\cdot)}(\cdot)$, taking into account the continuity of $F^{\Delta}_{\eta(\cdot)}(\cdot)$, we have

$$F^{2^{-d}}_{\eta\left(t_d^{2k-1}\right)}\left(x_1 \mid \xi\left(t_{d-1}^{k-1}\right),\ \xi\left(t_{d-1}^{k}\right)\right) = F^{2^{-d}}_{\eta\left(t_d^{2k-1}\right)}\left(x_2 \mid \eta\left(t_{d-1}^{k-1}\right),\ \eta\left(t_{d-1}^{k}\right)\right).$$

By the formula for finite increments

$$\begin{aligned} 0 = {} & F^{2^{-d}}_{\eta\left(t_d^{2k-1}\right)}\left(x_1 \mid \xi\left(t_{d-1}^{k-1}\right),\ \xi\left(t_{d-1}^{k}\right)\right) - \\ & - F^{2^{-d}}_{\eta\left(t_d^{2k-1}\right)}\left(x_2 \mid \eta\left(t_{d-1}^{k-1}\right),\ \eta\left(t_{d-1}^{k}\right)\right) = \frac{\partial}{\partial x} F_{\eta(\cdot)}(\theta_1 \mid \theta_2, \theta_3)(x_1 - x_2) + \\ & + \frac{\partial}{\partial u} F_{\eta(\cdot)}(\theta_1 \mid \theta_2, \theta_3)\left(\xi\left(t_{d-1}^{k-1}\right) - \eta\left(t_{d-1}^{k-1}\right)\right) + \\ & + \frac{\partial}{\partial v} F_{\eta(\cdot)}(\theta_1 \mid \theta_2, \theta_3)\left(\xi\left(t_{d-1}^{k}\right) - \eta\left(t_{d-1}^{k}\right)\right). \end{aligned}$$

It remains only to use inequality for the modulus of the sum.//

From (7), using Lemma 2, we obtain the following recurrence relationship:

$$\begin{aligned} \left|\xi\left(t_d^{2k-1}\right) - \eta\left(t_d^{2k-1}\right)\right| \leqslant {} & \left|\Delta_1\left(t_d^{2k-1}\right)\right| + B_1\left(t_d^{2k-1}\right)\left|\xi\left(t_{d-1}^{k-1}\right) - \right. \\ & \left. - \eta\left(t_{d-1}^{k-1}\right)\right| + B_2\left(t_d^{2k-1}\right)\left|\xi\left(t_{d-1}^{k}\right) - \eta\left(t_{d-1}^{k}\right)\right|. \end{aligned} \tag{8}$$

The inequality (8) allows us to obtain the upper estimate for the difference $|\xi(t_d^{2k-1}) - \eta(t_d^{2k-1})|$ in terms of R.V. $|\Delta_1(\cdot)|$.

<u>LEMMA 3</u>. For any t_d^{2k-1}, $d \geq 0$, $k \geq 1$, we have the inequality

$$\left|\xi\left(t_d^{2k-1}\right) - \eta\left(t_d^{2k-1}\right)\right| \leqslant \exp\left\{\sum_{s=0}^{d} h\left(2^{-s}\right)\right\} \sum_{t_j^{2i-1} \in \Gamma\left(t_d^{2k-1}\right)} \left|\Delta_1\left(t_j^{2i-1}\right)\right|. \tag{9}$$

Proof. We prove the Lemma by induction over d. For d=0 the assertion is obvious since

$$|\xi(1)-\eta(1)| \leqslant |\Delta_1(1)| + B_1(1)|\Delta_1(0)| \leqslant (1+h(1))(|\Delta_1(1)|+|\Delta_1(0)|) \leqslant$$
$$\leqslant e^{h(1)}(|\Delta_1(1)|+|\Delta_1(0)|).$$

Now, let (9) be satisfied for all $d \leq m$. Then for $d=m+1$, the inequality (8), by induction, yields the inequality

$$|\xi(t_{m+1}^{2k-1}) - \eta(t_{m+1}^{2k-1})| \leqslant |\Delta_1(t_{m+1}^{2k-1})| + B_1(t_{m+1}^{2k-1}) \exp\left\{\sum_{s=0}^{m_1} h(2^{-s})\right\} \times$$
$$\times \sum_{t_j^{2i-1}\in\Gamma(t_m^{k-1})} |\Delta_1(t_j^{2i-1})| + B_2(t_{m+1}^{2k-1}) \times$$

$$\times \exp\left\{\sum_{s=0}^{m_2} h(2^{-s})\right\} \sum_{t_j^{2i-1}\in\Gamma(t_m^{k})} |\Delta_1(t_j^{2i-1})| \leqslant \exp\left\{\sum_{s=0}^{\max(m_1,m_2)} h(2^{-s})\right\} \times$$
$$\times \left\{ |\Delta_1(t_{m+1}^{2k-1})| + B_1(\cdot) \sum_{t_j^{2i-1}\in\Gamma(t_m^{k-1})} |\Delta_1(t_j^{2i-1})| + B_2(\cdot) \sum_{t_j^{2i-1}\in\Gamma(t_m^{k})} |\Delta_1(t_j^{2i-1})| \right\},$$

where m_1 and m_2 are uniquely determined from the equalities $t_m^{k-1}=t_{m_1}^{2p-1}$ and $t_m^k=t_{m_2}^{2\ell-1}$. It is not hard to see that $\max(m_1,m_2)=m$. Collecting terms on the right-hand side of the above inequality (taking into account the condition $B_1(t_{m+1}^{2k-1})+B_2(t_{m+1}^{2k-1}) \leq 1+h(2^{-(m+1)})$, the inequality $1+x \leq e^x$ and the properties (1) of the sets $\Gamma(\cdot)$), we prove the Lemma.//

By Lemma 3 and the representation (6) we have

$$\mathbf{P}(\rho(\xi_N,\eta_N) > e^H\delta) \leqslant \sum_{k=0}^{2^N} \mathbf{P}\left(\sum_{t_j^{2i-1}\in\Gamma(t_N^k)} |\Delta_1(t_j^{2i-1})| > \delta \right). \tag{10}$$

By the definition of the functions $Q^{\Delta}_{\eta(\cdot)}(\cdot)$ from the inequality

$$F^{2^{-d}}_{\eta\left(t_d^{2k-1}\right)}(x-\gamma\,|\,u,v) \leqslant F^{2^{-d}}_{\xi\left(t_d^{2k-1}\right)}(x\,|\,u,v) \leqslant F^{2^{-d}}_{\eta\left(t_d^{2k-1}\right)}(x+\gamma\,|\,u,v)$$

we obtain

$$-\gamma \leqslant x - Q^{2^{-d}}_{\eta\left(t_d^{2k-1}\right)}\left(F^{2^{-d}}_{\xi\left(t_d^{2k-1}\right)}(x\,|\,u,v)\,|\,u,v\right) \leqslant \gamma. \tag{11}$$

Let

$$x = Q^{2^{-d}}_{\xi\left(t_d^{2k-1}\right)}\left(z_{k,d}\,|\,\xi\left(t_{d-1}^{k-1}\right), \xi\left(t_{d-1}^{k}\right)\right); \quad u = \xi\left(t_{d-1}^{k-1}\right); \quad v = \xi\left(t_{d-1}^{k}\right).$$

Then it is not hard to see that

$$Q^{2^{-d}}_{\eta\left(t_d^{2k-1}\right)}\left(F^{2^{-d}}_{\xi\left(t_d^{2k-1}\right)}(x\,|\,u,v)\,|\,u,v\right) = Q^{2^{-d}}_{\eta\left(t_d^{2k-1}\right)}(z_{k,d}\,|\,u,v). \tag{12}$$

Thus, from (11) and (12) it follows that

$$\left|\Delta_1\left(t_d^{2k-1}\right)\right| \leqslant \gamma\left(2^{-d}, t_d^{2k-1}, \xi\left(t_d^{2k-1}\right), \xi\left(t_{d-1}^{k-1}\right), \xi\left(t_{d-1}^{k}\right)\right). \tag{13}$$

Finally, from (10), (13) and condition 3 of the Theorem we obtain

$$\mathbf{P}\left(\rho\left(\xi_N, \eta_N\right) > e^H\delta\right) \leqslant C_2\delta.$$

//

The proof of Theorem 2 differs from the above proof only in its final part. Namely, using Lemma 3 and the Chebyshev inequality, we obtain

$$\mathbf{P}\left(\rho\left(\xi_N, \eta_N\right) > e^H\delta\right) \leqslant \sum_{k=0}^{2^N} \mathbf{P}\left(\sum_{t_j^{2i-1}\in\Gamma\left(t_N^k\right)} \left|\Delta_1\left(t_j^{2i-1}\right)\right| > \delta\right) \leqslant$$

$$\leqslant \delta^{-1} \sum_{k=0}^{2^N} \sum_{t_j^{2i-1}\in\Gamma\left(t_N^k\right)} \mathbf{E}\left|\Delta_1\left(t_j^{2i-1}\right)\right|.$$

It remains only to note (see [1]) that for any point t_d^{2k-1}

$$\mathbf{E}|\Delta_1(t_d^{2k-1})| = \int\int \mathbf{P}(\xi(t_{d-1}^{k-1}) \in du;\ \xi(t_{d-1}^{k}) \in dv) \times$$
$$\times \int_0^1 dz \left| Q^{2^{-d}}_{\xi(t_d^{2k-1})}(z\,|\,u,v) - Q^{2^{-d}}_{\eta(t_d^{2k-1})}(z\,|\,u,v) \right| = \mu(t_d^{2k-1}), \tag{14}$$

We also need to use condition 3. //

Proof of Theorem 3. Let $\{z_j;\ j\geq 0\}$ be a sequence of independent R.V. distributed uniformly on $[0,1]$.

We begin constructing the processes $\xi(t)$ and $\eta(t)$ on a common probability space by specifying the values $\xi(0)$ and $\eta(0)$ by the formulas $\ell^*(0) = Q_\ell^{0,N}(z_0|0)$, where $\ell(t)$ denotes, as before, any process among $\xi(t)$ or $\eta(t)$, $Q_\ell^{k,N}(x|u)=\sup\{y:F_\ell^{k,N}(y|u)<x\}$.

Now, using the recurrence relationship

$$l^*(t_{k+1}) = l^*(t_k) + Q_l^{k+1,N}(z_{k+1}\,|\,l^*(t_k)) = \sum_{j=0}^{k+1} Q_l^{j,N}(z_j\,|\,l^*(t_{j-1})),$$

we define the values of the processes $\xi(t)$ and $\eta(t)$ at points t_k, $1\leq k\leq N$.

LEMMA 4. The joint distributions of the families $\{\ell^*(t_j);\ 0\leqslant j\leqslant N\}$ and $\{\ell(t_j);\ 0\leqslant j\leqslant N\}$ coincide.

Proof. We compare the distribution of the family $\{\ell^*(t_j);\ 0\leqslant j\leqslant M\}$ and that of the family $\{\ell(t_j);\ 1\leqslant j\leqslant M\}$ for $M=0,1,2,\dots,N$. We prove the Lemma by induction over M.

For $M=0$ our assertion is, obviously, satisfied. Now let out assertion hold for $M=k\geq 0$. Then, for $M=k+1$, taking into account the Markovianness of $\ell(t)$, we have

$$\mathbf{P}\left(\bigcap_{j=0}^{k+1}\{l^*(t_j)<x_j\}\right)=\int_{-\infty}^{x_k}\mathbf{P}\left(\bigcap_{j=0}^{k-1}\{l^*(t_j)<x_j\};\ l^*(t_k)\in dy\right)\times$$

$$\times\mathbf{P}\left(Q_l^{k+1,N}(z_{k+1}\mid y)<x_{k+1}-y\right)=\int_{-\infty}^{x_k}\mathbf{P}\left(\bigcap_{j=0}^{k-1}\{l(t_j)<x_j\};\ l(t_k)\in dy\right)\times$$

$$\times F_l^{k+1,N}(x_{k+1}-y\mid y)=\mathbf{P}\left(\bigcap_{j=0}^{k+1}\{l(t_j)<x_j\}\right),$$

where $\bigcap_{j=0}^{k-1}\{\dots\}$ for $k=0$ is, by definition, the entire sample space.//

Thus, using Lemma 4 we can specify the random processes $\xi(t)$ and $\eta(t)$ on the extended probability space $R^{N+1}\times\times F[0,1]\times F[0,1]$ (for more detail, see the proof of Theorem 1).

Proceeding in the same way as in proving Theorem 1, we reduce the estimate $P(\rho(\xi,\eta)>2(1+a)\delta)$ to the estimate $P(\rho(\xi_N,\eta_N)>2a\delta)$, where $a>0$; $\ell_N(t)=\ell([tN]N^{-1})$; $t\in[0,1]$.

Let

$$\Delta_k^{(1)}=Q_\xi^{k,N}(z_k\mid\xi(t_{k-1}))-Q_\eta^{k,N}(z_k\mid\xi(t_{k-1}));$$
$$\Delta_k^{(2)}=Q_\eta^{k,N}(z_k\mid\xi(t_{k-1}))-Q_\eta^{k,N}(z_k\mid\eta(t_{k-1})).$$

Now, applying the Chebyshev inequality for the first moments, we obtain

$$\mathbf{P}(\rho(\xi_N,\eta_N)>2a\delta)\leqslant\mathbf{P}\left(\max_{0\leqslant s\leqslant N}\left|\sum_{k=0}^{s}\Delta_k^{(1)}\right|>a\delta\right)+$$
$$+\mathbf{P}\left(\max_{1\leqslant s\leqslant N}\left|\sum_{k=0}^{s}\Delta_k^{(2)}\right|>a\delta\right)\leqslant(a\delta)^{-1}\left\{\sum_{k=0}^{N}\mathbf{E}\left|\Delta_k^{(1)}\right|+\sum_{k=0}^{N}\mathbf{E}\left|\Delta_k^{(2)}\right|\right\}. \tag{15}$$

Taking into account the remark made while deriving (14), we have

$$\mathbf{E}\left|\Delta_k^{(1)}\right|=\mathbf{E}\nu_{k,N}(\xi(t_{k-1})),\quad 0\leqslant k\leqslant N.$$

LEMMA 5. For any $k \geq 0$

$$\mathbf{E}|\Delta_k^{(2)}| \leqslant \gamma(1+\varepsilon)^k,$$

where $\varepsilon = C_1/N$; $\gamma = C_2\delta^2/N$; C_1, C_2 are defined in Theorem 3.

Proof. We denote by u and v the quantities $Q_\eta^{k,N}(z_k|\xi(t_{k-1}))$ and $Q_\eta^{k,N}(z_k|\eta(t_{k-1}))$. Then, arguing in the same way as in proving Lemma 2, we obtain

$$F_\eta^{k,N}(u\,|\,\xi(t_{k-1})) = F_\eta^{k,N}(v\,|\,\eta(t_{k-1})).$$

This plus the formula for finite increments and condition 2 of the Theorem yield the inequality

$$|\Delta_k^{(2)}| = |u-v| \leqslant \varepsilon\,|\xi(t_{k-1}) - \eta(t_{k-1})|, \tag{17}$$

where $k=1,2,\ldots,N$. Since for any $k \geq 0$

$$\xi(t_k) - \eta(t_k) = \xi(t_{k-1}) - \eta(t_{k-1}) + \Delta_k^{(1)} + \Delta_k^{(2)},$$

from (17) for $k \geq 1$ we obtain

$$\begin{aligned}\mathbf{E}|\Delta_k^{(2)}| &\leqslant \varepsilon\mathbf{E}|\xi(t_{k-1}) - \eta(t_{k-1})| \leqslant \varepsilon\{\mathbf{E}|\xi(t_{k-2}) - \eta(t_{k-2})| + \\ &+ \mathbf{E}|\Delta_k^{(1)}| + \mathbf{E}|\Delta_k^{(2)}|\} \leqslant \varepsilon\{(1+\varepsilon)\mathbf{E}|\xi(t_{k-2}) - \eta(t_{k-2})| + \gamma\} \leqslant \\ &\leqslant \ldots \leqslant \varepsilon\gamma\sum_{i=0}^{k-1}(1+\varepsilon)^i \leqslant \gamma(1+\varepsilon)^k.\end{aligned}$$

For $k=0$ the estimate (16) is trivial ($\Delta_0^{(2)}=0$).//

Next, we have

$$\begin{aligned}\sum_{k=0}^{N}\mathbf{E}|\Delta_k^{(2)}| &\leqslant \gamma\sum_{k=0}^{N}(1+\varepsilon)^k = \gamma\varepsilon^{-1}[(1+\varepsilon)^{N+1} - 1] \leqslant \\ &\leqslant \gamma\varepsilon^{-1}\left(e^{2C_1} - 1\right) = \delta^2\left(e^{2C_1} - 1\right)C_2/C_1.\end{aligned}$$

Furthermore, by condition 3

$$\sum_{k=0}^{N}\mathbf{E}|\Delta_k^{(1)}| \leqslant \gamma(N+1) \leqslant 2C_2\delta^2.$$

Therefore

$$\mathbf{P}(\rho(\xi,\eta) > 2(1+a)\delta) \leqslant \left[2C_0 + a^{-1}\left(2C_2 + (e^{2C_1}-1)\,C_2/C_1\right)\right]\delta.$$

It is clear that as an optimal a we need to take the positive root of the equation

$$2(1+a) = 2C_0 + a^{-1}\left[2C_2 + (e^{2C_1}-1)\,C_2/C_1\right],$$

whence

$$a = \frac{1}{2}\left(C_0 - 1 + \sqrt{(1+C_0)^2 + 4(e^{2C_1}-1)\,C_2/C_1}\right).$$

//

3. SOME PARTICULAR CASES.

1. We shall make condition 2 of the above Theorems more specific in the case where an arbitrary Gaussian Markov process $\eta(t)$ plays the role of a limit process. But first we consider condition 2 of Theorems 1, 2. Let $t_1<t_2<t_3$, $t_j \in [0,1]$. Then for any Gaussian Markov process $\eta(t)$ with correlation function $R(t,s)$ we have the representation

$$\mathbf{P}(\eta(t_2) < x \mid \eta(t_1) = u,\ \eta(t_3) = v) = \mathbf{P}(\eta(t_2) - \\ - c_1(t_1,t_2,t_3)\eta(t_1) - c_2(t_1,t_2,t_3)\eta(t_3) < x - c_1(t_1,t_2,t_3)u - c_2(t_1,t_2,t_3)v), \tag{18}$$

where $c_1(\cdot)$, $c_2(\cdot)$ satisfy the system of equations

$$\begin{cases} R(t_1,t_2) = c_1 R(t_1,t_1) + c_2 R(t_1,t_3), \\ R(t_2,t_3) = c_1 R(t_1,t_3) + c_2 R(t_3,t_3). \end{cases} \tag{19}$$

This follows from the fact that the R.V. $\eta(t_2)-c_1\eta(t_1)-c_2\eta(t_3)$ does not depend on the pair of the R.V. $\eta(t_1)$ and $\eta(t_3)$ (since the R.V. $\eta(t_2)-c_1\eta(t_1)-c_2\eta(t_3)$ and $\alpha\eta(t_1)+\beta\eta(t_3)$ for any $\alpha,\beta \in R$ are uncorrelated and have the normal joint distribution).

If $D=R(t_1,t_1)R(t_3,t_3)-R^2(t_1,t_3)\neq 0$, the system (19) has a unique solution

$$c_1(t_1, t_2, t_3) = (R(t_1, t_2)R(t_3, t_3) - R(t_2, t_3)R(t_1, t_3))D^{-1};$$
$$c_2(t_1, t_2, t_3) = (R(t_1, t_1)R(t_2, t_3) - R(t_1, t_3)R(t_1, t_2))D^{-1}.$$

We note that using Lemma 6 (formulated below) it is not hard to show that $c_i(\cdot)\geq 0$, $i=1,2$.

If $D=0$, then for some $\alpha,\beta\in R$ $\alpha\eta(t_1)+\beta\eta(t_3)=0$ with probability 1. This means that the system (19) has infinitely many solutions; therefore, as c_1, c_2 we take any solution of (19) (for instance, $c_1=R(t_1,t_2)/R(t_1,t_1)$, $c_2=0$). In what follows we assume for certainty that $D\neq 0$.

Thus, it follows from (19) that

$$F^{\Delta}_{\eta(t)}(x \mid u, v) = \Phi_{t,\Delta}(x - c_1(\cdot)u - c_2(\cdot)v),$$

where $\Phi_{t,\Delta}(\cdot)$ is the distribution function of the Gaussian random variable $\eta(t)-c_1(\cdot)\eta(t-\Delta)-c_2(\cdot)\eta(t+\Delta)$. Therefore, condition 2 of Theorem 1 becomes in this case

$$c_1(t-\Delta,\ t,\ t+\Delta) + c_2(t-\Delta,\ t,\ t+\Delta) \leqslant 1 + h(\Delta) \tag{20}$$

for all $t,\Delta\in[0,1]$ such that $t-\Delta, t+\Delta\in[0,1]$, with $h(\cdot)\geq 0$ and $\sum_{s=0}^{\infty} h(2^{-s})<\infty$. For example, when $\eta(t)=w(t)$ is the standard Wiener process, or $\eta(t)=\mathring{w}(t)$ is a "Brownian bridge," we have $c_1(\cdot)\equiv c_2(\cdot)\equiv 1/2$. Therefore the condition (20) is satisfied.

Let K be the class of Gaussian processes on $[0,1]$ whose correlation functions are representable as $R(t,s)=T(\min(t,s))L(\max(t,s))$. This is a sufficiently wide class which includes, for instance, Gaussian processes with independent increments $(L(\cdot)\equiv\text{const})$ and stationary Gaussian Markov processes $(T(t)=Be^{\alpha t},\ L(t)=e^{-\alpha t},\ \alpha\geq 0)$.

We note that all random processes from K are Markov. To make sure this is the case, it suffices to take advantage of the following criterion for Gaussian processes to be Markov.

LEMMA 6 [8]. The Gaussian process $\eta(t)$ is Markov iff

$$R(t_1,t_2)R(t_2,t_3) = R(t_2,t_2)R(t_1,t_3)$$

for any $t_1<t_2<t_3$ in the domain of definition of $\eta(t)$.

The following useful fact is also worthy of note.

LEMMA 7. For any functions $T(\cdot)$ and $L(\cdot)$ satisfying the conditions

$$T(s)/L(s) \leq T(t)/L(t) \qquad \text{for } s<t;\ T(s)L(s)\geq 0\ , \tag{21}$$

the function $R(t,s)=T(\min(t,s))L(\max(t,s))$ is a correlation function.

Proof. We need only to show that for any $t_1<t_2<\cdots<t_r$ the quadratic form

$$Q=\sum_{i,j=1}^{r} R(t_i,t_j)\,z_iz_j, \qquad z_k\in R,$$

is nonnegative. We assume without loss of generality that $L(t_j)\neq 0$ for all $j\leq r$. Then

$$Q=\sum_{i=1}^{r} T(t_i)\,L(t_i)\;z_i^2+2\sum_{i=1}^{r-1} T(t_i)\,z_i\sum_{j=i+1}^{r} L(t_j)\,z_j=$$

$$= T(t_r)L(t_r)z_r^2 + \sum_{i=1}^{r-1} \frac{T(t_i)}{L(t_i)}\left[\sum_{j=i}^{r} L(t_j)z_j\right]^2 - \sum_{i=1}^{r-1} \frac{T(t_i)}{L(t_i)}\left[\sum_{j=i+1}^{r} L(t_j)z_j\right]^2 =$$

$$= \frac{T(t_1)}{L(t_1)}\left[\sum_{j=1}^{r} L(t_j)z_j\right]^2 + \sum_{i=1}^{r-1}\left[\frac{T(t_{i+1})}{L(t_{i+1})} - \frac{T(t_i)}{L(t_i)}\right]\left[\sum_{j=i+1}^{r} L(t_j)z_j\right]^2 \geqslant 0.$$

The conditions (21) of the Lemma in terms of $R(t,s)$ are $R^2(t,s) \leq R(t,t)R(s,s)$ and $R(s,s) \geq 0$, i.e., they are necessary for $R(t,s)=T(\min(t,s))L(\max(t,s))$ to be a correlation function.

Therefore, for the representatives of the class K the condition (20) becomes

$$\frac{L(t+\Delta)[T(t+2\Delta)-T(t)] - T(t+\Delta)[L(t+2\Delta)-L(t)]}{L(t)T(t+2\Delta) - T(t)L(t+2\Delta)} \leqslant 1 + h(\Delta)$$

or

$$[L(t+\Delta)-L(t)][T(t+2\Delta)-T(t)] - [T(t+\Delta)-T(t)][L(t+2\Delta)-L(t)] \leqslant$$
$$\leqslant h(\Delta)\{L(t)[T(t+2\Delta)-T(t)] - T(t)[L(t+2\Delta)-L(t)]\} \qquad (22)$$

for all $\Delta, t \in [0,1]$ and a certain $h(\cdot) \geq 0$ satisfying the condition $\sum_{s=0}^{\infty} h(2^{-s}) < \infty$.

It is easy to check that Gaussian processes with independent increments and stationary Gaussian Markov processes satisfy (22) with the function $h(\cdot) \equiv 0$.

The relation (22) somewhat clarifies the probabilistic sense of condition 2 of Theorems 1, 2, this condition being understood as the regularity condition for trajectories of the process $\eta(t)$.

We specify now condition 2 of Theorem 3. As before, let

$\eta(t)$ be an arbitrary Gaussian Markov process. Then, by analogy with (18), we obtain

$$F_{\eta(t_j)}(x\,|\,u) = \mathbf{P}(\eta(t_j) - \eta(t_{j-1}) < x\,|\,\eta(t_{j-1}) = u) = \mathbf{P}(\eta(t_j) - \\ - c(t_{j-1}, t_j)\,\eta(t_{j-1}) < x + (1 - c(t_{j-1}, t_j))\,u), \qquad (23)$$

where $c(\cdot)$ satisfies the equation $R(t_{j-1}, t_j) - cR(t_{j-1}, t_{j-1}) = 0$.

For simplicity, we assume that the R.V. $\eta(t_{j-1})$ is not degenerate (i.e., $R(t_{j-1}, t_{j-1}) \neq 0$). Then

$$c(t_{j-1}, t_j) = \frac{R(t_{j-1}, t_j)}{R(t_{j-1}, t_{j-1})}.$$

Therefore, from (23) we obtain

$$F_{\eta(t_j)}(x\,|\,u) = \Phi_j(x + (1 - c(t_{j-1}, t_j))\,u),$$

where $\Phi_j(\cdot)$ is the distribution function of the Gaussian random variable $\eta(t_j) - c(t_{j-1}, t_j)\eta(t_{j-1})$. Condition 2 of the Theorem is reduced in this case to the relation

$$\left| 1 - \frac{R(t_{j-1}, t_j)}{R(t_{j-1}, t_{j-1})} \right| \leqslant c/N$$

uniformly in $j \leq N$.

For representatives of the class K the condition (24) is

$$\left| \frac{L(t_j) - L(t_{j-1})}{L(t_{j-1})} \right| \leqslant c/N$$

uniformly in $j \leq N$. In turn, this condition is satisfied if

$$\max_{j<N} \frac{\sup_{t_{j-1}<t<t_j} |L'(t)|}{L(t_{j-1})} \leqslant C < \infty.$$

2. We illustrate one of the techniques for verification of condition 3 of Theorem 1 by concrete examples.

First we consider the classical invariance principle of

Donsker-Prokhorov. In this case

$$\xi(t) = S_n(t) = \sum_{k \leqslant [nt]} \xi_k / \sqrt{n}, \quad \eta(t) = w(t),$$

where $\{\xi_k\}$ are independent identically distributed R.V. with zero mean and unit dispersion. To simplify the notation, we assume that $n=2^{N_0}$ (N_0 is a natural number).

Suppose that the conditions of Theorem 1 given in [4] are satisfied. Then the following lemma holds.

LEMMA 8. There exists $\varepsilon>0$ such that in the condition

$$F^{2^{-d}}_{w(t_d^{2k-1})}(x-\gamma \mid u, v) \leqslant F^{2^{-d}}_{S_n(t_d^{2k-1})}(x \mid u, v) \leqslant F^{2^{-d}}_{w(t_d^{2k-1})}(x+\gamma \mid u, v) \quad (25)$$

we can take for $\gamma(\cdot)$ the function

$$\gamma(2^{-d}, t_d^{2k-1}, x, u, v) = C_1 2^{d-N_0/2}[(u-x)^2+(x-v)^2] + C_2/\sqrt{n}, \quad (26)$$

only if

$$|u-x| \leqslant \varepsilon 2^{N_0/2-d}, \quad |v-x| \leqslant \varepsilon 2^{N_0/2-d}, \quad (27)$$

and $\gamma(\cdot)=\infty$ otherwise.

Proof. We note that in our case

$$F^{\Delta}_{l(t)}(x \mid u, v) = \mathbf{P}(l(t)-l(t-\Delta)-[l(t+\Delta)-l(t)] < 2x-u- \\ -v \mid l(t+\Delta)-l(t-\Delta) = v-u) = \tilde{F}^{\Delta}_{l(t)}(2x-u-v \mid v-u).$$

For $\ell(t)=S_n(t)$ the asymptotics $\tilde{F}^{\Delta}_{S_n(\cdot)}(\cdot)$ in the domain of large deviations has been obtained in [4], implying thus the validity of (25) satisfying (27) for a function $\gamma(\cdot)$ of the form (26).//

Next, it is easy to see, from the fact that the R.V. $\{\xi_k\}$ are independent and identically distributed, and the manner of construction of the sets $\Gamma(\cdot)$, that the R.V.

$$\Lambda\left(t_d^{2k-1}\right) = \sum_{t_j^{2i-1} \in \Gamma\left(t_d^{2k-1}\right)} \gamma\left(2^{-j}, t_j^{2i-1} S_n\left(t_j^{2i-1}\right), S_n\left(t_{j-1}^{i-1}\right), S_n\left(t_{j-1}^{i}\right)\right)$$

has the same distribution as the R.V. $\Lambda(t_d^1)$. In turn, on the set $A_d = \bigcap_{j=0}^{d} \{|S_n(t_j^1)| \leqslant \varepsilon 2^{N_0/2-j};\ |S_n(t_j^1) - S_n(t_{j+1}^1)| \leqslant \varepsilon 2^{N_0/2-j}\}$

$$\Lambda\left(t_d^1\right) = C_1 \sum_{j=0}^{d} \frac{2^j}{\sqrt{n}} S_n^2\left(t_j^1\right) + C_1 \sum_{j=0}^{d-1} \frac{2^j}{\sqrt{n}} \left[S_n\left(t_j^1\right) - S_n\left(t_{j+1}^1\right)\right]^2 + \\ + C_2 (d+1)/\sqrt{n}. \quad (28)$$

Next, to the expression

$$S_n^2\left(t_j^1\right) = \left[S_n\left(t_d^1\right) + \left(S_n\left(t_{d-1}^1\right) - S_n\left(t_d^1\right)\right) + \ldots + \left(S_n\left(t_j^1\right) - S_n\left(t_{j+1}^1\right)\right)\right]^2$$

we apply the Hölder inequality in the form

$$\left(\sum_{i=1}^{M} a_i\right)^2 \leqslant \left(\sum_{i=1}^{M} q^i a_i^2\right)\left(\sum_{i=1}^{M} q^{-i}\right) \quad \text{for} \quad q = \sqrt{2}.$$

Then the first sum in (28) is estimated from the above by the quantity

$$C_1 \sum_{j=0}^{d} \frac{2^j}{\sqrt{n}} \sum_{i=0}^{d-j-1} 2^{i/2} \left[S_n\left(t_{j+i}^1\right) - S_n\left(t_{j+i+1}^1\right)\right]^2 \left(\sum_{i=0}^{d-j+1} 2^{-i/2}\right) \leqslant \\ \leqslant C_2 \sum_{j=0}^{d} \frac{2^j}{\sqrt{n}} \sum_{k=j}^{d-1} 2^{(k-j)/2} \left[S_n\left(t_k^1\right) - S_n\left(t_{k+1}^1\right)\right]^2 = \frac{C_2}{\sqrt{n}} \sum_{k=0}^{d-1} 2^{k/2} \left[\left(S_n\left(t_k^1\right) - \right.\right. \\ \left. - S_n\left(t_{k+1}^1\right)\right]^2 \sum_{j=0}^{k} 2^{j/2} \leqslant C_3 \sqrt{n} \sum_{k=0}^{d-1} 2^{k-N_0} \left[S_n\left(t_k^1\right) - S_n\left(t_{k+1}^1\right)\right]^2.$$

Thus, on the set A_d

$$\Lambda\left(t_d^1\right) \leqslant C_4 \sqrt{n} \sum_{k=0}^{d-1} 2^{k-N_0} \left[S_n\left(t_k^1\right) - S_n\left(t_{k+1}^1\right)\right]^2 + C_2 (d+1)/\sqrt{n}.$$

Now we set $\delta = Kn^{-\frac{1}{2}} \log n$, $N = N_0 - \log N_0$. In [4] it has been shown that for a sufficiently large K

$$\mathbf{P}(\rho(l_N, l) > \delta) \leqslant C_5\delta;$$

$$n\mathbf{P}\left(C_4\sqrt{n}\sum_{k=0}^{N-1} 2^{k-N_0}\left[S_n(t_k^1) - S_n(t_{k+1}^1)\right]^2 + C_2 N_0/\sqrt{n} > \delta\right) \leqslant C_6\delta;$$

$$\mathbf{P}(A_N) \geqslant 1 - C_7/n^2.$$

Therefore, Theorem 1 asserts that

$$\varkappa_\rho(S_n, w) \leqslant Cn^{-1/2}\log n, \ n > 1.$$

A similar result holds for the process $S_n^0(t)$ as well, whose distribution is given by the formula

$$\mathbf{P}\left(S_n^0(\cdot) \in B\right) = \mathbf{P}\left(S_n(\cdot) \in B \mid S_n(1) = 0\right)$$

provided, of course, that the assumptions of Theorem 2 given in [9] are satisfied.

For $\delta = Kn^{-\frac{1}{2}}\log n$ the conditions of the Theorem will be satisfied also for empirical (classical and inverse) processes (see [9]).

REFERENCES

[1] Yu.V. Prokhorov. "Convergence of Random Processes and Limit Theorems in Probability Theory." *Theory Prob. Applications*, 2 (1956): 157-214.

[2] A.V. Skorokhod. *Studies in the Theory of Random Processes*. Reading, Mass.: Addison-Wesley, 1965.

[3] A.A. Borovkov. "On the Rate of Convergence for the Invariance Principle." *Theory Prob. Applications*, 2 (1973): 207-224.

[4] J. Komlos, P. Major, and G. Tusnady. "An Approximation of Partial Sums of Independent RV'-s and Sample DF." I. *Z. Wahrscheinlichkeitstheorie verw. Gebiete*, 32, 1/2 (1975): 111-133.

[5] J. Komlos, P. Major, and G. Tusnady. "An Approximation of Partial Sums of Independent RV'-s and Sample DF." II. *Z. Wahrscheinlichkeitstheorie verw. Gebiete*, 34, 1 (1976): 33-58.

[6] P. Major. "The Approximation of Partial Sums of Independent RV'-s." *Z. Wahrscheinlichkeitstheorie verw. Gebiete*, 35, 3 (1976): 213-220.

[7] P. Major. "The Approximation of Partial Sums of I.I.D. R.V. When the Summands Have Only Two Moments." *Z. Wahrscheinlichkeitstheorie verw. Gebiete*, 35, 3 (1976): 221-229.

[8] W. Feller. *An Introduction to Probability Theory and Its Application*. Vol. 2. 3d ed. New York: John Wiley, 1968.

[9] I.S. Borisov. "On the Rate of Convergence in the 'Conditional' Invariance Principle." *Theory Prob. Applications*, 1 (1978): 63-76.

ON THE ASYMPTOTICS OF DISTRIBUTIONS CONNECTED WITH THE EXIT OF A NON-DISCRETE RANDOM WALK FROM AN INTERVAL

V.I. Lotov

We consider the random walk $S_0, S_1, S_2, \ldots,$ generated by the sequence of sums of independent identically distributed random variables $\xi_1, \xi_2, \ldots, S_0=0,$ $S_n=\sum_{i=1}^{n} \xi_i,$ $F(x)=P(\xi_i<x)$. Throughout this article we assume that $F(x)$ satisfies two Cramer-type conditions:

1) $F(x)$ has a nonzero absolutely continuous component:

2) $\inf\{\lambda: Me^{-\lambda\xi_1}<\infty\}<0,$ $\sup\{\lambda: Me^{-\lambda\xi_1}<\infty\}>0$. Let, in addition, $M\xi_1=0$. For $a>0$ and $b>0$ let $N=N(a,b)=$ $=\min\{k: S_k \notin [-a,b)\}$ and the functions

$$F_n^{(0)}(x) = \mathbf{P}(S_n < x, N > n); \quad F_n^{(1)}(x) = \mathbf{P}(S_N < x, N = n). \tag{1}$$

The results obtained in this article enable us to obtain complete asymptotic expansions of the distributions (1) in the case where $a=a(n)\to\infty$, $b=b(n)\to\infty$ as $n\to\infty$ in the entire spectrum of deviations $a+b=o(n)$, $a+b\geq C\sqrt{n}$. We note that the same problem has been solved by the author in [1,2] under the additional assumptions concerning the rationality of the function $M(e^{i\lambda\xi_1}; \xi_1<0)$.

The distributions of boundary value functionals connected with the exit of a random walk from an interval, have been investigated in many works (an extensive bibliography, although far from complete, can be found in [2]). The statements of problems, leading to investigations of this type are well known. Recall, for example, the classical problem of gambler's ruin. In our article we apply the asymptotic analysis technique developed in [2,3] to treat a similar problem for random walks on the lattice of integers, as well as for deducing the distributions for our case. The discussion is brief; in [3] the reader can find the detailed description of the application of the method. We focus our attention on obtaining asymptotic representations for generating functions of the form $\sum z^n P(S_N \in A, N=n)$ and $\sum z^n P(S_n \in B, N>n)$ (Theorems 1-3 and Corollaries). To the results obtained we apply directly the asymptotic analysis [3] (the latter including a modification of the saddle-point method) which enables us to derive immediately the complete asymptotic expansions. Some of the expansions are given in Theorems 4 and 5.

Let $r_z(\lambda) = 1 - zMe^{i\lambda\xi_1}$;

$$Q_0(z, \lambda) = 1 + \sum_{n=1}^{\infty} z^n \int_{-a}^{b} e^{i\lambda x} dF_n^{(0)}(x), \quad \mathrm{Im}\,\lambda = 0, \quad |z| < 1;$$

$$Q_1(z, \lambda) = \sum_{n=1}^{\infty} z^n \int_{-\infty}^{a} e^{i\lambda x} dF_n^{(1)}(x);$$

$$Q_2(z, \lambda) = \sum_{n=1}^{\infty} z^n \int_{b}^{\infty} e^{i\lambda x} dF_n^{(1)}(x), \quad \mathrm{Im}\,\lambda = 0, \quad |z| \leqslant 1.$$

From [5] we know that for $|z|<1$, $\mathrm{Im}\lambda=0$

$$r_z(\lambda) Q_0(z, \lambda) = 1 - Q_1(z, \lambda) - Q_2(z, \lambda) \tag{2}$$

and if $r_z(\lambda)=r_{z-}(\lambda)r_{z+}(\lambda)$ is a canonical V-factorization in the strip $|\mathrm{Im}\lambda|<\delta_1$ [4], then

$$Q_2(z,\lambda)=r_{z+}(\lambda)\left[r_{z+}^{-1}(\lambda)(1-Q_1(z,\lambda))\right]^{[b,\infty)} \tag{3}$$

and similarly

$$Q_1(z,\lambda)=r_{z-}(\lambda)\left[r_{z-}^{-1}(\lambda)(1-Q_2(z,\lambda))\right]^{(-\infty,-a)}. \tag{4}$$

Here, by definition,

$$\left[\int_{-\infty}^{\infty} e^{i\lambda x}dv(x)\right]^{A}=\int_{A} e^{i\lambda x}dv(x) \quad (\mathrm{var}\, v<\infty,\ \mathrm{Im}\,\lambda=0).$$

Substituting (4) into (3) and setting for $f(\lambda)=\int_{-\infty}^{\infty} e^{i\lambda x}dv(x)$ (var v $<\infty$, $\mathrm{Im}\lambda=0$)

$$Af(\lambda)=r_{z-}(\lambda)\left[r_{z-}^{-1}(\lambda)f(\lambda)\right]^{(-\infty,-a)};\quad Bf(\lambda)=r_{z+}\lambda\left[r_{z+}^{-1}(\lambda)f(\lambda)\right]^{[b,\infty)},$$

we obtain

$$Q_2(z,\lambda)=Be(\lambda)-BAe(\lambda)+BAQ_2(z,\lambda), \tag{5}$$

where $e(\lambda)\equiv 1$.

In the sequel we shall use the following [4]. Let $\lambda_1(z)$ and $\lambda(z)$ be zeros of the function $r_z(\lambda)$, defined in some δ-neighborhood of the point $z=1$ cut along the ray $z>1$, such that $\lambda_1(1)=\lambda(1)=0$ and for $1-\delta<z<1$, $\lambda_1(z)$ and $\lambda(z)$ lie on the imaginary axis, $\mathrm{Im}\lambda_1(z)>0>\mathrm{Im}\lambda(z)$. For $z\in\mathcal{D}_\delta=\{|z-1|<\delta\}\setminus\{z\geq 1\}$ we can write down the expansions (we mean everywhere the principal values $(z-1)^{\frac{1}{2}}$):

$$i\lambda(z)=-\psi_1 i(z-1)^{1/2}+\psi_2(z-1)-\ldots;\ i\lambda_1(z)=\psi_1 i(z-1)^{1/2}+$$
$$+\psi_2(z-1)+\ldots;\ \psi_1=\sqrt{2/\mathbf{D}\xi_1}>0.$$

There exist $\delta_1>0$ such that the positive component $r_{z+}(\lambda)$ is analytic in the half-plane $\mathrm{Im}\lambda>-\delta_1$ and continuous on the boundary. The functions

$$v_z^{\pm 1}(\lambda) = \left(\frac{r_{z+}(\lambda)}{\lambda - \lambda(z)}\right)^{\pm 1}$$

are analytic for $\text{Im}\lambda > -\delta_1$ and continuous on the boundary, $|z-1|<\delta$. Furthermore, the functions $v_z(\lambda)$ and $(v_z(\lambda)(\lambda+i))^{-1}$ are Fourier Stieltjes transforms of functions of bounded variation with variation concentrated on $[0,\infty)$. Similarly, the negative component $r_{z-}(\lambda)$ is analytic for $\text{Im}\lambda<\delta_1$, the functions

$$u_z^{\pm 1}(\lambda) = \left(\frac{r_{z-}(\lambda)}{\lambda - \lambda_1(z)}\right)^{\pm 1}$$

are analytic for $\text{Im}\lambda<\delta_1$, continuous on the boundary, and the functions $u_z(\lambda)$, $(u_z(\lambda)(\lambda-i))^{-1}$ are Fourier Stieltjes transforms of functions of bounded variation with variation concentrated on $(-\infty,0)$.

The letters γ, δ, δ_1 denote throughout sufficiently small positive integers, possibly different in different formulations.

<u>THEOREM 1</u>. There exist δ and γ such that for $z \in L_\delta = \{|z|<1, |z-1|<\delta\}$, Im =0 and sufficiently large a and b

$$Q_2(z,\lambda) = \frac{e^{i\lambda b} v_z(\lambda)\left(1 - H_1(z)\mu^a(z)\right)}{e^{i\lambda(z)b} v_z(\lambda(z))\left(1 - H_2(z)\mu^{a+b}(z)\right)} + \int_b^\infty e^{i\lambda x} d\rho_z(x),$$

where uniformly in L_δ

$$\left|\int_x^\infty d\rho_z(y)\right| = O\left(e^{-\gamma x}\right), \qquad x \geqslant b;$$

$$\mu(z) = e^{i(\lambda_1(z)-\lambda(z))}; \; H_1(z) = \frac{u_z(\lambda(z))}{u_z(\lambda_1(z))}; \; H_2(z) = \frac{v_z(\lambda_1(z))\,H_1(z)}{v_z(\lambda(z))}.$$

To prove the Theorem, we need two lemmas.

LEMMA 1. Let for $\operatorname{Im}\lambda=0$ the function $f(\lambda)=\int_{-\infty}^{\infty} e^{i\lambda x}dv(x)$, $\operatorname{var} v < \infty$, be defined; furthermore, for some $\delta_1>0$ let $f(\lambda)$ be analytic in the strip $\delta_1<\operatorname{Im}\lambda<0$ and continuous on the boundary. Then we can find $\gamma>0$ and $\delta>0$ such that

$$Af(\lambda) = \frac{u_z(\lambda)\, e^{i\lambda_1(z)a} f(\lambda_1(z))}{e^{i\lambda a} u_z(\lambda_1(z))} + (\lambda - \lambda_1(z)) \int_{-\infty}^{-a} e^{i\lambda x} d\theta_z(x),\ \operatorname{Im}\lambda = 0,\ z \in \mathcal{L}_\delta,$$

where uniformly in $x<-a$ and $z\in L_\delta$

$$\left| \int_{-\infty}^{x} d\theta_z(y) \right| \leqslant Ce^{\gamma x}. \tag{6}$$

Proof. Let

$$\frac{f(\lambda)\, u_z^{-1}(\lambda)}{\lambda - i} = \int_{-\infty}^{\infty} e^{i\lambda x} dw_z(x)$$

for real λ. Then

$$\left[r_{z-}^{-1}(\lambda) f(\lambda)\right]^{(-\infty,-a)} = \left[\frac{\lambda - i}{\lambda - \lambda_1(z)} \int_{-\infty}^{\infty} e^{i\lambda x} dw_z(x)\right]^{(-\infty,-a)} =$$

$$= \int_{-\infty}^{-a} e^{i\lambda x} dw_z(x) + i(\lambda_1(z) - i) \left[\int_{-\infty}^{0} e^{i(\lambda-\lambda_1(z))x} dx \int_{-\infty}^{\infty} e^{i\lambda x} dw_z(x)\right]^{(-\infty,-a)} =$$

$$= \int_{-\infty}^{-a} e^{i\lambda x} dw_z(x) + i(\lambda_1(z) - i) \int_{-\infty}^{-a} e^{i(\lambda-\lambda_1(z))x} \left[\frac{u_z^{-1}(\lambda_1(z))\, f(\lambda_1(z))}{\lambda_i(z) - i} - \right.$$

$$\left. - \int_{-\infty}^{x} e^{i\lambda_1(z)t} dw_z(t)\right] dx = \frac{e^{i\lambda_1(z)a} f(\lambda_1(z))}{e^{i\lambda a} u_z(\lambda_1(z))\,(\lambda - \lambda_1(z))} +$$

$$+ \int_{-\infty}^{-a} e^{i\lambda x} \left(dw_z(x) - i(\lambda_1(z) - i) \int_{-\infty}^{x} e^{i\lambda_1(z)(t-x)} dw_z(t)\, dx \right).$$

If for $\operatorname{Im}\lambda=0$

$$u_z(\lambda)\int_{-\infty}^{-a} e^{i\lambda x}\left(dw_z(x) - i(\lambda_1(z)-i)\int_{-\infty}^{x} e^{i\lambda_1(z)(t-x)}dw_z(t)\,dx\right) = \int_{-\infty}^{-a} e^{i\lambda x}d\theta_z(x),$$

then

$$Af(\lambda) = \frac{u_z(\lambda)\,e^{i\lambda_1(z)a} f(\lambda_1(z))}{e^{i\lambda a}u_z(\lambda_1(z))} + (\lambda-\lambda_1(z))\int_{-\infty}^{-a} e^{i\lambda x}d\theta_z(x).$$

The estimate (6) follows from the analytic property of $u_z(\lambda)$ and $(u_z(\lambda)(\lambda-i))^{-1}f(\lambda)$ for $0<\mathrm{Im}\lambda<\delta_1$ and Lemma 2 of [4].//

Similarly we prove

LEMMA 2. Let for $\mathrm{Im}\lambda=0$ the function $f(\lambda)=\int_{-\infty}^{\infty} c^{i\lambda x}dv(x)$, $\mathrm{var}\, v<\infty$, be defined; furthermore, for some $\delta_1>0$ let $f(\lambda)$ be analytic in the strip $-\delta_1<\mathrm{Im}\lambda<0$ and continuous on the boundary. Then we can find $\gamma>0$ and $\delta>0$ such that

$$Bf(\lambda) = \frac{v_z(\lambda)\,e^{i(\lambda-\lambda(z))b} f(\lambda(z))}{v_z(\lambda(z))} + (\lambda-\lambda(z))\int_{b}^{\infty} e^{i\lambda x}d\varphi_z(x),\ \mathrm{Im}\,\lambda=0,\ z\in\mathscr{L}_\delta,$$

and uniformly in $x\geq b$ and $z\in L_\delta$

$$\left|\int_{x}^{\infty} d\varphi_z(y)\right| = O(e^{-\gamma x}).$$

Proof of Theorem 1. From (5) and Lemmas 1, 2 we derive

$$Q_2(z,\lambda) = \frac{e^{i(\lambda-\lambda(z))b}v_z(\lambda)}{v_z(\lambda(z))}\left(1-\mu^a(z)H_1(z)+\mu^a(z)H_1(z)Q_2(z,\lambda_1(z))\right) +$$
$$+(\lambda-\lambda(z))\int_{b}^{\infty} e^{i\lambda x}d\varphi_z^{(1)}(x) + (\lambda(z)-\lambda_1(z))\int_{b}^{\infty} e^{i\lambda x}d\varphi_z^{(2)}(x), \quad (7)$$

where uniformly in $x\geq b$ and $z\in L_\delta$

$$\left|\int_x^\infty d\varphi_z^{(1)}(y)\right| = O\left(e^{-\gamma x}\right); \quad \left|\int_x^\infty d\varphi_z^{(2)}(y)\right| = O\left(e^{-\gamma(x-b+a)}\right).$$

The value of $Q_2(z,\lambda_1(z))$ can be found from (7) for $\lambda=\lambda_1(z)$. We have, uniformly in L_δ

$$Q_2(z,\lambda_1(z)) = \frac{\mu^b(z)\,v_z(\lambda_1(z))\left(1-\mu^a(z)H_1(z)\right)}{v_z(\lambda(z))\left(1-\mu^{a+b}(z)H_2(z)\right)} + O\left(e^{-\gamma a}+e^{-\gamma b}\right). \qquad (8)$$

The boundedness in L_δ of the function $(\lambda_1(z)-\lambda(z))(1-\mu^{a+b}(z)H_2(z))^{-1}$ necessary in obtaining (8), can be proved for some $\delta>0$ and sufficiently large $a+b$ in the same way as in [2]. The substitution of (8) into (7) completes the proof of the Theorem.

Let for $A\subset[0,\infty)$

$$T(z,A) = \sum_{n=1}^\infty z^n P(S_N - b\in A, N=n).$$

COROLLARY 1. There exist $\delta>0$ and $\gamma>0$ such that uniformly in $z\in L_\delta$ for sufficiently large a and b:

$$T(z,A) = \frac{b(z,A)\left(1-H_1(z)\mu^a(z)\right)}{e^{i\lambda(z)b}\left(1-H_2(z)\mu^{a+b}(z)\right)} + O\left(e^{-\gamma a}+e^{-\gamma b}\right)\int_A e^{-\gamma x}dx,$$

where

$$\int_0^\infty e^{i\lambda x} b(z,dx) = v_z(\lambda)\,v_z^{-1}(\lambda(z)), \quad \operatorname{Im}\lambda = 0.$$

COROLLARY 2.

$$P(S_N - b\in A) = \frac{b(1,A)(a+\nu_1/2\psi_1)}{a+b+\eta_1/2\psi_1} + O\left(e^{-\gamma a}+e^{-\gamma b}\right)\int_A e^{-\gamma x}dx,$$

where

$$\nu_1 = \left[\frac{H_1(z)-1}{i(z-1)^{1/2}}\right]_{z=1}; \quad \eta_1 = \left[\frac{H_2(z)-1}{i(z-1)^{1/2}}\right]_{z=1}.$$

From Theorem 1 and Eq. (4) we have

<u>THEOREM 2</u>. For $z\in L_\delta$ for some δ, $\mathrm{Im}\lambda=0$ and sufficiently large a and b

$$Q_1(z,\lambda) = \frac{u_z(\lambda)\,e^{i(\lambda_1(z)-\lambda)a}\left(1-H_3(z)\mu^b(z)\right)}{u_z(\lambda_1(z))\left(1-H_2(z)\mu^{a+b}(z)\right)} + \int_{-\infty}^{-a} e^{i\lambda x}\,d\varepsilon_z(x),$$

where uniformly in $x\le -a$ and $z\in L_\delta$

$$\left|\int_{-\infty}^{x} d\varepsilon_z(y)\right| = O\left(e^{\nu x}\right), \quad H_3(z) = \frac{H_2(z)}{H_1(z)} = \frac{v_z(\lambda_1(z))}{v_z(\lambda(z))}.$$

<u>COROLLARY 3</u>. For $A\subset(-\infty,0)$, let

$$T(z,A) = \sum_{n=1}^{\infty} z^n P(S_N + a \in A,\ N=n),$$

then, uniformly in $z\in L_\delta$, for sufficiently large a and b

$$T(z,A) = \frac{\beta(z,A)\,e^{i\lambda_1(z)a}\left(1-H_3(z)\mu^b(z)\right)}{1-H_2(z)\mu^{a+b}(z)} + O\left(e^{-\gamma a}+e^{-\gamma b}\right)\int_A e^{\gamma x}\,dx,$$

where

$$\int_{-\infty}^{0} e^{i\lambda x}\beta(z,dx) = u_z(\lambda)\,u_z^{-1}(\lambda_1(z)), \quad \mathrm{Im}\,\lambda = 0.$$

From Theorems 1, 2 and Eq. (2) it readily follows for $z\in L_\delta$ and $\mathrm{Im}\lambda=0$:

$$Q_0(z,\lambda)=\frac{1}{r_z(\lambda)}-\frac{e^{i(\lambda_1(z)-\lambda)a}\left(1-H_3(z)\mu^b(z)\right)}{r_{z+}(\lambda)(\lambda-\lambda_1(z))\,u_z(\lambda_1(z))\left(1-H_2(z)\mu^{a+b}(z)\right)}-$$
$$-\frac{e^{i(\lambda-\lambda(z))b}\left(1-H_1(z)\mu^a(z)\right)}{r_{z-}(\lambda)(\lambda-\lambda(z))\,v_z(\lambda(z))\left(1-H_2(z)\mu^{a+b}(z)\right)}-$$
$$-\frac{1}{r_z(\lambda)}\left(\int_b^{\infty}e^{i\lambda x}d\rho_z(x)+\int_{-\infty}^{-a}e^{i\lambda x}d\varepsilon_z(x)\right). \quad (9)$$

For these z and λ, let

$$V_z^{(1)}(\lambda)=r_z^{-1}(\lambda)=\int_{-\infty}^{\infty}e^{i\lambda x}dP_z^{(1)}(x);\quad V_z^{(2)}(\lambda)=\frac{r_{z+}^{-1}(\lambda)}{\lambda-\lambda_1(z)}=\int_{-\infty}^{\infty}e^{i\lambda x}dP_z^{(2)}(x);$$

$$V_z^{(3)}(\lambda)=\frac{r_{z-}^{-1}(\lambda)}{\lambda-\lambda(z)}=\int_{-\infty}^{\infty}e^{i\lambda x}dP_z^{(3)}(x);$$

$$\frac{1}{r_z(\lambda)}\left(\int_b^{\infty}e^{i\lambda x}d\rho_z(x)+\int_{-\infty}^{-a}e^{i\lambda x}d\varepsilon_z(x)\right)=\int_{-\infty}^{\infty}e^{i\lambda x}d\psi_z(x).$$

Then for any Borel $A\subset[-a,b)$ we have from (9)

$$S(z,A)\equiv\sum_{n=0}^{\infty}z^nP(S_n\in A,\,N>n)=\int_A dP_z^{(1)}(x)-$$
$$-\frac{\int_A dP_z^{(2)}(x+a)\,e^{i\lambda_1(z)a}\left(1-H_3(z)\mu^b(z)\right)}{u_z(\lambda_1(z))\left(1-H_2(z)\mu^{a+b}(z)\right)}-$$
$$-\frac{\int_A dP_z^{(3)}(x-b)\,e^{-i\lambda(z)b}\left(1-H_1(z)\mu^a(z)\right)}{v_z(\lambda(z))\left(1-H_2(z)\mu^{a+b}(z)\right)}-\int_A d\psi_z(x). \quad (10)$$

We isolate the singularities of the functions $V_z^{(i)}(\lambda)$ and use Lemma 2 of [4]; and we easily obtain then uniformly in $z\in L_\delta$ the following asymptotic representations (i=1,2,3):

$$P_z^{(i)}(x) = p_1^{(i)}(z) + \frac{e^{-i\lambda(z)x}}{\lambda(z)\left(\frac{1}{V_z^{(i)}}\right)'(\lambda(z))} + O\left(e^{-\gamma x}\right), \quad x \to \infty;$$
$$P_z^{(i)}(x) = p_2^{(i)}(z) - \frac{e^{-i\lambda_1(z)x}}{\lambda_1(z)\left(\frac{1}{V_z^{(i)}}\right)'(\lambda_1(z))} + O\left(e^{\gamma x}\right), \quad x \to -\infty; \tag{11}$$

moreover, uniformly in the sets $A \subset [-a,b)$ and $z \in L_\delta$

$$\left|(\lambda_1(z) - \lambda(z)) \int_A d\psi_z(x)\right| = O\left(e^{-\gamma a} + e^{-\gamma b}\right). \tag{12}$$

The functions $p_1^{(i)}(z)$ and $p_2^{(i)}(z)$ do not depend on x. The relations (10)-(12) enable us to obtain asymptotic representations of the generating functions $S(z,A)$ for any constraints on $A \subset [-a,b)$.

THEOREM 3. There exist $\delta > 0$ and $\gamma > 0$ such that uniformly in $A \subset [-a,b)$ and $z \in L_\delta$ for sufficiently large a and b

$$S(z, A) = \sum_{n=0}^{\infty} z^n P(S_n \in A) - \frac{i\int_A e^{-i\lambda(z)x} dx \mu^a(z)\left(1 - \Pi_3(z)\mu^b(z)\right)}{u_z(\lambda_1(z)) v_z(\lambda(z))(\lambda_1(z) - \lambda(z))\left(1 - \Pi_2(z)\mu^{a+b}(z)\right)} -$$
$$- \frac{i\int_A e^{-i\lambda_1(z)x} dx \mu^b(z)\left(1 - \Pi_1(z)\mu^a(z)\right)}{u_z(\lambda_1(z)) v_z(\lambda(z))(\lambda_1(z) - \lambda(z))\left(1 - \Pi_2(z)\mu^{a+b}(z)\right)} + \frac{O\left(e^{-\gamma a} + e^{-\gamma b}\right)}{\lambda_1(z) - \lambda(z)}, \quad z \in \mathcal{L}_\delta. \tag{13}$$

If $\sup_{x \in A}|x|$ grows together with $a+b$, for further analysis it is expedient to replace the summand $\sum_{n=0}^{\infty} z^n P(S_n \in A) = \int_A dP_z^{(1)}(x)$ in (13) by its asymptotic representation in accord with (11).

The asymptotic expansions of the distributions (1) obtain now by the immediate application of Theorem 4 of [3] and its corollaries to the asymptotic representations of generating functions contained in the formulations of Corollaries 1 and 3 and Theorem 3. As an illustration, we give some asymptotic expansions.

THEOREM 4. Let $a=x_1\sqrt{n}$, $b=x_2\sqrt{n}$, $0<C_1\le x_1$, $x_2\le C_2<\infty$, $A_1\subset[0,\infty)$, $A_2\subset(-\infty,0)$; then for any integer $q\ge 1$

$$\mathbf{P}(S_N-b\in A_1,\ N=n)=\sum_{i=0}^{q-1}\frac{1}{n^{i+1/2}}\sum_{r=0}^{\infty}\frac{1}{n^{r/2}}\times$$
$$\times P_{ri}(1,0,0,0,b(.,A_1))+O\left(\frac{1}{n^{q+1/2}}\right),$$

$$\mathbf{P}(S_N+a\in A_2,\ N=n)=\sum_{i=0}^{q-1}\frac{1}{n^{i+1/2}}\sum_{r=0}^{\infty}\frac{1}{n^{r/2}}\times$$
$$\times P'_{ri}(0,0,1,0,\beta(.,A_2))+O\left(\frac{1}{n^{q+1/2}}\right);$$

the functions $P_{ri}(k_1,k_2,k_3,k_4,F)$ and $P'_{ri}(k_1,k_2,k_3,k_4,F)$ have been introduced in [3];

$$P_{00}(1,0,0,0,b(.,A_1))=P'_{00}(0,0,1,0,\beta(.,A_2))=0;$$

$$P_{10}(1,0,0,0,b(.,A_1))=b(1,A_1)\sqrt{\frac{2}{\pi\sigma^2}}\sum_{s=0}^{\infty}\left\{\left[x_1\left(s+\frac{1}{2}\right)+x_2 s\right]\times\right.$$
$$\left.\times e^{-\frac{2}{\sigma^2}\left[x_1\left(s+\frac{1}{2}\right)+x_2 s\right]^2}-\left[x_1\left(s+\frac{1}{2}\right)+x_2(s+1)\right]e^{-2/\sigma^2[x_1(s+1/2)+x_2(s+1)]^2}\right\},$$

$$P'_{10}(0,0,1,0,\beta(.,A_2))=\beta(1,A_2))=\sqrt{\frac{2}{\pi\sigma^2}}\sum_{s=0}^{\infty}\left\{\left[x_1 s+x_2\left(s+\frac{1}{2}\right)\right]\times\right.$$
$$\left.\times e^{-2/\sigma^2[x_1 s+x_2(s+1/2)]^2}-\left[x_1(s+1)+x_2\left(s+\frac{1}{2}\right)\right]e^{-2/\sigma^2[x_1(s+1)+x_2(s+1/2)]2}\right\},$$

$\sigma^2=\mathbf{D}\xi_1$.

THEOREM 5. Let $a=x_1\sqrt{n}$, $b=x_2\sqrt{n}$, $0<C_1\le x_1$, $x_2\le C_2<\infty$, $A\subset[-a,b)$, $\sup_A|x|<C<\infty$ uniformly in n; then for any integer $q\ge 1$

$$\mathbf{P}(S_n \in A, N \leqslant n) = \sum_{i=0}^{q-1} \frac{1}{n^{i+1/2}} \sum_{r=0}^{\infty} \frac{1}{n^{r/2}} \left(P_{ri}(1, 0, 1, 0, F_A^{(1)}) + \right.$$

$$\left. + P'_{ri}(1, 0, 1, 0, F_A^{(2)})\right) + O\left(\frac{1}{n^{q+1/2}}\right);$$

$$F_A^{(1)}(z) = \frac{i \int_A e^{-i\lambda_1(z)x} dx}{(\lambda_1(z) - \lambda(z))\, u_z(\lambda_1(z))\, v_z(\lambda(z))}; \quad F_A^{(2)}(z) = F_A^{(1)}(z) \frac{\int_A e^{-i\lambda(z)x} dx}{\int_A e^{-i\lambda_1(z)x} dx};$$

$$P_{00}(1, 0, 1, 0, F_A^{(1)}) + P'_{00}(1, 0, 1, 0, F_A^{(2)}) =$$

$$= \frac{1}{\sigma\sqrt{2\pi}} \sum_{s=0}^{\infty} \left(e^{-2/\sigma^2\,[x_1(s+1)+x_2 s]^2} + e^{-2/\sigma^2\,[x_1 s+x_2(s+1)]^2} - \right.$$

$$\left. - 2e^{-2/\sigma^2\,[x_1(s+1)+x_2(s+1)]^2}\right), \quad \sigma^2 = \mathbf{D}\xi_1.$$

REFERENCES

[1] V.I. Lotov. "Asymptotic Analysis in the Scheme of Random Walk With a Two-sided Boundary." In *The Second Vilnius Conference on Probability Theory and Mathematical Statistics, Summary of Reports,* 244-245. Vol. 1. Vilnius, 1977. (In Russian.)

[2] V.I. Lotov. *Asymptotic Expansions in Two-sided Boundary value Problems for Random Walks.* Doctoral Dissertation. Moscow, 1977. (In Russian.)

[3] V.I. Lotov. "Asymptotic Analysis of Distributions in Problems for Sums of Independent Terms." I, II. *Theory Prob. Applications,* 3; 4 (1979): 480-491; 873-879.

[4] A.A. Borovkov. "New Limit Theorems in Boundary value Problems for Sums of Independent Terms." *Sibirskij matemat. zh.*, 3, 5 (1962): 645-694. (In Russian.)

[5] J.H.B. Kemperman. "A Wiener-Hopf Type Method for a General Random Walk With a Two-sided Boundary." *Ann. Math. Stat.*, 34, 4 (1963): 1163-1193.

LARGE DEVIATIONS OF THE WIENER PROCESS

A.A. Mogul'skij

1. INTRODUCTION

Let $C(\kappa,1)$ be the space of continuous functions $f(t)$; $0\leq\kappa\leq t\leq 1$, with norm $\|f\| = \sup_{\kappa\leq t\leq 1} |f(t)|$; $C_0\subset C(0,1)$ be a family of absolutely continuous functions $f(t)$ with $f(0)=0$. For $f\in C_0$ we take

$$v(f) = 1/2\cdot\int_0^1 \dot{f}^2(t)\,dt; \tag{1}$$

for any set $G\subseteq C(0,1)$ let

$$V(G) = \inf_{f\in G\cap C_0} v(f);$$

if $G\cap C_0=\emptyset$, we assume that, by definition, $V(G)=\infty$.

We introduce the following condition:

$$0 < V(\bar{G}) = V(G)-V(G_0) < \infty , \tag{2}$$

where $\bar{G}$ is the closure of the set G; G_0 is the set of interior points of the set G. We denote by $\omega(t)$, $0\leq t\leq 1$, the standard Wiener process. It is well known [1,2] that as $r\to\infty$ we have the relation defining the "rough" asymptotics of the probability $P(\omega\in r\cdot G)$: for any set G satisfying (2),

$$\ln P(\omega\in r\cdot G) \sim -r^2 V(G) . \tag{3}$$

In particular, we consider sets G of the form

$$G=G_\varkappa(g,\ h)=\{f\in C(0,\ 1);\ g(t)>f(t)>h(t);\ \varkappa\leqslant t\leqslant 1\}, \tag{4}$$

where $g,h\in C(\kappa,1)$; $\inf\{g(t)-h(t);\ \kappa\leq t\leq 1\}>0$; $0\leq\kappa\leq 1$. It is clear enough that if the set G of the form (4) satisfies the condition $V(G)>0$ (i.e., the closure $\bar{G}$ does not contain $f(t)\equiv 0$), the conditions (2) are satisfied and the relation (3) holds.

Now we consider the probability

$$P(\omega\in r\cdot G) \tag{5}$$

for the set $G=G_\kappa(f,g)$ of the form (4). As $r\to\infty$ this probability tends to 1 when the set G contains the function $h(t)\equiv 0$, and to zero when the set $\bar{G}$ does not contain $h(t)\equiv 0$. In the first case we have the problem of investigating the behavior of $1-P(\omega\in r\cdot G)=P(\omega\notin r\cdot G)$ as $r\to\infty$; this problem has been exhaustively studied by A.A. Borovkov in [3] for a Wiener process and for the broken lines generated by sums of independent identically distributed random variables.

This paper deals with the latter case, that is, the investigation of accurate asymptotic behavior of the probability (5) as $r\to\infty$ for a set G of the form (4) satisfying the condition $V(G)>0$. We shall give later on a qualitative description of the behavior of this probability.

The functional $v(f)$ is semi-continuous in the following sense: if $\|f_n - f_0\| \to 0$ as $n \to \infty$, then $\liminf_{n\to\infty} v(f_n) \geq v(f_0)$ (see, for example, [2]). It is not hard to verify next that the set $\{f \in C_0 : v(f) \leq A\}$ is compact in $C(0,1)$ for any $0<A<\infty$. Hence for any open set G satisfying (2) we can find at least one function $f \in \bar{G} \cap C_0$ such that $v(f)=V(G)$. It is not hard to show that due to the inequalities of Cauchy and Jensen we have

$$v(pf_1 + qf_2) \leqslant (p\sqrt{v(f_1)} + q\sqrt{v(f_2)})^2 \leqslant pv(f_1) + qv(f_2), \tag{6}$$

where $p>0$; $q>0$; $p+q=1$; $f_1, f_2 \in C_0$; here equality in the first inequality is possible iff $af_1+bf_2=0$ for some a,b, $|a|+|b|>0$; in the second inequality of (6) equality is possible iff $v(f_1)=v(f_2)$. It follows from (6) that for convex open sets satisfying (2) there will be a unique function $f_0 \in \bar{G} \cap C_0$ such that $v(f_0)=V(G)$. In particular, for a curvilinear strip $G=G_K(g,h)$ such unique function f_0 exists: we call it the shortest path lying in $\bar{G}$.

Thus, let $G=G_K(g,h)$ of the form (4), $V(G)>0$ and let f_0 be the shortest path of G. The function f_0 is representable meaningfully in the following way. Imagine on a plane a "tube" formed by functions $g(t)$ and $h(t)$; any function f of the set $G \cap C_0$ can be interpretable as a "thread" fixed at the origin and lying entirely in our "tube." Then the shortest path f_0 is in correspondence with the stretched thread fixed at the origin and lying entirely in our "tube."

This interpretation makes the ensuing considerations more obvious.

Let

$$A_+ = \{t \in [\varkappa,\ 1] : f_0(t) = g(t)\}, \quad A_- = \{t \in [\varkappa,\ 1] : f_0(t) = h(t)\},$$
$$A_0 = [0,\ 1] \setminus (A_+ \cup A_-).$$

Obviously, for $t \in A_+$ the function f_0 is convex downward (i.e., $\dot{f}_0(t)$ does not decrease for $t \in A_+$), for $t \in A_-$ the function f_0 is convex upward (i.e., $f_0(t)$ does not increase for $t \in A_-$), for $t \in A_0$ the function f_0 changes linearly (i.e., $\ddot{f}_0(t)=0$ for $t \in A_0$). In this case for any set $G=G_\varkappa(g,h)$ of the form (4) the segment $[0,1]$ can be partitioned into finitely many intervals on which $f_0(t)$ is either convex upward or convex downward.

In this article we present a method enabling one to determine the asymptotics of the probability (5) for the class M of sets G of the form (4). This method consists in the following. The strip $G_\varkappa(g,h)$ belongs to M if the segment $[0,1]$ can be partitioned into finitely many non-intersecting intervals $(0,t_1), (t_1,t_2),\dots, (t_k,1)$ so that the shortest path $f_0(t)$ for $t \in [t_i,t_{i+1}]$ has three continuous derivatives,† with either $\ddot{f}_0(t)=0$ for $t \in [t_i,t_{i+1}]$ or $0<a \le |\ddot{f}_0(t)| \le A < \infty$ for all $t \in [t_i,t_{i+1}]$. Therefore, the functions $\dot{f}_0(t)$, $\ddot{f}_0(t)$, $\dddot{f}_0(t)$ on $[0,1]$ have no more than a finite number of discontinuities of the first kind and discon-

---

† On the end-points of the interval (t_i,t_{i+1}) derivatives are defined by the continuity inside the interval.

tinuity points fall on the points $t_1,\ldots,t_k$. Furthermore, in the main Theorem it is assumed that the third derivative $\dddot{f}(t)$ changes sign finitely many times; this technical restriction can be excluded (it will, however, complicate the proof).

Under these assumptions, the required asymptotics is

$$\mathbf{P}(\omega \in r\cdot G) \sim \frac{C}{r^{h/3}} \exp\left\{-r^2/2\int_0^1 \dot{f}_0^2(t)\,dt - \lambda r^{2/3}\int_0^1 |\dddot{f}_0(t)|^{2/3}\,dt\right\},$$

where λ is an absolute constant; the constant C and the exponent h are determined mainly by the form of the function f_0 and of the set A_0 (by rather complicated calculations). The power term $Cr^{-h/3}$ obtains as the product of power terms corresponding to each interval (t_i,t_{i+1}) and depends on such characteristics as the values of discontinuities of the functions $\dot{f}_0(t)$ and $\ddot{f}_0(t)$ at points t_i, $i=1,\ldots,k$. Since the formula is rather cumbersome, we restrict ourselves to the case, for simplicity, where for some $0<\kappa\leq 1$, $\ddot{f}_0(t)=0$, $0\leq t\leq\kappa$; and $0<a\leq-\ddot{f}_0(t)\leq A<\infty$ for $\kappa\leq t\leq 1$. In this case we consider the probability of the event $\{\omega(t)>rf_0(t);\ \kappa\leq t\leq 1\}$. The main difficulties of the problem are resolved exactly in this case; and the move to a situation when intervals of the type indicated alternate is sufficiently simple.

2. THE FORMULATION OF THE BASIC RESULT

Thus, let us consider a set $G=G_0(g,h)$ of the form (4), such that for some $0<\kappa<1$ the function $f_0(t)$ is linear for

$0 \le t \le \kappa$ and convex upward for $\kappa \le t \le 1$; in this case, obviously, $h(t)=f_0(t)$ holds for $\kappa \le t \le 1$. It is not hard to see then that for any function $g_1(t)$, $\kappa \le t \le 1$, $g_1(t) \ge g(t)$, $g_1(t) \not\equiv g(t)$, we have

$$V(G_\kappa(g_1,\ h)) = V(G_\kappa(g,\ h)) < V(G_\kappa(g_1,\ h) \backslash G_\kappa(g,\ h)),$$

hence by (3) we have

$$\mathbf{P}(\omega \in r \cdot G_\kappa(g,\ h)) \sim \mathbf{P}(\omega \in r \cdot G_\kappa(g_1,\ h)), \tag{7}$$

i.e., the upper boundary g of the strip $G_\kappa(g,h)$ influences only slightly the character of the asymptotics (5). It is possible, in particular, to set $g(t)=+\infty$, going over to a one-point boundary problem.

By the Airy function $W(x)$ we mean the unique solution of the equation

$$\ddot{W}(x) = xW(x) \tag{8}$$

satisfying the conditions

$$W(0) = \sqrt{\pi}/3^{1/3}\Gamma(2/3);\ \dot{W}(0) = -\sqrt{\pi}/3^{1/3}\Gamma(1/3).$$

This function has been studied exhaustively (see, for instance, [3]); the Airy function will play an essential role in solving our problem.

<u>THEOREM</u>. Let there be a strip $G_0(g,h)$ of the form (4) and let a shortest path satisfy the following relations:

1) $\ddot{f}_0(t)=0$ for $0 \le t \le \kappa$; 2) $f_0(t)=h(t)$ for $\kappa \le t \le 1$;

3) $\dot{f}_0(t)$ is continuous for $0 \le t \le 1$ and $\dot{f}_0(1)=0$;

4) $0 < a \le -\ddot{f}_0(t) \le A < \infty$ for $\kappa \le t \le 1$; 5) $\dddot{h}(t)$ is continuous on $[\kappa,1]$ and alternates sign finitely many times. Then as $r \to \infty$

$$\mathbf{P}(\omega(t) > rh(t);\ \varkappa \leqslant t \leqslant 1) \sim \frac{\Lambda^2 \exp\left\{-r^2/2 \int_0^1 \dot{f}_0^2(t)\,dt - \lambda \cdot r^{2/3} \int_\varkappa^1 |\ddot{h}(t)|^{2/3}\,dt\right\}}{\sqrt{2\pi\varkappa}\, r^{1/3} |\ddot{h}(\varkappa)\ddot{h}(1)|^{1/6}},$$

where $\lambda=-2^{-1/3}\lambda_0$; $\lambda_0=-2,3\ldots$ is the maximum zero of the Airy function $W(x)$;

$$\Lambda^2 = \left(\int_{\lambda_0}^{\infty} W(x)\,dx\right)^2 \Big/ \int_{\lambda_0}^{\infty} W^2(x)\,dx.$$

REMARK 1. Let $g(t)$ be a continuous function and let $g(t)>f_0(t)$ be satisfied for all $x\leq t\leq 1$. Then in the same way as we have done for (7), we prove that as $r\to\infty$

$$\mathbf{P}(\omega(t) > r \cdot h(t);\ \varkappa \leqslant t \leqslant 1) \sim \mathbf{P}(\omega \in r \cdot G_\varkappa(g, h)).$$

REMARK 2. It follows from the proof of the Theorem that if for $0\leq t<\kappa$ $f_0(t)>h(t)$ and the functions $\ddot{h}(t)$ and $\dddot{h}$ are continuous in a neighborhood of the point κ, then we have

$$\mathbf{P}(\omega(t) > r \cdot h(t);\ 0 \leqslant t \leqslant 1) \sim c \cdot \mathbf{P}(\omega(t) > r \cdot h(t);\ \varkappa \leqslant t \leqslant 1),$$

where $0<c<\infty$ is an absolute constant.

A few words about the ideas on which the proof of the Theorem rests are in order. The proof is based on the consecutive application of two absolutely continuous transformations of the measure corresponding to the Wiener process in $C(0,1)$. The first transformation (see Section 3) converts, roughly speaking, the Wiener process $\omega(t)$ into the process $\omega(t)-rf_0(t)$; this transformation or its modifications are used in most of the works dealing with large deviations for various spaces (see, for instance, [1,5-8]). The second absolutely continuous transformation (see Section 4) converts the Wiener process into a Markov (nonhomogeneous) process with continuous trajectories with a phase space $[0,\infty)$. To the best of the author's knowledge, this transformation together with the Airy function has not appeared in the literature. We need to note, nevertheless, that similar applications can be found in [9], where a rough asymptotics of probabilities of large deviations has been studied for local times.

3. THE FIRST ABSOLUTELY CONTINUOUS TRANSFORMATION

We recall that C_0 designates the set of absolutely continuous "rooted" functions $f(t)$ (i.e., $f(0)=0$). For the Wiener process $\omega(t)$ for $f \in C_0$ we define as a stochastic integral the functional

$$\langle \omega, f \rangle = \int_0^1 \dot{f}(t)\, d\omega(t) = \omega(1) - \dot{f}(1) \cdot \int_0^1 \omega(t)\, d\dot{f}(t). \qquad (9)$$

For a nonrandom function $g \in C(0,1)$ the functional (9) is the standard Lebesgue-Stieltjes integral $\langle g, f \rangle = \int_0^1 \dot{f}(t) dg(t)$.

It is seen that for $f \in C_0$

$$S(f) \equiv \ln \mathbf{M} \exp\{\langle \omega, f \rangle\} = 1/2 \cdot \langle f, f \rangle. \qquad (10)$$

We define the function of deviations ([1]) for a Wiener process as

$$v(g) = \sup_{f \in C_0} \{\langle g, f \rangle - S(f)\}, \quad g \in C(0,1). \qquad (11)$$

If $g \in C_0$, then (see also [1]) $v(g)$ is easily determined:

$$v(g) = \sup_{f \in C_0} \{\langle g, f \rangle - 1/2 \cdot \langle f, f \rangle\} = 1/2 \cdot \langle g, g \rangle.$$

If we denote by f_g the function in C_0, on which sup is attained in (11), then, obviously, for $g \in C_0$ we obtain $f_g = g$.

Let the set $G = \{f \in C(0,1): f(t) > h(t);\ \varkappa \le t \le 1\}$ be as in the Theorem and let $f_0 \in \bar{G}$ be the shortest path to G. For each $r > 0$ we define the absolutely continuous transformation of the Wiener measure on C(0,1), setting for any measurable set $H \subset C(0,1)$

$$\tilde{\mathbf{P}}(\omega \in H) = \mathbf{M}(\exp\{\langle\omega, rf_0\rangle\};\ \omega \in H)/\mathbf{M}\exp\{\langle\omega, rf_0\rangle\} = \\ = \exp\{-r^2/2 \cdot \langle f_0, f_0\rangle\} \cdot \mathbf{M}(\exp\{\langle\omega, rf_0\rangle\};\ \omega \in H).$$

Using this transformation, we express the probability sought in terms of

$$\mathbf{P}(\omega \in r \cdot G) = \exp\{-r^2/2 \cdot \langle f_0, f_0\rangle\}\mathbf{M}(\exp\{-\langle\omega - rf_0, rf_0\rangle\};\ \omega \in r \cdot G) = \\ = \exp\{-r^2 v(f_0)\}\mathbf{M}(\exp\{-\langle\omega - rf_0, rf_0\rangle\};\ \omega - rf_0 \in r \cdot (G - f_0)). \quad (12)$$

For any arbitrary function $f \in C_0$, it is not hard to see that

$$\ln \mathbf{M}\exp\{\langle\omega - rf_0, f\rangle\} = 1/2\,\langle f, f\rangle = \ln \mathbf{M}\exp\{\langle\omega, f\rangle\};$$

hence the process $\omega(t) - rf_0(t)$, $0 \le t \le 1$, has distribution $\tilde{P}$ coinciding with the distribution of the Wiener process: for any measurable $H \subseteq C(0,1)$

$$\tilde{\mathbf{P}}(\omega - rf_0 \in H) = \mathbf{P}(\omega \in H).$$

It is easy to see that the set $G' = G - f_0$ will be of the form:

$$G' = \{f \in C(0,1): f(t) > h(t) - f_0(t) = 0;\ \varkappa \leqslant t \leqslant 1\}. \quad (13)$$

Therefore, we can rewrite (12) as

$$\mathbf{P}(\omega \in r \cdot G) = \exp\{-r^2 v(f_0)\}\mathbf{M}(\exp\{-r\langle\omega, f_0\rangle\};\ \omega \in G'). \quad (14)$$

Next, by the fact that for $\varkappa \le t \le 1$ we have $h(t) = f_0(t)$, and also because $\dot{h}(t) = \ddot{f}_0(t) = 0$ for $0 \le t \le \varkappa$, we obtain

$$\langle \omega, f_0 \rangle \equiv \int_0^1 \dot{f}_0(t)\, d\omega(t) = -\int_0^1 \ddot{f}_0(t)\, \omega(t)\, dt = -\int_\varkappa^1 \ddot{h}(t)\, \omega(t)\, dt.$$

Hence (14) becomes

$$\mathbf{P}(\omega \in rG) = \exp\{-r^2 v(f_0)\}\, \mathbf{M}\left(\exp\left\{-r\int_\varkappa^1 |\ddot{h}(t)|\, \omega(t)\, dt\right\};\quad \omega \in G'\right). \tag{15}$$

Using the explicit form of the set G' (see (13)), we express (15) in terms of

$$\mathbf{P}(\omega \in rG = \exp\{-r^2 v(f_0)\}\cdot\int_0^\infty p(x) N_1(x)\, dx, \tag{16}$$

where

$$p(x) = \frac{1}{\sqrt{2\pi\varkappa}} \exp\{-x^2/(2\varkappa)\};$$

$$N_1(x) = \mathbf{M}\left(\exp\left\{-r\int_\varkappa^1 |\ddot{h}(t)|\,(x+\omega(t))\, dt\right\};\quad \inf_{\varkappa\leqslant t\leqslant 1}(x+\omega(t)) > 0\right).$$

Next, we set $\rho(t)=-h(t+\kappa)$ for $0\le t\le 1-\kappa$; we denote the Wiener process equal to x at time t by $\omega_t^x(u)=\omega(u-t)+x$ for $u\ge t$. In the formula defining the function $N_1(x)$ we make the following simple transformation:

$$N_1(x) \equiv \mathbf{M}\left(\exp\left\{-r\int_0^{1-\varkappa} \rho(t)\, \omega_0^x(t)\, dt\right\};\quad \inf_{0\leqslant t\leqslant 1-\varkappa} \omega_0^x(t) > 0\right) =$$

$$= \mathbf{M}\left(\exp\left\{-\int_0^{(1-\varkappa)T} \rho(t/T)\, \omega_0^{\sqrt{T}x}(t)\, dt\right\};\quad \inf_{0\leqslant t\leqslant (1-\varkappa)T} \omega_0^{\sqrt{T}x}(t) > 0\right),$$

where $T=r^{2/3}$. Letting $T_1=(1-\kappa)T$;

$$N_2(x) = \mathbf{M}\left(\exp\left\{-\int_0^{T_1} \rho(t/T)\, \omega_0^x(t)\, dt\right\};\quad \inf_{0\leqslant t\leqslant T_1} \omega_0^x(t) > 0\right),$$

we rewrite (16) as

$$\mathbf{P}(\omega \in rG) = T^{-1/2} \exp\{-r^2 v(f_0)\} \cdot \int_0^\infty p(x/\sqrt{T}) N_2(x)\, dx. \qquad (17)$$

4. THE SECOND ABSOLUTELY CONTINUOUS TRANSFORMATION

From [4] we know that as $x \to \infty$ the Airy function $W(x)$ (see (8)) can be approximated

$$W(x) \sim 1/2 \cdot x^{-1/4} \exp\{-s\}; \quad \dot{W}(x) \sim -1/2 \cdot x^{1/4} \exp\{-s\}, \qquad (18)$$

where $s = 2/3 \cdot x^{3/2}$; the largest zero of the function $W(x)$, denoted by $-\lambda_0$, equals $-2.338107...$ Let for $x \geq 0$

$$z(x) = W(2^{1/3}x - \lambda_0) \left(\int_0^\infty W^2 (2^{1/3}x - \lambda_0)\, dx \right)^{-1/2}; \qquad (19)$$

the function $z(x)$ (see (8)) satisfies the equation

$$1/2 \cdot \ddot{z}(x) = xz(x) - \lambda z(x), \; \lambda = 2^{-1/3} \cdot \lambda_0, \qquad (20)$$

and the conditions

$$z(0) = z(\infty) = 0; \quad \int_0^\infty z^2(x)\, dx = 1. \qquad (21)$$

Let, next,

$$z(x, \rho) = \rho^{1/6} z(x\rho^{1/3}); \; p(x, \rho) = z^2(x, \rho). \qquad (22)$$

It is not hard to see that by (20) the function $z(x,\rho)$ satisfies the equation

$$\frac{1}{2} \cdot \frac{\partial^2 z(x, \rho)}{\partial x^2} = \rho \cdot x \cdot z(x, \rho) - \lambda(\rho) z(x, \rho), \qquad (23)$$

where $\lambda(\rho) = \lambda \rho^{2/3}$, and the conditions (21).

Let $F(\omega)$ be a functional of the Wiener process; let $H \subseteq C(0,1)$ be a measurable set; we denote by $M'(F(\omega_0^x);\ \omega_0^x \in H,$

$\omega_0^x(t)=y)$ the density

$$\frac{d}{dy}\mathbf{M}\left(F\left(\omega_0^x\right);\ \omega_0^x\in H,\ \omega_0^x(t)<y\right)$$

which we shall use later in the cases where it is obvious (or proved) that such a continuous density exists; if $F=1$, let

$$\mathbf{P}\left(\omega_0^x\in H,\ \omega_0^x(t)=y\right)=\mathbf{M}\left(1;\ \omega_0^x\in H,\ \omega_0^x(t)=y\right).$$

Thus, for $0<\rho<\infty$ let

$$\mathbf{M}^\rho(t;x,y)=\mathbf{M}'\left(\exp\left\{-\int_0^t\rho\omega_0^x(u)\,du+\lambda\rho^{2/3}t\right\};\ \inf_{0\le u\le t}\omega_0^x(u)>0,\ \omega_0^x(t)=y\right),$$

$$\widetilde{M}^\rho(t)=z^{-1}(x,\rho)M^\rho(t;x,y)z(y,\rho). \qquad (24)$$

By Lemma 7.1 the functions $M^\rho(t;x,y)$ and $\tilde{M}^\rho(t;x,y)$ are continuous for $x>0$, $y>0$. These functions can be interpreted as transition densities (see Chapter 2 of [10]), for example:

$$\widetilde{M}^\rho(t+u;x,y)=\int_0^\infty\widetilde{M}^\rho(t;x,x')\widetilde{M}^\rho(u;x',y)\,dx'.$$

The infinitesimal operator corresponding to the transition density $\tilde{M}^\rho(t;x,y)$, has the form:

$$A^\rho f(x)=\left[-\rho x-\lambda\rho^{2/3}+\frac{1}{2}\cdot\frac{\partial^2 z(x,\rho)}{\partial x^2}\cdot\frac{1}{z(x,\rho)}\right]f(x)+$$
$$+\frac{1}{2}\ddot{f}(x)+\dot{f}(x)\frac{\partial\ln z(x,\rho)}{\partial x}.$$

Since the function $f(x)\equiv 1$ belongs to the domain of definition A^ρ and by (23) $A^\rho 1\equiv 0$, then (see Lemma 2.3 in Chapter 2 of [10]) the transition density $\tilde{M}^\rho(t;x,y)$ is conservative, i.e., for $t>0$, $x>0$

$$\int_0^\infty\widetilde{M}^\rho(t;x,y)\,dy\equiv 1. \qquad (25)$$

We note that $M^{\rho}(t;\ x,y)=M^{\rho}(t;\ y,x)$; from the last expression and (25) we obtain immediately that for $y>0$, $t>0$

$$p(y,\rho)=\int_0^\infty p(x,\rho)\,\widetilde{M}^{\rho}(t;\ x,y)\,dx. \tag{26}$$

Therefore, the family $\widetilde{M}^{\rho}$ is the transition density of a homogeneous Markov process $X^{\rho}(t)$ with continuous trajectories and stationary density $p(y,\rho)$.

As before, let $\rho(t)=-\dot{h}(t+\kappa)$ for $0\leq t\leq 1-\kappa$. We set for $x>0$, $T>0$, $T_1=(1-\kappa)T$, $0\leq t\leq T_1$

$$A(x,t)=\frac{1}{6}\cdot\frac{\dot{\rho}(t)}{\rho(t)}+\frac{1}{3}\cdot\frac{\dot{\rho}(t)}{\rho^{2/3}(t)}\cdot x\cdot\frac{\dot{z}(x\rho^{1/3}(t))}{z(x\rho^{1/3}(t))}. \tag{27}$$

On the set of trajectories of the Wiener process $\omega_x^t(u)$, $t\leq u\leq s\leq T_1$, which leave the point $x>0$ at an instant of time t and do not go below zero till an instant of time s, we define the functional

$$\gamma^x(t,s)=1/T\cdot\int_t^s A\left(\omega_t^x(u),\,u/T\right)du. \tag{28}$$

By (18) there exists $C<\infty$ such that for all $x>0$, $0\leq t\leq T_1$ we have $|A(x,t/T)|\leq Cx^{3/2}$.

We denote next for λ from (20) $x>0$, $0\leq t\leq s\leq T_1$

$$\alpha^x(t,s)=-\int_t^s\rho(u/T)\,\omega_t^x(u)du,\quad \beta(t,s)=\lambda\cdot\int_t^s\rho^{2/3}(u/T)\,du; \tag{29}$$

and for $x>0$, $y>0$, $0\leq t\leq s\leq T_1$, $-1\leq\mu\leq 1$ we introduce the family of functions

$$\widetilde{M}(t,s;\,x,y;\,\mu)=z^{-1}(x,\rho(t,\rho(t/T))\cdot z(y,\rho(s/T))\,\mathrm{M}'(\exp\{\alpha^x(t,s)+$$
$$+\beta(t,s)+\mu\gamma^x(t,s)\};\ \inf_{t<u<s}\omega_t^x(u)>0,\ \omega_t^x(s)=y). \tag{30}$$

Due to the fact that $|A(x,t)|\leq Cx^{3/2}$, for all sufficiently large T the conditions of Lemma 15 are satisfied for the family $\tilde{M}$; hence the function $\tilde{M}(t,s;x,y;\mu)$ is continuous in $x>0$, $y>0$. The family (30) can be viewed as transition (nonhomogeneous) densities:

$$\tilde{M}(t, s; x, y; \mu) = \int_0^\infty \tilde{M}(t, r; x, x'; \mu)\, \tilde{M}(r, s; x', y; \mu)\, dx'$$

for $t<r<s$. The following lemma is an analog of (24) and (25) for the nonhomogeneous case where $\rho(t)\neq\rho$.

LEMMA 1. For $T_1=(1-\kappa)T$, $0\leq t\leq s\leq T_1$, $x>0$, $y>0$

$$\int_0^\infty \tilde{M}(t, s; x, y; -1)\, dy = 1; \tag{31}$$

$$\int_0^\infty p(x, \rho\ (t/T))\, \tilde{M}(t, s; x, y; 1)\, dx = p(y, \rho(s/T)). \tag{32}$$

Proof. Let (see also Section 7)

$$\overset{\circ}{\omega}(u) = \overset{\circ}{\omega}{}_{t,s}^{x,y}(u) \equiv \omega_t^x(u) - \left(\omega_t^x(s) - y\right)\cdot(u-t)/(s-t), \quad t\leqslant u\leqslant s,$$

be a conditional Wiener process originating at x at an instant of time t and equal to y at an instant of time s. The function $\tilde{M}(t,s;x,y,\mu)$ (30) can be rewritten as (see Section 7)

$$\tilde{M}(t, s; x, y; \mu) = \tilde{M}^0(t, s; x, y; \mu)\cdot \mathbf{P}'\left(\omega_t^x(s) = y\right),$$

where $\tilde{M}^0$ is defined in the same way as $\tilde{M}$ but by replacing, however, in (30) (and respectively in (28) and (29)) the process ω by the process $\overset{\circ}{\omega}$. Let $\rho_n(u)=\rho(k/n)$ for $k/n\leq u<(k+1)/n$. Next we put

$$\alpha_n^x(t,s) = -\int_t^s \rho_n(u/T)\,\omega_t^x(u)\,du,$$
$$\beta_n(t,s) = \lambda\cdot\int_t^s \rho_n^{2/3}(u/T)\,du; \tag{33}$$
$$\gamma_n^x(t,s) = \sum_{k=k_1}^{k_2-1} \ln\frac{z\left(\omega_t^x(k/n),\ \rho((k+1)/n)\right)}{z\left(\omega_t^x\left(\frac{k}{n}\right),\ \rho\left(\frac{k}{n}\right)\right)},$$

where $k_1=\sup\{k:k/n\le tT\}$; $k_2=\sup\{k:k/n<sT\}$.

Let the functions $\tilde{M}_n(t,s;x,y;\mu)$ be defined in the same way as $M(t,s;x,y;\mu)$, by replacing, however, in (30) the functionals α, β, γ by α_n, β_n, γ_n respectively; if, next, in the definition of $\tilde{M}_n(t,s;x,y;\mu)$ (and in (33)) we replace the process ω by the process $\mathring{\omega}$, we obtain the definition of the functions $\tilde{M}_n^0(t,s;x,y,\mu)$. In this case, obviously,

$$\tilde{M}_n(t,s;x,y;\mu) = \tilde{M}_n^0(t,s;x,y;\mu)\,\mathbf{P}'\left(\omega_t^x(s)=y\right).$$

If in the definitions (28), (29) and (33) of the functionals α, γ; α_n, γ_n we replace the process ω by the process $\mathring{\omega}$, we obtain the definitions of the functionals $\mathring{\alpha}$, $\mathring{\gamma}$; $\mathring{\alpha}_n$, $\mathring{\gamma}_n$. Therefore,

$$\tilde{M}^0(t,s;x,y;\mu) = z^{-1}(x,\rho(t/T))\cdot z(y,\rho(s/T))\,\mathbf{M}(\exp\{\mathring{\alpha}^x(t,s)+\beta(t,s)+ \\ +\mu\mathring{\gamma}^x(t,s)\};\ \inf_{t\le u\le s}\mathring{\omega}_{t,s}^{x,y}(u)>0); \tag{34}$$

$$\tilde{M}_n^0(t,s;x,y;\mu) = z^{-1}(x,\rho(t/T))\cdot z(y,\rho(s/T))\,\mathbf{M}(\exp\{\mathring{\alpha}_n^x(t,s)+ \\ +\beta_n(t,s)+\mu\cdot\mathring{\gamma}_n^x(t,s)\};\ \inf_{t\le u\le s}\mathring{\omega}_{t,s}^{x,y}(u)>0), \tag{35}$$

where, we recall, the functions $z(x,\rho)$ are given by (22).

The functionals $\mathring{\alpha}_n$ and $\mathring{\gamma}_n$ given on the continuous processes $\mathring{\omega}(u)=\mathring{\omega}^{x,y}_{t,s}(u)$, $t\leq u\leq s$, obviously converge with probability 1 as $n\to\infty$ to the functionals $\mathring{\alpha}$ and $\mathring{\gamma}$; obviously, next, $\beta_n\to\beta$ as $n\to\infty$. Hence the random variables under the sign of mathematical expectation in the definition (35) of the function $\tilde{M}^0_n$ converge with probability 1 to the random variables under the sign of mathematical expectation in the definition (34) of the function $\tilde{M}^0$. For convergence

$$\widetilde{M}^0_n(t, s; x, y; \mu)\to\widetilde{M}^0(t, s; x, y; \mu) \tag{36}$$

as $n\to\infty$ it is sufficient (by the Lebesgue theorem) to have coarser uniform (in n) boundedness. We wish to establish that we can find $C_1, C_2<\infty$ such that for $|\mu|\leq 1$, $n=1,2,\ldots$

$$\exp\{\mathring{\alpha}^x_n(t, s)+\beta_n(t, s)+\mu\mathring{\gamma}^x_n(t, s)\}\leqslant$$
$$\leqslant C_1\exp\left\{C_2\sup_{t\leqslant u\leqslant s}\left|\mathring{\omega}^{x,y}_{t,s}(u)\right|^{3/2}\right\}\equiv\eta(x, y). \tag{37}$$

It is obvious that only the estimate $\exp\{\mu\mathring{\gamma}^x_n(t,s)\}$ needs to be clarified. It follows from (18) that as $x\to\infty$

$$\ln z(x)\sim -2/3\cdot x^{3/2},\ (\ln z(x))'\sim -x^{3/2}.$$

This plus the fact that as $x\to 0$ $z(x)\sim cx$, $0<c<\infty$, easily yield the required inequality for $\exp\{\mu\mathring{\gamma}^x_n(t,s)\}$.

Since, obviously, $\eta(x,y)$ on the right-hand side of the inequality (37) has a finite mathematical expectation, the required boundedness results; the convergence (36) is proved.

Recalling how the functions $\tilde{M}_n$ and $\tilde{M}^0_n$, $\tilde{M}$ and $\tilde{M}^0$ are interrelated, we obtain that for any $x>0$, $y>0$, $|\mu|\leq 1$ as $n\to\infty$

$$\widetilde{M}_n(t, s; x, y; \mu)\to\widetilde{M}(t, s; x, y; \mu). \tag{38}$$

Next, immediately from (25) it follows that

$$\int_0^\infty \widetilde{M}_n(t, s; x, y; -1)\,dy \equiv 1, \tag{39}$$

hence by the convergence (38) for proving (31) it suffices to make sure that it is possible to pass to the limit under the sign of the integral in (39). By (37) we have

$$z(x, \rho_n(t/T))\,\widetilde{M}_n(t, s; x, y; -1) \leqslant \mathbf{M}\eta(x, y)\cdot z(y, \rho_n(s/T))\cdot \mathbf{P}'\left(\omega_t^x(s) = y\right),$$

and therefore it suffices to prove that the right-hand side of the last inequality is integrable over y in the range from 0 to ∞ uniformly in n. It is easy to see that

$$\int_0^\infty \mathbf{M}\eta(x, y)\, z(y, \rho_n(s/T))\, \mathbf{P}'\left(\omega_t^x(s) = y\right) dy \leqslant \mathbf{M}C_1 \exp\left\{\sup_{t\leqslant u\leqslant s} C_2 \left|\omega_t^x(u)\right|^{3/2}\right\}\times$$
$$\times z\left(\left|\omega_t^x(s)\right|,\ \rho_n(s/T)\right) \leqslant A\cdot C_1\mathbf{M}\exp\left\{\sup_{t\leqslant u\leqslant s} C_2\left|\omega_t^x(u)\right|^{3/2}\right\},$$

where

$$A = \sup_{0<x<\infty}\ \sup_{0\leqslant s\leqslant T_1}\ \sup_{n=1,2,\ldots} z(x, \rho_n(s/T)) < \infty.$$

Hence the required uniform integrability in (39) is established, the relation (31) is proved. Since (32) can be proved in the similar way, Lemma 1 is proved.//

The second absolutely continuous transformation in accord with Lemma 1 is defined as follows:

$$\widetilde{\mathbf{P}}'\left(\omega_t^x(s) = y\right) = \widetilde{M}(t, s; x, y; -1), \tag{40}$$

where, we recall, the function $\tilde{M}$ is given by (30). In these terms the function $N_2(x)$ from (17) is expressible as

$$N_2(x) = \exp\left\{-\lambda \int_0^{T_1} \rho^{2/3}(u/T)\,du\right\} \times$$
$$\times \int_0^\infty z(x, \rho(0))\,\widetilde{M}(0, T_1; x, y; 0)\,z^{-1}(y, \rho(1-\varkappa))\,dy,$$

where $T=z^{2/3}$, $T_1=(1-\kappa)T$. The relation (17) becomes

$$\mathbf{P}(\omega \in r\cdot G) = \exp\left\{-r^2 v(f_0) - \lambda \int_0^{T_1} \rho^{2/3}(u/T)\,du\right\} \times$$
$$\times 1/\sqrt{T}\cdot \int_0^\infty\int_0^\infty p\left(\frac{x}{\sqrt{T}}\right) z(x, \rho(0))\,\widetilde{M}(0, T_1; x, y; 0)\,z^{-1}(y, \rho(1-\varkappa))\,dxdy. \tag{41}$$

The following non-rigorous arguments mark the trail we shall follow. It seems obvious that $p(x/\sqrt{T}) \sim 1/\sqrt{2\pi\kappa}$ as $T \to \infty$. Next, by (32) it is reasonable to expect that

$$\widetilde{M}(0, T_1; x, y; 0) \sim p(y; \rho(1-\varkappa)), \tag{42}$$

hence the integrand in (41) converges to

$$1/\sqrt{2\pi\varkappa}\cdot z(x, \rho(0))\cdot z(y, \rho(1-\varkappa)).$$

The rest of this paper deals with the proof of (42) and justification of the passage to the limit in (41). We prove (42) in several stages. We assume without loss of generality that for $\kappa \le t \le 1$ either $\dddot{h}(t) \le 0$ or $\dddot{h}(t) \ge 0$; the situation when $\dddot{h}(t)$ alternates the sign on the interval $[\kappa, 1]$ reduces to the preceding situation through partitioning $[\kappa, 1]$ into intervals where the third derivative is sign-definite.

We turn now to the obvious relationship

$$\widetilde{M}(0, T_1; x, y; 0) = \int_{x'=0}^{\infty}\int_{y'=0}^{\infty} \widetilde{M}(0, T_1^{1/3}; x, x'; 0)\, \widetilde{M}(T_1^{1/3}, T_1 - T_1^{1/3}; x', y'; 0)\, \widetilde{M}(T_1 - T_1^{1/3}, T_1; y', y; 0)\, dx'dy'. \quad (43)$$

In Section 5 we shall prove that as $T \to \infty$

$$\int_{x'=0}^{\infty}\int_{y'=0}^{\infty} p(x', \rho(T_1^{1/3}/T))\, \widetilde{M}(T_1^{1/3}, T_1 - T_1^{1/3}; x', y'; 0)\, dx'dy' \sim 1. \quad (44)$$

We note that in proving (44) (only in this case) we shall use the technical restriction consisting in the fact that $\dddot{h}(t)$ alternates its sign a finite number of times. In Section 5 we shall also establish (Lemma 2)

$$\begin{aligned} &\widetilde{M}(0, T_1^{1/3}; x, x'; 0) \sim p(x', \rho(T_1^{1/3}/T)); \\ &\widetilde{M}(T_1 - T_1^{1/3}; T_1; y', y; 0) \sim p(y, \rho(T_1/T)); \end{aligned} \quad (45)$$

in proving (45) we use Lemma 16' from Section 7. Applying next formally (44) and (45) to the formula (43), it is possible to obtain (42). Section 6 concerns the justification of passages to the limit in both (43) and (41).

5. PROOF OF THE RELATION (44)

Let

$$\widetilde{\mathrm{M}}^t \exp\{\mu\gamma(t, s)\} \equiv \int_0^{\infty} p(x, \rho(t/T)) \int_0^{\infty} \widetilde{M}(t, s; x, y; \mu - 1)\, dxdy.$$

The superscript t means that at an instant of time t the Wiener process $\omega_t^a(u)$ starts at a random point a the distribution of which is given by the density $p(x, \rho(t/T))$; ~ over the sign of mathematical expectation indicates the second absolutely continuous transformation. According to the assertions of Lemma 1 we have

$$\tilde{M}^t \exp\{\mu\gamma(t,s)\} = 1$$

for $\mu=0$ and $\mu=2$. In this notation the relation (44) is then written as

$$\lim_{T_1\to\infty} \tilde{\mathbf{M}}^{T_1^{1/3}} \exp\{\gamma(T_1^{1/3}, T_1 - T_1^{1/3})\} = 1, \tag{46}$$

where, as before, $T_1=(1-\kappa)T$. We take

$$N(u, v) = \left.\frac{\partial\tilde{\mathbf{M}}^u \exp\{\mu\gamma(u, v)\}}{\partial\mu}\right|_{\mu=2}; \tag{47}$$

we make sure that the relation

$$\lim_{T_1\to\infty} N(T_1^{1/3}, T_1 - T_1^{1/3}) = 0 \tag{48}$$

implies (46). Indeed, the function

$$\tilde{\mathbf{M}}^{T_1^{1/3}} \exp\{\mu\gamma(T_1^{1/3}, T_1 - T_1^{1/3})\}$$

is convex in μ (this is the Laplace transform of the random functional $\gamma(T_1^{1/3}, T_1-T_1^{1/3})$, defined on trajectories of the process $\omega^a_{T_1^{1/3}}(u)$, where the random variable a has density $p(x,\rho(T_1^{1/3}/T))$, hence it lies entirely above the tangent drawn at the point $\mu=2$; on the other hand, for $0\leq\mu\leq 2$ this function does not exceed 1. Since (48) implies that the derivative at $\mu=2$ tends to zero, i.e., the tangent at this point approaches the function $f(\mu)=1$, (46) follows.

We prove (48) with the aid of several lemmas. As was noted at the end of Section 4, it is possible to assume without loss of generality that for $0 \leq t \leq 1-\kappa$ the function $\rho(t)$ does not alternate the sign; for certainty we assume that $\rho(t) \geq 0$ for $0 \leq t \leq 1-\kappa$. The case where $\rho(t) \leq 0$ reduces to the preceding case, if we consider the converted process $\omega^*(t)=\omega(T_1-t)$, to which the function $\rho^*(t)=\rho(1-\kappa-t)$ with $\rho^*(t) \geq 0$ corresponds.

LEMMA 2. For $0<a \leq A<\infty$, $\varepsilon>0$, $0 \leq u \leq T_1-T_1^{1/3}$ we can find T_0 such that for all $T_1 \geq T_0$, $a \leq x$, $y \leq A$, $|\mu| \leq 1$ we have

$$\left|\widetilde{M}\left(u, u+T_1^{1/3}; x, y; \mu\right)-p(y, \rho(u/T))\right|<\varepsilon,$$

where the functions $\tilde{M}$ and p are defined respectively by (30) and (22).

Proof. By (18) we can find $0<c \leq C<\infty$ such that for all $x>0$ we have

$$\sup_{0 \leqslant t \leqslant 1-\varkappa}|A(x, t)| \leqslant c x^{3/2}+C,$$

where the function $A(x,t)$ is defined by (27). Hence we can find $\rho_+=\rho_+(u,T_1)$ and $\rho_-=\rho_-(u,T_1)$ such that for some $C_1<\infty$ we have

$$\delta \equiv\left|\rho_{+}^{2/3}-\rho_{-}^{2/3}\right| \leqslant C_1/T_1^{2/3}$$

and (see (28) and (29)) for $|\mu| \leq 1$

$$-C/T_1^{1/6}-\rho_{+} \int_u^{u+T_1^{1/3}} \omega_u^x(t)\, dt \leqslant \alpha^x\left(u, u+T_1^{1/3}\right)+\mu \gamma^x\left(u, u+T_1^{1/3}\right) \leqslant$$

$$\leqslant-\rho_{-} \int_u^{u+T_1^{1/3}} \omega_u^x(t)\, dt+C/T_1^{1/6}$$

on the set $\{0<\omega_u^x(t)<T_1^{1/3};\ u \leq t \leq u+T_1^{1/3}\}$. Hence we have the

inequalities

$$A_+ e^{-\delta T_1^{1/3}} \widetilde{M}^{\rho+}(T_1^{1/3}; x, y) \leqslant \widetilde{M}(u, u+T_1^{1/3}; x, y; \mu) \leqslant$$

$$\leqslant A_- e^{\delta T_1^{1/3}} \widetilde{M}^{\rho-}(T_1^{1/3}; x, y),$$

where the function $\tilde{M}^{\rho}$ is defined by (24)

$$A_\pm = z(x, \rho(u/T)) \cdot z^{-1}(x, \rho_\pm) \cdot z^{-1}(y, \rho(u+T_1^{1/3}/T)) z(y, \rho_\pm).$$

Since, as $T_1 \to \infty$ $A_\pm \to 1$ uniformly in $x, y \in [a, A]$, and also by Lemma 16' as $T_1 \to \infty$ we have

$$\tilde{M}^{\rho_\pm}(T_1^{1/3}; x, y) \sim p(y, \rho_\pm)$$

uniformly in $x, y \in [a, A]$ and, finally,

$$p(y, \rho_\pm) \sim p(y, \rho(u/T))$$

uniformly in $y \in [a, A]$, then Lemma 2 is proved.

<u>LEMMA 3</u>. For any $0 \leq u < v \leq T_1$

$$N(u, v) = 1/T \cdot \int_{t=u}^{v} \int_0^\infty \int_0^\infty q(x, t/T) \widetilde{M}(t, v; x, y; 1)\, dx dy dt,$$

where the function N is defined by (47); $q(y, t/T) = p(y, \rho(t/T)) A(y, t/T)$; the function A is defined by (27).

<u>Proof</u>. Starting from the definition (47) and changing the order of integration, we obtain

$$N(u,v) \equiv \int_0^\infty p(x,\rho(u/T))\,\widetilde{\mathbf{M}}\left(\exp\left\{2/T\cdot\int_u^v A(\omega_u^x(t),\,t/T)\,dt\right\}\times\right.$$

$$\left.\times 1/T\cdot\int_u^v A(\omega_u^x(t'),\,t'/T)\,dt'\right)dx = 1/T\cdot\int_u^v\int_0^\infty\int_0^\infty\int_0^\infty p(x,\rho(u/T))\times$$

$$\times\widetilde{\mathbf{M}}'\left(\exp\left\{2/T\cdot\int_u^{t'} A(\omega_u^x(t),\,t/T)\,dt\right\};\quad \omega_u^x(t')=x'\right)A(x',\,t'/T)\times$$

$$\times\widetilde{\mathbf{M}}'\left(\exp\left\{2/T\cdot\int_{t'}^v A(\omega_{t'}^{x'}(t),\,t/T)\,dt\right\};\quad \omega_{t'}^{x'}(v)=y\right)dydx'dxdt' =$$

$$= 1/T\cdot\int_u^v\int_0^\infty\int_0^\infty\int_0^\infty p(x,\rho(u/T))\,\widetilde{M}(u,\,t';\,x,\,x';\,1)\times$$

$$\times A(x',\,t'/T)\,\widetilde{M}(t',\,v;\,x',\,y;\,1)\,dydx'dxdt.$$

Next, by (32) the integral in x is given by $p(x',\rho(t'/T))$.//

<u>LEMMA 4</u>. For any $0\leq t\leq T_1$ we have $\int_0^\infty q(x,t/T)dx=0$, where the function q has been defined in Lemma 3.

<u>Proof</u>. We turn now to defining the function q. To prove Lemma 4, it suffices to make sure that for any $\rho>0$

$$\int_0^\infty z^2(y\rho)\,dy + 2\rho\int_0^\infty y\dot{z}(y\rho)\,z(y\rho)\,dy = 0\,,$$

which can be done by integrating by parts.//

<u>LEMMA 5</u>. For any $\varepsilon>0$ we can find $a'>0$ such that for all $0\leq t\leq T_1$, $0<a\leq a'$

$$\int_{x=0}^a q(x,\,t/T)\int_{y=0}^\infty \widetilde{M}\left(t,\,T_1-T_1^{1/3};\,x,\,y;\,1\right)dxdy<\varepsilon.$$

<u>Proof</u>. By the condition $\rho(u)\geq 0$ for $0\leq u\leq 1-\kappa$ we can find $C<\infty$ such that for all $x>0$

$$\gamma^x\left(t,\,T_1-T_1^{1/3}\right)\leqslant C. \tag{49}$$

Hence

$$\int_{x=0}^{a'} q(x, t/T) \int_{y=0}^{\infty} \widetilde{M}\left(t, T_1 - T_1^{1/3}; x, y; 1\right) dxdy \leqslant e^{2C} \int_{x=0}^{a'} q(x, t/T) \times$$
$$\times \int_{y=0}^{\infty} \widetilde{M}\left(t, T_1 - T_1^{1/3}; x, y; -1\right) dxdy = e^{2C} \int_{0}^{a'} q(x, t/T)\, dx,$$

where the last equality holds by Lemma 1. Taking next $a'>0$ sufficiently small, we obtain what was to be proved.

LEMMA 6. For any $\varepsilon>0$ we can find $0<a'\leq A'<\infty$ and T_0 such that for all $T_1\geq T_0$, $0<a\leq a'\leq A'\leq A<\infty$

$$N\left(T_1^{1/3}, T_1 - T_1^{1/3}\right) \leqslant \varepsilon + 1/T \cdot \int_{T_1^{1/3}}^{T_1 - T_1^{1/3}} \int_{x=a}^{A} \int_{y=0}^{\infty} q(x, t/T)\, \widetilde{M}\left(t, T_1 - T_1^{1/3}; x, y; 1\right) dydxdt.$$

Proof. By Lemma 5 we can find $a'>0$ such that for all sufficiently large T_1, $0<a\leq a'$

$$N\left(T_1^{1/3}, T_1 - T_1^{1/3}\right) \leqslant \varepsilon + 1/T \cdot \int_{T_1^{1/3}}^{T_1 - T_1^{1/3}} \int_{x=a}^{\infty} \int_{y=0}^{\infty} q(x, t/T)\, \widetilde{M}\left(t, T_1 - T_1^{1/3}; x, y; 1\right) dydxdt. \quad (50)$$

It follows from the condition $\rho(u)\geq 0$ for $0\leq u\leq 1-\kappa$ that we can find $A'<\infty$ such that $q(x,t/T)\leq 0$ holds for all $x\geq A'$, $0\leq t\leq T_1$; the function $\widetilde{M}(t, T_1-T_1^{1/3}; x,y; 1)$ is positive, therefore, diminishing in (50) the domain of integration over x down to $[a,A]$, $A'\leq A<\infty$, we increase thereby the right-hand side of (50).//

LEMMA 7. For any $\varepsilon>0$, $0<a\leq A<\infty$ we can find $0<b\leq B<\infty$ such that

$$K \equiv \int_{x=a}^{A} q(x, t/T)\left(\int_{y'=0}^{b} + \int_{y'=B}^{\infty}\right)\int_{y=0}^{\infty} \widetilde{M}(t, t+T_1^{1/3}; x, y'; 1) \times$$
$$\times \widetilde{M}(t+T_1^{1/3}, T_1 - T_1^{1/3}; y', y; 1)\, dy dy' dx < \varepsilon$$

is satisfied for all $0 \le t \le T_1 - 2T_1^{1/3}$.

Proof. Let

$$C_1 = C_1(a, A) \equiv \sup_{0 \le t \le T_1}\left[\sup_{a \le x \le A} q(x, t/T) \Big/ \inf_{a \le x \le A} p(x, \rho(t/T))\right]. \qquad (51)$$

It is easy to see that $C_1 < \infty$ and

$$K \le C_1 \int_{x=a}^{A} p(x, \rho(t/T))\left(\int_{y'=0}^{b} + \int_{y'=B}^{\infty}\right)\int_{y=0}^{\infty} \widetilde{M}(t, t+$$
$$+ T_1^{1/3}; x, y'; 1)\, \widetilde{M}(t+T_1^{1/3}, T_1 - T_1^{1/3}; y', y; 1)\, dy dy' dx.$$

By Lemma 1 the integral over x does not exceed $p\left(y', \rho\left(\frac{t+T_1^{1/3}}{T}\right)\right)$, hence by (49)

$$K \le C_1\left(\int_{y'=0}^{b} + \int_{y'=B}^{\infty}\right) p(y', \rho(t+T_1^{1/3}/T)) \times$$
$$\times \int_{y=0}^{\infty} \widetilde{M}(t+T_1^{1/3}, T_1 - T_1^{1/3}; y', y; 1)\, dy dy' \le$$
$$\le C_1 e^{2C}\left(\int_{y'=0}^{b} + \int_{y'=B}^{\infty}\right) p(y', \rho(t+T_1^{1/3}/T))\, dy'.$$

Choosing b and B, we have proved the Lemma.

Now we proceed directly to prove (48). It follows from Lemmas 6 and 7 that for any $\varepsilon > 0$ we can find $0 < a' \le A' < \infty$ and for any $0 < a \le a' \le A' \le A < \infty$ we can find $0 < b \le B < \infty$, $T_0 < \infty$ such that for all $T_1 \ge T_0$ we have

$$N(T_1^{1/3}, T_1 - T_1^{1/3}) \le \varepsilon + I_1 + I_2,$$

where

$$I_1 = 1/T \cdot \int_{T_1-2T_1^{1/3}}^{T_1-T_1^{1/3}} \int_{x=a}^{A} \int_{x=0}^{\infty} q(x, t/T)\, \widetilde{M}(t, T_1 - T_1^{1/3}; x, y; 1)\, dydxdt;$$

$$I_2 = 1/T \cdot \int_{0}^{T_1-2T_1^{1/3}} \int_{x=a}^{A} \int_{y'=b}^{B} \int_{y=0}^{\infty} q(x, t/T)\, \widetilde{M}(t, t + T_1^{1/3}; x, y'; 1) \times$$
$$\times \widetilde{M}(t + T_1^{1/3}, T_1 - T_1^{1/3}; y', y; 1)\, dydy'dxdt.$$

Using the notation of (51), we estimate I_1:

$$I_1 \leqslant C_1(a, A)/T \cdot \int_{T_1-3T_1^{1/3}}^{T_1-T_1^{1/3}} \int_{x=a}^{A} p(x, \rho(t/T)) \int_{y=0}^{\infty} \widetilde{M}(t, T_1 - T_1^{1/3}; x, y; 1)\, dydxdt.$$

Since the integral over x by Lemma 1 does not exceed $p(y, \rho(T_1 - T_1^{1/3}/T))$, then

$$I_1 \leqslant C_1(a, A)/T \cdot \int_{T_1-2T_1^{1/3}}^{T_1-T_1^{1/3}} dt = \frac{C_1(a, A) \cdot T_1^{1/3}}{T} = C_1(a, A)/(1 - \varkappa) \cdot T_1^{-2/3}.$$

Now we turn to the third summand. By Lemma 2

$$\delta = \delta(a, A) \equiv \sup_{\substack{a \leqslant x \leqslant A \\ b \leqslant y' \leqslant B}} \left| \widetilde{M}(t, t + T_1^{1/3}; x, y'; 1) - p\left(y', \rho\left(\frac{t + T_1^{1/3}}{T}\right)\right) \right| \to 0$$

as $T_1 \to \infty$. Hence $I_2 \leq I_{2,1} + I_{2,2}$, where

$$I_{2,1} = 1/T \cdot \int_0^{T_1 - 2T_1^{1/3}} \int_{x=a}^{A} \int_{y'=b}^{B} |q(x, t/T)| \, \delta \int_{y=0}^{\infty} \widetilde{M}(t + T_1^{1/3}, T_1 - T_1^{1/3}; y', y; 1) \, dt dy dy' dx;$$

$$I_{2,2} = 1/T \cdot \int_0^{T_1 - 2T_1^{1/3}} \int_{x=a}^{A} \int_{y'=b}^{B} q(x, t/T) \, p(y', \rho(t + T_1^{1/3}/T)) \times \int_{y=0}^{\infty} \widetilde{M}(t + T_1^{1/3}, T_1 - T_1^{1/3}; y', y; 1) \, dy dy' dx dt.$$

By the inequality (49) the integral over y' and y in the definition of $I_{2,1}$ does not exceed e^{2C}, hence

$$I_{2,1} \leqslant 1/T \, e^{2C} \delta \int_0^{T_1 - 2T_1^{1/3}} \int_{x=a}^{A} |q(x, t/T)| \, dx dt \leqslant \delta e^{2C} D,$$

where

$$D = \sup_{0 \leqslant t \leqslant T_1} \int_0^{\infty} |q(x, t/T)| \, dx < \infty.$$

Next, by Lemma 1 the integral over y, y' in the definition of $I_{2,2}$ does not exceed 1, which leads to

$$I_{2,2} \leqslant 1/T \int_0^{T_1 - 2T_1^{1/3}} \int_{x=a}^{A} q(x, t/T) \, dx dt.$$

Therefore,

$$N(T_1^{1/3}, T_1 - T_1^{1/3}) \leqslant \varepsilon + I_1 + I_{2,1} + I_{2,2} \leqslant \varepsilon + \frac{C_1(a, A)}{1 - \varkappa} T_1^{-2/3} + e^{2C} \cdot \delta(a, A) \cdot D + 1/T \int_0^{T_1} \int_{x=a}^{A} q(x, t/T) \, dx dt.$$

By an appropriate choice of a ($\leq a'$) and A ($\geq A'$)

(see Lemma 4) it is possible to make the fourth summand on the right-hand side of the last inequality smaller than ε; for these a and A the second and third summands can be made arbitrarily small by choosing an appropriate T_1. We have thus proved (48) (that is, (44)).

6. THE JUSTIFICATION OF PASSAGES TO THE LIMIT

Initially we formulate two assertions.

LEMMA 8. For any $\varepsilon>0$ we can find $0<a'<A'<\infty$ and $T_0<\infty$ such that for all $T_1\geq T_0$, $0<a\leq a'\leq A'\leq A<\infty$ we have

$$\left|\int_a^A\int_a^A p\left(x,\rho\left(T_1^{1/3}/T\right)\right)\widetilde{M}\left(T_1^{1/3},T_1-T_1^{1/3};x,y;0\right)dxdy-1\right|<\varepsilon. \quad (52)$$

LEMMA 9. For any $\varepsilon>0$ we can find $0<a\leq A<\infty$, $T_0<\infty$ such that for all $T_1\geq T_0$

$$\left|\left(\int_0^\infty\int_0^\infty\int_0^\infty\int_0^\infty-\int_a^A\int_a^A\int_a^A\int_a^A\right)p\left(x/\sqrt{T}\right)M_1(x,x')M_2(x',y')M_3(y',y)\,dxdx'dydy'\right|<$$
$$<\varepsilon, \quad (53)$$

where

$$M_1(x,y)=M\left(0,T_1^{1/3};x,y;0\right);\quad M_2(x,y)=M\left(T_1^{1/3},T_1-T_1^{1/3};x,y;0\right);$$
$$M_3(x,y)=M\left(T_1-T_1^{1/3},T_1;x,y;0\right),$$

where the function $M(t,s;x,y;\mu)$ is

$$M(t,s;x,y;\mu)=z(t,\rho(t/T))\widetilde{M}(t,s;x,y;\mu)z^{-1}(y,\rho(s/T)), \quad (54)$$

where $\widetilde{M}$ and z are defined respectively by (30) and (22).

Using these two Lemmas, we complete proving the Theorem. First we denote the constant in (41):

$$K\equiv\int_0^\infty\int_0^\infty p\left(x/\sqrt{T}\right)\widetilde{M}(0,T_1;x,y;0)\,z(x,\rho(0))\cdot z^{-1}(y,\rho(1-\varkappa))\,dxdy.$$

By Lemma 9 for any $\varepsilon>0$ we can find $0<a\le A<\infty$ such that for sufficiently large T we have

$$\left| K - \int_a^A\int_a^A\int_a^A\int_a^A p(x/\sqrt{T})\, M_1(x, x')\, M_2(x', y')\, M_3(y', y)\, dx dx' dy' dy \right| < \varepsilon,$$

where the notation of Lemma 9 has been used. Next, by Lemma 2 we have the relations

$$M_1(x, x') \sim z(x, \rho(0))\, z\left(x', \rho\left(T_1^{1/3}/T\right)\right),$$
$$M_3(y', y) \sim z\left(y', \rho\left(T_1 - T_1^{1/3}/T\right)\right) \cdot z(y, \rho(1-\varkappa))$$

uniformly in x, x' and $y, y' \in [a, A]$. It is obvious, further, that

$$p(x/\sqrt{T}) \sim 1/\sqrt{2\pi\kappa}$$

uniformly in $x \in [a, A]$. Hence we can assert that for any $\varepsilon>0$ we can find $0<a'\le A'<\infty$ and $T_0<\infty$ such that for all $0<a\le a'\le A'\le A<\infty$, $T_1\ge T_0$ we have

$$\Bigg| K - 1/\sqrt{2\pi\varkappa} \cdot \int_a^A\int_a^A\int_a^A\int_a^A z(x, \rho(0))\, p\left(x', \rho\left(T_1^{1/3}/T\right)\right) \times$$
$$\times \widetilde{M}\left(T_1^{1/3}, T_1 - T_1^{1/3}; x', y'; 0\right) z(y, \rho(1-\varkappa))\, dx dx' dy' dy \Bigg| < \varepsilon.$$

Applying Lemma 8 to the last relation completes the proof.

Thus, it remains only to prove Lemmas 8 and 9. To this end, we shall use the inequality

$$M(t, s; x, y; 0) \le M^{1/2}(t, s; x, y; 1) \cdot M^{1/2}(t, s; x, y; -1), \qquad (55)$$

which follows from the Cauchy inequality, and also two relations which follow from Lemma 1:

$$\int_{x=0}^{\infty} z(x, \rho(t/T)) M(t, s; x, y; 1)\, dx = z(y, \rho(s, T));$$
$$\int_{y=0}^{\infty} M(t, s; x, y; -1)\, z(y, \rho(s/T))\, dy = z(x, \rho(t/T)). \tag{56}$$

Proof of Lemma 8. By the relation (44) proved it suffices to show that for any $\varepsilon>0$ we can find $0<a'\leq A'<\infty$ such that for all T_1, $0<a\leq a'\leq A'\leq A<\infty$

$$\left[\left(\int_{x=0}^{a} + \int_{x=A}^{\infty}\right)\int_{y=0}^{\infty} + \int_{x=0}^{\infty}\left(\int_{y=0}^{a} + \int_{y=A}^{a}\right)\right] p(x, \rho(T_1^{1/3}/T)) \times$$
$$\times \widetilde{M}(T_1^{1/3}, T_1 - T_1^{1/3}; x, y; 0)\, dx dy < \varepsilon. \tag{57}$$

Let us estimate the first summand in (57). The definition (30) and the inequality (55) imply that the first summand in (57) is not greater than

$$\left[\left(\int_{x=0}^{a} + \int_{x=A}^{\infty}\right)\int_{y=0}^{\infty} z(x, \rho(t/T)) M(t, s; x, y; 1)\, z(y, \rho(s/T))\, dx dy\right]^{1/2} \times$$
$$\times \left[\left(\int_{x=0}^{a} + \int_{x=A}^{\infty}\right)\int_{y=0}^{\infty} z(x, \rho(t/T)) M(t, s; x, y; -1)\, z(y, \rho(s/T))\, dx dy\right]^{1/2},$$

where $t=T_1^{1/3}$; $s=T_1-T_1^{1/3}$. By (56) the first factor is not greater than 1, the second factor is equal to

$$\left[\left(\int_{x=0}^{a} + \int_{x=A}^{A}\right) z^2(x, \rho(t/T))\, dx\right]^{1/2}.$$

Thus, the first summand in (57) is estimated properly. Since the second summand can be estimated similarly, Lemma 8 is proved.//

LEMMA 10.

A. We can find $C<\infty$ such that for all T_1,

$0\le t\le t+1\le s\le T_1$, $x>0$, $y>0$

$$\int_{x=0}^{\infty} M(t,s;x,y;0)\,dx+\int_{y=0}^{\infty} M(t,s;x,y;0)\,dxdy<C.$$

B. For any $\varepsilon>0$ we can find $0<a\le A<\infty$ and $T_0<\infty$ such that for all $T_1\ge T_0$, $0\le t\le s\le T_1$

$$\left[\left(\int_{x=0}^{a}+\int_{x=A}^{\infty}\right)\int_{y=0}^{\infty}+\int_{x=0}^{\infty}\left(\int_{y=0}^{a}+\int_{y=A}^{\infty}\right)\right]M(t,s;x,y;0)\,dxdy<\varepsilon.$$

Proof of Lemma 9. It suffices to show that for any $\varepsilon>0$ we can find $T_0<\infty$, $0<a\le A<\infty$ such that for $T_1\ge T_0$

$$\Bigg[\int_{x=0}^{\infty}\int_{x'=0}^{\infty}\int_{y'=0}^{\infty}\left(\int_{y=0}^{a}+\int_{y=A}^{\infty}\right)+\int_{x=0}^{\infty}\int_{x'=0}^{\infty}\left(\int_{y'=0}^{a}+\int_{y'=A}^{\infty}\right)\int_{y=0}^{\infty}+$$
$$+\int_{x=0}^{\infty}\left(\int_{x'=0}^{a}+\int_{x'=A}^{\infty}\right)\int_{y'=0}^{\infty}\int_{y=0}^{\infty}+\left(\int_{x=0}^{a}+\int_{x=A}^{\infty}\right)\int_{x'=0}^{\infty}\int_{y'=0}^{\infty}\int_{y=0}^{\infty}\Bigg]\times$$
$$\times M_1(x,x')M_2(x',y')M_3(y',y)\,dxdx'dy'dy<\varepsilon,$$

where the functions M_1, M_2, M_3 have been used in the statement of Lemma 9. The smallness of each of the four summands follows from Lemma 10. For example, the first summand is

$$\int_{x=0}^{\infty}\left(\int_{y=0}^{a}+\int_{y=A}^{\infty}\right)M(0,T_1;x,y;0)\,dxdy,$$

as can be estimated through the second summand in Lemma 10-B. Since the other three summands can be estimated similarly through Lemma 10, Lemma 9 is proved. It remains only to prove Lemma 10.

Proof of Lemma 10. Let $\rho^*(t)=\rho(1-\kappa-t)$, $0\le t\le 1-\kappa$ and let the family of functions $M^*(t,s;x,y;\mu)$ be the family of functions $M(t,s;x,y;\mu)$ (see (54)) for the function ρ^*. It is obvious that

$$M(t,s;\ x,y;\ 0) = M^*(t^*,s^*;\ x,y\ 0)\ ,$$

where $t^*=T_1-s$; $s^*=T_1-t$, with $0\le t^*<s^*\le T_1$. Hence, if one summand has been properly estimated in Lemma 10-A, the second summand has been estimated as well therein; the same is true for Lemma 10-B. Therefore, it is sufficient to estimate one summand in Lemma 10-A and in Lemma 10-B.

LEMMA 11. We can find a continuous function $C(u)$, $u>0$, such that for all T_1, $0\le t<t+2\le s\le T_1$, $x>0$, $y>0$

$$M(t,s;\ x,y;\ 0)\ \le\ C(x)C(y)\ . \tag{58}$$

Proof. It is easy to see that

$$M(t,s;x,y;0)=\int\limits_{x'=0}^{\infty}\int\limits_{y'=0}^{\infty} M(t,t+1;x,x';0)\times \\ \times M(t+1,s-1;x',y';0)\,M(s-1,s;y',y;0)\,dx'dy'.$$

By (18) we can find a continuous function $C(u)$, $u>0$, such that

$$M(t,t+1;x,x';0)\le C(x)z(x',\rho((t+1)/T)), \\ M(s-1,s;y',y;0)\le C(y)z(y',\rho((s-1)/T)).$$

Hence

$$M(t,s;x,y;0)\le C(x)\cdot C(y)\int\limits_{x'=0}^{\infty}\int\limits_{y'=0}^{\infty} z(x',\rho((t+1)/T))\,M(t+1,s-1; \\ x',y';0)\,z(y',\rho((s-1)/T))\,dx'dy';$$

the integral on the right-hand side of the last relationship is not greater than 1 (which can be proved in the same way as Lemma 8). //

LEMMA 12. For any $\varepsilon>0$ we can find $T_0<\infty$, $A<\infty$, $t_0<\infty$ such for all $T_1 \geq T_0$, $0 \leq t \leq t+t_0 \leq s \leq T_0$, $y>0$

$$\int_A^\infty M(t, s; x, y; 0)\, dx \leqslant C(y) \cdot \varepsilon,$$

where the function $C(y)$ is defined in Lemma 11.

<u>Proof</u>. Let $A_0 = 2\lambda\rho_+^{2/3}/\rho_-$, where $\rho_+ = \max_{0 \leq t \leq 1-\kappa} \rho(t)$; $\rho_- = \min_{0 \leq t \leq 1-\kappa} \rho(t)$. For $x>A_0$ let $\eta_t^x = \inf\{u \geq t: \omega_t^x(u) = A_0\}$ be the first time that $\omega_t^x = A_0$. It is obvious that for $x>A_0$, $\delta>0$

$$\mathbf{M} \exp\{-\delta\eta_t^x\} = \exp\{-\psi(\delta)(x-A_0)\}, \tag{59}$$

where $\psi(\delta)>0$. Let $A>A_0$, $0 \leq t \leq t+2 \leq s \leq T_1$. It is easy to see that

$$\int_A^\infty M(t, s; x, y; 0)\, dx \leqslant \mathbf{P}'\left(\omega_t^x(s) = y\right) \exp\{-1/2A_0\rho_-(s-t)\} +$$

$$+ \int_{x=A}^\infty \int_{u=t}^{s-2} \exp\{-1/2A_0\rho_- u\} \mathbf{P}\left(\eta_t^x \in du\right) M(u, s; A_0, y; 0)\, dx.$$

Using Lemma 11 and Eq. (59), we obtain

$$\int_A^\infty M(t, s; x, y; 0)\, dx \leqslant \mathbf{P}'\left(\omega_t^x(s) = y\right) \exp\{-1/2A_0\rho_-(s-t)\} +$$

$$+ C(A_0) C(y) \int_{x=A}^\infty \mathbf{M} \exp\{-1/2A_0\rho_-\eta_t^x\}\, dt \leqslant \mathbf{P}'\left(\omega_t^x(s) = y\right) \times$$

$$\times \exp\{-1/2A_0\rho_-(s-t)\} + C(A_0) C(y) \int_A^\infty \exp\{-\psi(1/2A^0\rho_-)(x-A_0)\}\, dx.$$

<u>LEMMA 13</u>. For any $\varepsilon>0$ we can find $a>0$, $T_0<\infty$, $t_0<\infty$ such that for all $T_1 \geq T_0$, $0 \leq t \leq t+t_0 \leq s \leq T_1$, $y>0$

$$\int_0^a M(t, s; x, y; 0)\, dx \leqslant \varepsilon \cdot C(y),$$

where the function $C(y)$ is defined in Lemma 11.

Proof. The proof of Lemma 13 is similar to that of Lemma 11. Here we need to use the functional

$$\xi_t^x = \inf\left\{u \geqslant t : \omega_t^x(u) = a_0, \inf_{t \leqslant s \leqslant u} \omega_t^x(s) > 0\right\}$$

for $0<x\leq a_0/2$, where $a_0>0$ is sufficiently small. In this case instead of (59) we have the inequality (see, for example, [11])

$$\mathbf{P}(\xi_t^x > u) \leqslant \mathbf{P}\left(\sup_{t \leqslant s \leqslant u} |\omega_t^0(s)| \leqslant 2a_0\right) \leqslant \exp\{-c(u-t)/a_0^2\},$$

where $c>0$ is an absolute constant.

Using the assertions of Lemmas 11-13, we obtain for some $C<\infty$ and for all T_1, $0\leq t<s\leq T_1$, $y>0$

$$\int_{x=0}^{\infty} M(t,s;\ x,y;\ 0)dx \leq CC(y) , \tag{60}$$

where the function $C(y)$ is defined in Lemma 11. Turning next to the function $M^*(t,s;x,y;0)$, we obtain from (60) that for all T_1, $0\leq t<s\leq T_1$,

$$\int_{y=0}^{\infty} M(t,s;\ x,y;\ 0)dy \leq CC(x) , \tag{61}$$

where C and $C(y)$ are defined in (60).

Using the relation (61) and repeating the proofs of Lemmas 12 and 13, we can prove the following lemma.

LEMMA 14. For any $\varepsilon>0$ we can find $0<a\leq A<\infty$, $T_1<\infty$ such that for all $T_1\geq T_0$

$$\left(\int_{x=0}^{a} + \int_{x=A}^{\infty}\right)\int_{y=0}^{\infty} M(t, s; x, y; 0)\,dx dy < \varepsilon. \qquad (62)$$

We continue proving Lemma 10. Since the left-hand side of (62) is exactly the first summand in Lemma 10-B, then assertion B is proved. To prove assertion A, we use the obvious relation

$$\int_{y=0}^{\infty} M(t, s; x, y; 0)\,dy = \int_{x'=0}^{\infty}\int_{y=0}^{\infty} M(t, t+1; x, x'; 0) \times$$
$$\times M(t+1, s; x', y; 0)\,dx' dy. \qquad (63)$$

Since from (62) and (61) it follows that for some $C_1<\infty$ we have

$$\int_{x'=0}^{\infty}\int_{y=0}^{\infty} M(t+1, s; x', y; 0)\,dx' dy < C_1,$$

where T_1 is arbitrary, $0\le t+1<s\le T_1$, and for all x, x', t we have

$$M(t, t+1; x, x'; 0) \leqslant \exp\{\lambda\rho_+^{2/3}\}/\sqrt{2\pi},$$

then the right-hand side of (63) is not greater than

$$C = C_1 \cdot \exp\{\lambda\rho_+^{2/3}\}/\sqrt{2\pi}.$$

Therefore, we have estimated the second summand in Lemma 10-A. //

7. AUXILIARY LEMMAS

Suppose a nonnegative function $\rho(t)$ is given, continuous on the interval [0,1]. Also, for $x>0$, $y>0$ we define the function

$$A(x, y) = \mathbf{M}\left(\exp\left\{-\int_0^1 \rho(t)\,\omega_0^x(t)\,dt\right\};\ \inf_{0<t\leq 1} \omega_0^x(t) > 0,\ \omega_0^x(1) < y\right).$$

LEMMA 15. The function $A(x,y)$ is absolutely continuous in y; the density

$$B(x,y) = \partial A(s,y)/\partial y$$

is continuous for all x, y and representable as

$$B(x,y) = B_0(x,y)P^t(\omega_0^x(1)=y)\ , \qquad (64)$$

where

$$B^0(x, y) = \mathbf{M}\left(\exp\left\{-\int_0^1 \rho(t)\,\overset{\circ}{\omega}{}_{0,1}^{x,y}(t)\,dt\right\};\ \inf_{0\leq u<t} \overset{\circ}{\omega}{}_{0,1}^{x,y}(u) > 0\right),$$

where

$$\overset{\circ}{\omega}{}_{0,1}^{x,y}(t) = \omega_0^x(t) - \left(\omega_0^x(1) - y\right)t,\ 0\leq t\leq 1$$

is a conditional Wiener process which equals x at $t=0$ and equals y at $t=1$.

Proof. For $0<y_1<y_2$

$$A(x, y_2) - A(x, y_1) \leq \mathbf{P}\left(\omega_0^x(1) \in [y_1, y_2)\right).$$

Hence by the Radon-Nikodym theorem the function $A(x,y)$ is absolutely continuous in y. Obviously, it is possible to take the right-hand side of (64) for the density $B(x,y)$; since the function $P'(\omega_0^x(1)-y)$ is continuous in (x,y), it remains only to prove that the function $B_0(x,y)$ is continuous in (x,y).

Suppose $x_0>0$ and $y_0>0$ are fixed. For any $\varepsilon>0$ and

and $\delta>0$, $\delta\leq\frac{1}{2}\min(x_0,y_0)$ we can find $0<a\leq A<\infty$ such that for all x, y such that $|x-x_0|+|y-y_0|<\delta$,

$$\mathbf{M}\left(\exp\left\{-\int_0^1 \rho(t)\,\overset{\circ}{\omega}{}^{x,y}_{0,1}(t)\,dt\right\};\ \inf_{0\leqslant t\leqslant 1}\overset{\circ}{\omega}{}^{x,y}_{0,1}(t)>0;\right.$$
$$\left.(A_+(A)\cup A_-(a)\vphantom{\int_0^1}\right)<\varepsilon, \qquad (65)$$

where the events $A_\pm$ are defined as

$$A_+(A)=\left\{\sup_{0\leqslant t\leqslant 1}\overset{\circ}{\omega}{}^{x,y}_{0,1}(t)>A\right\};$$
$$A_-(a)=\left\{\inf_{0\leqslant t\leqslant 1}\overset{\circ}{\omega}{}^{x,y}_{0,1}(t)<a\right\}.$$

Next, for any $\varepsilon>0$, $0<a\leq A<\infty$ we can find $\delta>0$ such that on the set $\overline{A_-(a)\cap A_+(A)}$ from $|x-x_0|<\delta/2$, $|y-y_0|<\delta/2$ there follows

$$\left|\exp\left\{-\int_0^1\rho(t)\,\overset{\circ}{\omega}{}^{x,y}_{0,1}(t)\,dt\right\}-\exp\left\{-\int_0^1\rho(t)\,\overset{\circ}{\omega}{}^{x_0y_0}_{0,1}(t)\,dt\right\}\right|<\varepsilon. \qquad (66)$$

Using (65) and (66), it is easy to obtain the required continuity of the function $B_0(x,y)$ at the point (x_0,y_0). //

Let the function $z(y)$ be defined by (19), and let $p(y)=z^2(y)$; also (see (24)) let

$$p(t;x,y)=\widetilde{M}^1(t;x,y)\equiv$$
$$\equiv z^{-1}(x)\,\mathbf{M}'\left(\exp\left\{-\int_0^t\omega_0^x(u)\,du+\lambda t\right\};\ \inf_{0\leqslant u\leqslant t}\omega_0^x(u)>0,\ \omega_0^x(1)=y\right)\cdot z(y).$$

By (25) the function $p(t;x,y)$ can be viewed as the transition density of a homogeneous Markov process $X_u^x(t)$ starting from x at time u, with phase space $(0,\infty)$ and continuous trajectories; by (26) the function $p(y)$ is the invariant density for $X_u^x(t)$:

$$\int_0^\infty p(x)p(t;x,y)\,dx = p(y) . \tag{67}$$

The next lemma asserts that the transition density $p(t;x,y)$ of the process $X_u^x(t)$ as $t\to\infty$ converges to the stationary density $p(y)$ uniformly in $0<x$, $y\le A<\infty$.

LEMMA 16. For any $\varepsilon>0$, $A<\infty$ we can find $t_0<\infty$ such that for all $t\ge t_0$, $0<x$, $y\le A$

$$|p(t;\, x,y)-p(y)| < \varepsilon .$$

Since the function $\tilde{M}^1$ passes to the function $\tilde{M}^\rho$ through an obvious substitution, from Lemma 16 there follows

LEMMA 16'. For any $\varepsilon>0$, $0<a\le A<\infty$ and any function $\rho(t)$, $0\le t<\infty$, such that $a\le\rho(t)\le A$, we can find $t_0<\infty$ such that for all $t\ge t_0$, $x\in[a,A]$, $y\in[a,A]$

$$\left|\frac{\widetilde{M}^{\rho(t)}(t;\, x,\, y)}{p(y,\, \rho(t))}-1\right|<\varepsilon.$$

Proof of Lemma 16. Let $\alpha=(1,2\cdot 2^{1/3}/\lambda_0)^{1/2}$; we introduce the functional

$$\tau=\inf\left\{t\geqslant 0: X_0^{2\alpha}(t)=2\alpha,\ \sup_{0\leqslant u\leqslant t}\left|X_0^{2\alpha}(u)-2\alpha\right|\geqslant\alpha\right\}$$

which is the moment of the first return of the process $X_0^{2\alpha}(t)$ to the point 2α after it has left the interval $(\alpha,3\alpha)$. We shall give the proof of the next lemma at the end of the proof of Lemma 16.

LEMMA 17. We can find $0<c\le C<\infty$ such that $P(\tau>t)\le Ce^{-ct}$.

Let

$$F(t) = \mathbf{P}(\tau < t);\ H(t) = \sum_{k=0}^{\infty} F^{(*k)}(t);$$
$$h(t, y) = \mathbf{P}'\left(X_0^{2\alpha}(t) = y,\ \tau > t\right).$$

Since

$$p(t;\ 2\alpha,\ y) = h(t,\ y) + \int_0^t p(t-u;\ 2\alpha,\ y)\, F(du),$$

then [12, p. 425]

$$p(t;\ 2\alpha,\ y) = \int_0^t h(t-u,\ y)\, H(du). \tag{68}$$

Let

$$\widetilde{p}(y) = 1/a \cdot \int_0^\infty h(t,\ y)\, dt,$$

where $a = M\tau < \infty$; changing the order of integration, it is possible to show that $\tilde{p}(y)$ is a density. By Lemma 17 and Theorem 3 (see [13, p. 323])

$$H(t) = t/a + M\tau^2/(2a^2) + R(t)\ ,$$

where the function $R(t)$ satisfies the condition $\operatorname*{var}_{[t,\infty]} R \leqslant Ce^{-ct}$ for some $0 < c \leq C < \infty$. Hence from (68) we obtain

$$|\,p(t;\ 2\alpha,\ y) - \widetilde{p}(y)\,| \leqslant 1/a \cdot \int_t^\infty h(u,\ y)\, du + \left|\int_0^t h(t-u,\ y)\, R(du)\right| \leqslant$$
$$\leqslant 1/a \cdot \int_t^\infty h(u,\ y)\, du + \sup_{t/2 < u < \infty} h(u,\ y) \operatorname*{var}_{[0,\infty)} R + \sup_{0 < u < \infty} h(u,\ y) \operatorname*{var}_{[t/2,\infty)} R. \tag{69}$$

For any set $S \subseteq [0,\infty)$ we consider two conditions:

$$1_S:\ \sup_{t \geqslant 0} \sup_{y \in S} p(t;\ 2\alpha,\ y) < \infty;$$
$$2_S:\ \lim_{t\to\infty} \sup_{y \in S} |\,p(t;\ 2\alpha,\ y) - \widetilde{p}(y)\,| = 0.$$

it follows from (69) that the conditions

$$3_S\colon \lim_{t\to\infty}\sup_{y\in S}\int_t^\infty h(u,y)\,du = 0;$$

$$4_S\colon \lim_{t\to\infty}\sup_{y\in S} h(t,y) = 0;$$

$$5_S\colon \sup_{0\leqslant t<\infty}\sup_{y\in S} h(t,y) < \infty$$

are sufficient for 1_S and 2_S. In turn, conditions 3_S-5_S to follow from 6_S: we can find a function $\Pi(t)$ such that

$$\lim_{t\to\infty}\Pi(t)=0,\quad \sup_{0\leqslant t<\infty}\Pi(t)<\infty,\ \int_0^\infty \Pi(t)\,dt<\infty,$$

and such that for $0\le t<\infty$

$$\sup_{y\in S} h(t,y) < \Pi(t) .$$

We prove now that for $S=[0,2\alpha-\varepsilon]\cup[2\alpha+\varepsilon,A]$ ($\varepsilon>0$, $A<\infty$ are arbitrary) condition 6_S is satisfied. From $t\ge 1$ we have

$$h(t,y)\equiv \mathbf{P}'\left(X_0^{2\alpha}(t)=y;\ \tau>t\right)\leqslant$$

$$\leqslant\int_0^\infty \mathbf{P}\left(X_0^{2\alpha}(t-1)\in dx,\ \tau>t-1\right)\cdot\mathbf{P}'\left(X_{t-1}^x(t)=y\right)\leqslant$$

$$\leqslant \sup_{\substack{0<x<\infty\\ y\in S}}\mathbf{P}'\left(X_{t-1}^x(t)=y\right)\cdot\mathbf{P}(\tau>t-1).$$

Since, obviously,

$$\sup_{x>0,\,y\in S}\mathbf{P}'\left(X_{t-1}^x(t)=y\right)<\infty,$$

then, by Lemma 17 for $t\ge 1$ $h(t,y)\le Ce^{-ct}$. For $0\le t\le 1$ we have

$$h(t,y)\equiv\mathbf{P}'\left(X_0^{2\alpha}(t)=y,\ \tau>t\right)\leqslant\mathbf{P}'\left(X_0^{2\alpha}(t)=y\right)\leqslant C\cdot\mathbf{P}'\left(\omega_0^{2\alpha}(t)=y\right),$$

hence for any $\varepsilon>0$ we can find $C_1<\infty$ such that for $|y-2\alpha|>\varepsilon$, $0\le t\le 1$ we have $h(t,y)\le C_1$.

Thus, we have proved that condition 6_S (therefore, conditions 1_S and 2_S as well) is satisfied for the set $S=[0,2\alpha-\varepsilon]\cup[2\alpha+\varepsilon,A]$ where $\varepsilon>0$, $A<\infty$ are arbitrary.

For any $x>0$ we introduce the functional

$$\eta_x = \inf\{t>0 : X_0^x(t) = 2\alpha\}.$$

Also, let

$$v(t;x,y) = \mathbf{P}'\left(X_0^x(t)=y;\ \eta_x>t\right),\quad F(t,x)=\mathbf{P}(\eta_x<t).$$

It is easily seen that

$$p(t;x,y) = v(t;x,y) + \int_0^t p(t-u;2\alpha,y)\,F(du,x);$$

hence

$$\begin{aligned}|p(t;x,y)-\tilde{p}(y)| &\leqslant v(t;x,y) + \tilde{p}(y)(1+F(t,x)) + \\ &+ \int_0^{t/2} |p(t-u;2\alpha,y)-\tilde{p}(y)|\,F(du,x) + \\ &+ \left[\tilde{p}(y) + \sup_{0\leqslant u<\infty} p(u;2\alpha,y)\right](1-F(t/2,x)),\end{aligned}$$

implying that in order that $|\tilde{p}(y)-p(t;x,y)|\to 0$ uniformly in $x,y\in S$, as $t\to\infty$, it is sufficient that conditions 1_S, 2_S, 7_S, 8_S be satisfied, where

$$7_S\colon \lim_{t\to\infty}\sup_{x,y\in S} v(t;x,y)=0;$$

$$8_S\colon \lim_{t\to\infty}\sup_{x\in S}(1-F(t,x))=0.$$

Since conditions 1_S and 2_S have been verified, we turn to conditions 7_S and 8_S. It is easily seen that

$$\begin{aligned}&v(t;x,y)\equiv \mathbf{P}'\left(X_0^x(t)=y,\ \eta_x>t\right)\leqslant \\ &\leqslant \int_{x'=0}^{\infty} \mathbf{P}\left(X_0^x(t-1)\in dx',\ \eta_x>t-1\right)\mathbf{P}'\left(X_{t-1}^{x'}(t)=y\right)\leqslant C\,\mathbf{P}(\eta_x>t-1).\end{aligned}$$

The last inequality obtains because

$$\sup_{\substack{0<x'<\infty \\ y\in S}} \mathbf{P}'\left(X_{t-0}^{x'}(t)=y\right)\leqslant C$$

for some $C<\infty$. Therefore, 7_S follows from 8_S; to prove 8_S, it is sufficient to repeat the arguments of the proof of Lemma 17, which we shall do later on.

Thus, we have proved that as $t\to\infty$

$$|p(t;\, x,\, y)-\tilde{p}(y)|\to 0 \tag{70}$$

uniformly in $x, y\in S=[0,2\alpha-\varepsilon]\cup[2\alpha+\varepsilon, A]$, where $\varepsilon>0$ is arbitrary. Since we can carry over this construction to the case $\alpha'\neq\alpha$, for $\delta=\alpha'-\alpha$ sufficiently small, we have proved thereby that (70) holds uniformly in $x, y\leq A$.

Now we make sure that $p(y)=\tilde{p}(y)$. It follows from (70) that for any $0<a\leq A<\infty$

$$\left|\int_a^A p(x)\,p(t;\, x,\, y)\,dx-\int_a^A p(x)\,dx\,\tilde{p}(y)\right|\to 0$$

as $t\to\infty$. From the last convergence and from (67) it follows that for all $y>0$ $p(y)\geq\tilde{p}(y)$ is satisfied. Since both functions are densities, they coincide.

It remains to prove the following.

<u>Proof of Lemma 17</u>. Let $\alpha=(1,2\cdot 2^{1/3}/\lambda_0)^{1/2}$, where α is such that $2\alpha<\lambda<3\alpha$, where $\lambda\leq 2^{-1/3}\lambda_0$ (see (20)). Furthermore, let

$$\mathbf{P}\left(\sup_{0\leqslant u\leqslant t}\omega_0^{\beta}(u)\leqslant\alpha,\ \inf_{0\leqslant u\leqslant t}\omega_0^{\beta}(u)>-\alpha\right)\leqslant Ce^{-ct}, \tag{71}$$

where $0\leq\beta\leq(3\alpha+\lambda)/2-2\alpha$ [11], is satisfied for some c, C such that $\lambda<c\leq C<\infty$.

The assertion of Lemma 17 is equivalent to the fact that for some $\delta>0$

$$Me^{\delta\tau} < \infty \tag{72}$$

is satisfied.

For $x>0$, $y>0$, the sets $S\subseteq R^1$ we introduce the random variable

$$\tau(x, y; S) = \inf\{t>0 : X_0^x(t) = y;\ X_0^x(u) \in S \text{ for } 0\leqslant u\leqslant t\},$$

which, as expected, is defined only on those trajectories of the process $X_0^x(u)$ which do not leave the set S until its entry into y. The distribution functions of these functionals can be degenerate and the Laplace transforms can be less than 1 at zero. Let

$$\tau_1 = \tau(2\alpha, 3\alpha; (\alpha, 3\alpha)), \qquad \tau_2 = \tau(2\alpha, \alpha; (\alpha, 3\alpha));$$
$$\tau_3 = \tau(\alpha, 2\alpha; (0, 2\alpha)), \qquad \tau_4 = \tau(3\alpha, 3\alpha+\lambda/2; (3\alpha+\lambda/2, \infty));$$
$$\tau_5 = \tau(3\alpha+\lambda/2, 3\alpha; (2\alpha, 3\alpha)), \quad \tau_6 = \tau(3\alpha+\lambda/2, 2\alpha; (2\alpha, 3\alpha)).$$

It is obvious that

$$Me^{\delta\tau} = Me^{\delta\tau_2}\cdot Me^{\delta\tau_3} + Me^{\delta\tau_1}\cdot \sum_{k=1}^{\infty}\left(Me^{\delta\tau_4}\cdot Me^{\delta\tau_5}\right)^k Me^{\delta\tau_6}. \tag{73}$$

The first summand corresponds to those trajectories of the process $X_0^{2\alpha}(u)$, which have not risen to the level 3α before crossing the level α; the kth term of the sum of the second summand in (73) corresponds to the fact that the process $X_0^{2\alpha}(t)$ reaches the point 3α passing by the point α, then moves from the point 3α to the point 3α stopping at the point $(3\alpha+\lambda)/2$ k times staying above the level 2α and, finally, enters the point 2α for the first time. Since

$$\mathrm{M}e^{\delta\tau_4}\cdot \mathrm{M}e^{\delta\tau_5}\Big|_{\delta=0} = \mathbf{P}(\tau_4<\infty)\cdot\mathbf{P}(\tau_5<\infty)<1,$$

(73) becomes

$$\mathrm{M}e^{\delta\tau}=\mathrm{M}e^{\delta\tau_2}\cdot\mathrm{M}e^{\delta\tau_3}+\frac{\mathrm{M}e^{\delta\tau_1}\cdot\mathrm{M}e^{\delta\tau_4}\cdot\mathrm{M}e^{\delta\tau_5}\cdot\mathrm{M}e^{\delta\tau_6}}{1-\mathrm{M}e^{\delta\tau_4}\cdot\mathrm{M}e^{\delta\tau_5}}.$$

The Lemma will be proved if we prove that (72) holds for τ_i, $i=1,\dots,6$. The functionals τ_i, $i=1,\dots,6$ are such that before the time $t=\tau_i$ the span of the trajectory of the process $X_0^x(t)$, i.e., the difference between the maximum and minimum on an interval $[0,\tau_i]$ is not greater than 2α; hence by (71)

$$\mathbf{P}(\tau_i>t)\leqslant Ce^{\lambda t}\mathbf{P}\Big(\sup_{0\leqslant u\leqslant t}\omega(u)-\inf_{0\leqslant u\leqslant t}\omega(u)\leqslant 2\alpha\Big)\leqslant Ce^{-\varepsilon t}$$

for some $\varepsilon>0$. Finally, the functional τ_4 is such that $X_0^{3\alpha}(u)\geq(3\alpha+\lambda)/2$ is satisfied for $0\leq t\leq\tau_4$; hence

$$P(\tau_4>t)\leq Ce^{-\varepsilon t},$$

where $\varepsilon>0$. //

REFERENCES

[1] A.A. Borovkov. "Boundary-value Problems for Random Walks and Large Deviations in Function Spaces." *Theory Prob. Applications*, 4 (1967): 575-595.

[2] A.A. Mogul'skii. "Large Deviations for Trajectories of Multi-dimensional Random Walks." *Theory Prob. Applications*, 2 (1976): 300-315.

[3] A.A. Borovkov. "The Analysis of Large Deviations in Boundary-value Problems With Arbitrary Boundaries, I." *Sibirskij matemat. zh.,* 5 (1964): 253-289. (In Russian.)

[4] G.D. Yakovleva. *Tables of Airy Functions and Their Derivatives*. Moscow: Nauka Publishers, 1969. (In Russian.)

[5] G. Kramer. "On a New Limit Theorem in Probability Theory." *Uspekhi matem. nauk,* 10 (1944): 166-184. (In Russian.)

[6] A.D. Ventsel'. "Rough Limit Theorems on Large Deviations for Markov Stochastic Processes, I, II." *Theory Prob. Applications,* 2 (1976): 227-242; 3 (1976): 499-512.

[7] Jurgen Gartner. "On Large Deviations From the Invariant Measure." *Theory Prob. Applications,* 1 (1977): 24-39.

[8] A.A. Borovkov and A.A. Mogul'skij. "On Probabilities of Large Deviations in Topological Spaces, I." *Sibirskij matemat. zh.,* 19, 4 (1978): 697-709. (In Russian.)

[9] M.D. Donsker, and S.R.S. Varadhan. "Asymptotical Evaluation of Certain Markov Process Expectations for Large Time. I." *Comm. Pure Appl. Math.,* 28, 6 (1975): 1-47.

[10] E.B. Dynkin. *Markov Processes*. Berlin-Heidelberg-New York: Springer-Verlag, 1965.

[11] A.A. Mogul'skii. "Small Deviations in a Space of Trajectories." *Theory Prob. Applications,* 4 (1974): 726-736.

[12] W. Feller. *An Introduction to Probability Theory and Its Application*. Vol. 2. 3rd ed. New York: John Wiley, 1968.

[13] A.A. Borovkov. *Random Processes in Queueing Theory*. Moscow: Nauka Publishers, 1972. (In Russian.)

THE RATE OF CONVERGENCE IN BOUNDARY VALUE PROBLEMS FOR DOMAINS WITH DISCONTINUITIES

I.F. Pinelis

1. INTRODUCTION AND FORMULATIONS OF THE BASIC RESULTS

The rate of convergence in boundary value problems has been the subject of investigation by many authors (see [1]). S.V. Nagaev [1] suggested a method for determining the rate of convergence in boundary value problems, to establish the rate of convergence of the true order $1/\sqrt{n}$ for domains of a general type.

Let $\xi_1, \xi_2, \ldots$ be independent random variables with a common distribution function F, $M\xi_1=0$, $M\xi_1^2=1$, $c_3 \equiv M|\xi_1|^3<\infty$.

Consider a random walk in R^2, defined such that if $(t,x)\in R^2$ is the location of the walking point at an instant of time m, then at the next instant of time $m+1$ this point is on a ray $\{t+1/n\}(-\infty,y)$ with probability $F(y-x)\sqrt{n}$. Let

$$\xi_n(t,x,m) = (\xi_n^{(1)}(t,x,m),\ \xi_n^{(2)}(t,x,m))$$

be the location of a walking point at an instant of time m provided walking began at an instant of time 0 at a point (t,x).

Let $[x]$ be the integer part of x, $]x[=[-x]$. Let $\xi(t)$ be the standard (continuous) Wiener process on $[0,\infty)$. For an arbitrary $D\subseteq \mathbf{R}^2$ let

$$D(t) = \{x:(t,x)\in D\} \quad (-\infty<t<\infty) .$$

Let

$$P_n(t, x, D, b) = \mathbf{P}(\xi_n(t, x, m) \in D(0 \leqslant m \leqslant [n(b-t)]));$$
$$P(t, x, D, b) = \mathbf{P}(\xi(u) + x \in D(t+u)(0 \leqslant u \leqslant b-t)),$$

assuming these probabilities are definable. In other words, $P_n(t,x,D,b)(P(t,x,D,b))$ is the probability of the event that a trajectory of a random walk (of a Wiener process) starting at time zero at the point (t,x), will lie in D in the interval $[0,b-t]$.

It follows from [1] that if

$$D = \{(t,x):\ 0 \le t \le 1,\ g_2(t) < x < g_1(t)\}\ ,$$

where g_1, g_2 satisfy the Lipschitz condition with constant K, then

$$(P_n - P)(0,0,D,1) = O((K+1)\ c_3^2/\sqrt{n})\ . \tag{1}$$

Here and in what follows the constant in $O(\cdot)$ is assumed to be absolute; as usual, the values of the difference between the functions equal the difference of the values of the functions.

A.I. Sahanenko [2] proved, using the method proposed in [1], that c_3^2 in (1) can be replaced by c_3. In what follows, we shall make use of the remarks given in [2] concerning the substitution of c_3 for c_3^2, 1 for c_3, without further comment.

We note that the estimate $O((K+1)c_3/\sqrt{n})$ is non-improvable in the sense that it is not possible to replace the right-hand side of (1) by $O(f(K)c_3/\sqrt{n})$ if $f(K)/K \to 0$ as $K \to \infty$. Indeed, arguing in the same way as in [2], we derive from the validity of $O(f(K)c_3/\sqrt{n})$ that

$$\mathbf{P}\Big(\sup_{0<m<Nn}\big(\xi_n^{(2)}(0,0,m)-m/n\big)\geqslant x\Big)=$$
$$=\mathbf{P}\Big(\sup_{0\leqslant t\leqslant N}(\xi(t)-t)\geqslant x\Big)+O\Big(f(\sqrt{N})\,c_3/\sqrt{Nn}\Big),$$

and, therefore,

$$\mathbf{P}\Big(\sup_{0\leqslant m<\infty}\big(\xi_n^{(2)}(0,0,m)-m/n\big)\geqslant x\Big)=\mathbf{P}\Big(\sup_{0\leqslant t<\infty}(\xi(t)-t)\geqslant x\Big),$$

which, generally speaking, is wrong (see, for instance, [3]).

We proceed to formulate the main results we have obtained. We say that $D\subseteq R^2$ is a domain with no more than s discontinuities on $[a,b]$ $(s\geq 2)$ (at points $a=t_0<t_1<\cdots<t_{s-1}=b$), if for any $t\in[a,b]$ (the section $D(t)$ is the union of a finite number $m(t)$ of non-intersecting non-empty (open, closed or half-open) intervals $\delta_i(t)$ with the end-points $g_2^{(i)}(t)<g_1^{(i)}(t)$ $(i=1,\ldots,m(t))$, with $m(t)\equiv m_j$ for $t\in\Delta_j=(t_{j-1},t_j)$ and the functions $g_1^{(i)}(t)$, $g_2^{(i)}(t)$ satisfy the Lipschitz condition in t with the constant K_j in the interval Δ_j $(i=1,\ldots,m_j,\ j=1,\ldots,s-1)$.

For example, all the domains obtained from domains of the form

$$D=\{(t,x):\ g_2(t)<x<g_1(t)\ (c\leq t\leq d)\}\ , \qquad (1')$$

where g_1, g_2 satisfy the Lipschitz condition, through the subtraction of s domains of the same form, belong to a class of domains with no more than $2s+2$ discontinuities.

Domains of the form (1') with the piecewise-Lipschitz boundaries g_1, g_2 also belong to the class considered; for them $m(t)\equiv 1$ for $D(t)\neq\emptyset$.

Separately, we consider domains of the form

$$D = \{(t,x):\ g_2(t)<x<g_1(t)\ (c\le t<d),\ x<y\ (t=d)\}\ ,$$

where g_1, g_2 are Lipschitz.

The following theorem is the central result of our paper.

THEOREM 1. Let

$$D_y = \{(t,x):\ g_2(t)<x<g_1(t)\ (0\le t<1),\ x<y\ (t=1)\}\ ,$$

where $g_2(t)<g_1(t)$, g_1, g_2 satisfy the Lipschitz condition on $[0,1]$ with the constant K. then

$$(P_n-P)(0,x,D_y,1) = O((K+1)c_3/\sqrt{n})\ .$$

Let $S_n=\sum_{i=1}^{n}\xi_i$ $(n=1,2,\dots)$.

COROLLARY 1.

$$\mathbf{P}\left(\max_{1\le i\le n} S_i<x,\ S_n<y\right)= \\ =\mathbf{P}\left(\max_{0\le t\le 1}\xi(t)<x/\sqrt{n},\ \xi(1)<y/\sqrt{n}\right)+O\left(c_3/\sqrt{n}\right).$$

We call the domain D having no more than s discontinuities on $[a,b]$, regular (on $[a,b]$), if

$D(t_j)\subseteq D(t_j-0)\cap D(t_j+0)$ $(1\le j\le s-2)$, $D(a)\subseteq D(a+0)$, $D(b)\subseteq D(b-0)$,

where

$$D(t\pm 0)=\bigcup_{i=1}^{m(t\pm 0)}\delta_i(t\pm 0);$$

$$\delta_i(t\pm 0)=\left(g_2^{(i)}(t\pm 0),\ g_1^{(i)}(t\pm 0)\right);\ (i=1,\dots,m(t\pm 0)).$$

Also, we introduce the "regular" probability $P_n^r(a,x,D,b)$ for the regular domain D in the following manner. For $(t,x)\in D$ we set $i(t,x)=i$, if $x\in\delta_i(t)$ $(i=1,\dots,m(t))$.

We assume that nt_j $(j=0,\ldots,s-1)$ are integers. Then, let $P_n^r(a,x,D,b)$ be the probability of the event that $\xi_n(a,x,m)\in D(0\le m\le n(b-a))$, with

$$i\left(\xi_n^{(1)}(m),\ \xi_n^{(2)}(m)\right)=i\left(\xi_n^{(1)}(n(t_{j-1}-a))+\right.$$
$$\left.+0,\ \xi_n^{(2)}(n(t_{j-1}-a))\right)=i\left(\xi_n^{(1)}(n(t_j-a))-0,\ \xi_n^{(2)}(n(t_j-a))\right)$$

for $n(t_{j-1}-a)<m<n(t_j-a)$ $(j=1,\ldots,s-1)$; here $\xi_n^{(\alpha)}(m)=\xi_n^{(\alpha)}(a,x,m)$ $(\alpha=1,2)$. In other words, $P_n^r(a,x,D,b)$ is the probability of the event that the trajectory $\xi_n(a,x,m)$ sojourns in D without intersecting the boundaries $g_i^{(\alpha)}(t)$.

THEOREM 2. Let D be a regular domain having no more than s discontinuities on $[a,b]$, nt_j $(j=0,\ldots,s-1)$ being integers. Then

$$|(P_n^r-P)(a,x,D,b)|\leqslant\Psi_s c_3/\sqrt{n},$$

where

$$\Psi_s=L\sum_{j=1}^{s-1}\lambda_j\prod_{i=j}^{s-1}\mu_i;$$

$$\lambda_j=K_j+1/\sqrt{|\Delta_j|};\ |\Delta_j|=t_j-t_{j-1};\ \mu_j=2m(t_j)\quad(j=1,\ldots,s-1);$$

L being an absolute constant.

COROLLARY 2. Let D be a domain of the form (1') with piecewise-Lipschitz boundaries; more precisely, let D be a (not necessarily regular) domain having no more than s discontinuities on $[a,b]$, $m(t)\equiv 1$ $(t\in[a,b])$ and nt_j $(j=0,\ldots,s-1)$ being integers. Then

$$|(P_n-P)(a,x,D,b)|\leqslant L\sum_{j=1}^{s-1}2^{s-j}\left(K_j+1/\sqrt{|\Delta_j|}\right)c_3/\sqrt{n}.$$

If, however, for example, the discontinuity points t_j

are irrational and, therefore, nt_j cannot be integers, it is more appropriate to use random processes with continuous time, constructed on the basis of the random walk $\xi_n(t,x,m)$.

Thus, let $\hat{s}_n(u)=\hat{s}_{n,a,x}(u)$ be a random polygon with vertices $\xi_n(t,x,m)$ $(m=0,1,2,\ldots)$, $\bar{s}_n(u)=\bar{s}_{n,a,x}(u)=$ $=\xi_n^{(2)}(a,x,[n(u-a)])$ is a "step" process. For the regular domain D having no more than s discontinuities on $[a,b]$, let $\bar{P}_n^r(a,x,D,b)$ be the probability of the event that $\bar{s}_n(u)\in D(u)$ $(u\in[a,b])$, with

$$i(u,\ \bar{s}_n(u))=i(t_{j-1}+0,\ \bar{s}_n(t_{j-1}))=i(t_j-0,\ \bar{s}_n(t_j))$$

for $t_{j-1}<u<t_j$ $(j=1,\ldots,s-1)$.

For $D\subseteq R^2$ we put

$$\bar{P}_n(a,\ x,\ D,\ b)=\mathbf{P}(\bar{s}_n(u)\in D(u)(u\in[a,\ b]));$$
$$\hat{P}_n(a,\ x,\ D,\ b)=\mathbf{P}(\hat{s}_n(u)\in D(u)(u\in[a,\ b])).$$

REMARK 1. We have introduced regular domains and "regular" probabilities because of the discontinuity of $\bar{s}_n(u)$ and $\xi_n(t,x,m)$; this enables us to avoid too cumbersome formulations.

To each domain D having no more than s discontinuities on $[a,b]$, we can set into correspondence a regular domain D^r, replacing the sections $D(t_j)$ by $D(t_j)\cap D(t_j-0)\cap D(t_j+0)$ $(j=1,\ldots,s-2)$, $D(a)$ by $D(a)\cap D(a+0)$, and $D(b)$ by $D(b)\cap D(b-0)$. Obviously,

$$\hat{P}_n(a,\ x,\ D^r,\ b)=\hat{P}_n(a,\ x,\ D,\ b);$$
$$P(a,\ x,\ D^r,\ b)=P(a,\ x,\ D,\ b).$$

Using Lemma 11 (see below), we can show that

$$\bar{P}_n(a, x, D^r, b) = \bar{P}_n(a, x, D, b) + O\left(\frac{c_3}{\sqrt{n}}\left(\sum_{j=1}^{s-1}\frac{m_j}{\sqrt{t_j - a}}\right)\right);$$

$$P_n(a, x, D^r, b) = P_n(a, x, D, b) + O\left(\frac{c_3}{\sqrt{n}}\sum_{j=1}^{s-2}\frac{m_j + m_{j+1}}{\sqrt{t_j - a}}\right).$$

While considering the "irregular" probabilities $P_n(a,x,D,b)$, $\bar{P}_n(a,x,D,b)$ instead of $P_n^r(a,x,D,b)$, $\bar{P}_n^r(a,x,D,b)$, we would have encountered a term of the form

$$O\left(\frac{c_3}{\sqrt{n}}\sum_{j=1}^{s-1}\frac{|\Delta_j|}{(d_j - 2K_j/n)^3}\right),$$

where

$$d_j = \inf\{g_2^{(i+1)}(t) - g_1^{(i)}(t) : i = 1, \ldots, m_j - 1, t \in \Delta_j\} \times \times (j = 1, \ldots, s-1),\ n > 2 \max_{1 \le j \le s-1} K_j/d_j.$$

THEOREM 1'. Let

$$D = \{(t, x) : g_2(t) < x < g_1(t)(c \le t < d),\ x < y(t = d)\},$$

where $g_2(t) < g_1(t)$, g_1, g_2 satisfy a Lipschitz condition on $[c,d]$ with a constant K. Then

$$(\hat{P}_n - P)(c, x, D, d) = O((K + 1/\sqrt{d-c})c_3/\sqrt{n}),$$

and the same estimate holds for $\bar{P}_n(c,x,D,d)$.

THEOREM 2. Let D be a regular domain having no more than s discontinuities on $[a,b]$. Then

$$|(\hat{P}_n - P)(a, x, D, b)| \le \hat{\Psi}_s c_3/\sqrt{n},$$

where

$$\hat{\Psi}_s = L\sum_{j=1}^{s-1}\hat{\lambda}_j\prod_{i=j}^{s-1}\mu_i;$$

$$\hat{\lambda_j} = \lambda_j + \Delta\hat{\lambda}_j;$$

$$\Delta\hat{\lambda}_j = \begin{cases} 1/\sqrt{t_{j-1}-a} + K_j/c_3\sqrt{n} & (nt_{j-1} \text{ is not an integer}),\\ & (nt_{j-1} \text{ is an integer})\end{cases}$$

(obviously, $\Delta\hat{\lambda}_j \le 1/\sqrt{t_{j-1}-a}+K_j$ for $j=1,\dots,s-1$; $\hat{\lambda}_1=\lambda_1$); L is an absolute constant.

Similarly

$$|(\bar{P}_n^r - P)(a, x, D, b)| \leq \hat{\Psi}_s c_3/\sqrt{n}.$$

<u>COROLLARY 2'</u>. Let D be a domain of the form (1') with piecewise-Lipschitz boundaries, more precisely, D is a (not necessarily regular) domain having no more than s discontinuities on $[a,b]$ with $m(t)\equiv 1$ $(t\in[a,b])$. Then

$$(\bar{P}_n - P)(a, x, D, b) = O\left(\sum_{j=1}^{s-1} 2^{s-j}(K_j + 1/\sqrt{|\Delta_j|})\,c_3/\sqrt{n}\right),$$

and the same estimate holds for $\hat{P}_n(a,x,D,b)$.

<u>REMARK 2</u>. If, for example, $g_2(t)\equiv-\infty$, then in Corollaries 2, 2' we have the estimate

$$O\left(\sum_{j=1}^{s-1}(K_j + 1/\sqrt{|\Delta_j|})\,c_3/\sqrt{n}\right).$$

This follows from the proof of Theorems 2, 2'.

We prove Theorem 1 in our longest section, Section 2. The most difficult part in proving Theorem 1 consists in estimating the stability of solution of the boundary value problem with respect to time, i.e., estimating the difference of the form $\Delta_n(h)=P_n(a+h,x,D,b) - P_n(a,x,D,b)$. In Section 2 we shall obtain the estimate

$\Delta_n(h)=O((K^2+1/(b-a))h+(K+1/\sqrt{(b-a)})(c_3/\sqrt{n}+h|\ln h|/\sqrt{b-a}))$.
This estimate, although not very precise, is still sufficient for proving Theorem 1. In turn, it is possible, using Theorem 1, to obtain a more precise estimate

$$\Delta_n(h)=O((K^2+1/(b-a))h+(K+1/\sqrt{b-a})c_3/\sqrt{n}). \qquad (1'')$$

This is the generalization of Lemma 9 from [1].

The derivation of the estimate (1") and of a similar estimate for a Wiener process is handled in Section 3. We note that the estimate for a Wiener process follows from (1") due to the invariance principle. We shall obtain, however, first the estimate of stability with respect to time for a Wiener process, and then, from this estimate we shall derive (1"), using Theorem 1. In Section 4 we shall prove the remaining results from those formulated above.

The author expresses his deep gratitude to S.V. Nagaev for the statements of the problems and his useful comments, and also to A.A. Borovkov for his interest in this paper and his suggestions which have substantially improved it.

2. THE PROOF OF THEOREM 1

The proof of Theorem 1 is based on the method used in [1]; we shall use, basically, the notation adopted in [1]. For the convenience of the reader we repeat a part of the notation here. Let

$$D = \{(t,x):\ c_1 \le t \le c_2,\ d_2(t) < x < d_1(t)\} ,$$

where d_1, d_2 are some functions on $[c_1,c_2]$;

$$\begin{gathered}\tau_n(t,\ x,\ D)=\inf\{m:\xi_n(t,\ x,\ m)\notin D,\ m>0\};\\ P_n(t,\ x,\ E_1,\ E_2,\ D)=\mathbf{P}(\xi_n(t,\ x,\ \tau_n(t,\ x,\ D))\in E_1\times E_2);\\ P_n(t,\ x,\ E_1,\ D)=P_n(t,\ x,\ E_1,\ \mathbf{R}^1,\ D),\end{gathered}\tag{2}$$

where E_1, E_2 are Borel sets on the line;

$P_n(t,x,D)=P_n(t,x,(c_2,\infty),D)$ for $(t,x)\in D$, 0 otherwise.

We assume (this causes no loss of generality) that F is continuous, x=0. Assume for the moment that

$$g_1(a)=K/n > y > g_2(1)+K/n \ . \tag{2'}$$

First we prove some auxiliary results (Lemmas 1-6). The basic technical tool for proving the Theorem is the following.

LEMMA 1. Let $0<b\leq 1$ and let there be given a nondecreasing function M_1 on $[0,b]$, with

$$M_1(t)=O\left(p_1\frac{h}{b+h-t}+p_2\frac{c_3\sqrt{K+1}}{\sqrt{n(b-t)}}+p_3\frac{(K+1)c_3}{\sqrt{nb}}+\right.$$
$$\left.+p_4\frac{c_3}{\sqrt{n(b-t)}}\ln\frac{b}{b-t}\right) \tag{3}$$

for any $t\in[0,b)$ and some $h\in(0,b)$. Next, let M_2 be a function given on $[0,b]$ with a variation $\mathrm{Var}\ M_2=O(1)$, with $M_2(b)=0$ and

$$M_2(t) = O((K+1)((b-t)/b+qc_3/\sqrt{nb})) \ . \tag{4}$$

In (3) and (4) $p_i\geq 0$ $(i=1,2,3,4)$, $q\geq 0$.

Let at least one of the following two conditions be satisfied: $\mathrm{Var}\ M_1=O(1)$ or in the condition (4) $q=0$. If $q+\sum_{i=2}^{4}p_i=O(1)$, then

$$\int_0^b M_1(t)\,dM_2(t)=O\big((K+1)\,(p_1h\,|\ln h|/b+c_3/\sqrt{nb})\big).$$

Proof of Lemma 1. It suffices to prove the Lemma for $b=1$, since otherwise it is possible to substitute bt', bh' for t, h, n' for nb, $M_i'(t)=M_i(t/b)$ for $M_i(t)$ $(i=1,2)$. Therefore, we take $b=1$. Then

$$\int_0^1 M_1(t)\,dM_2(t) = -M_1(0)M_2(0) - \int_0^1 M_2(t)\,dM_1(t), \tag{5}$$

since $M_2(1)=0$. Taking into account that $\operatorname{Var} M_2=O(1)$, we obtain

$$M_1(0)M_2(0) = O(M_1(0)) , \tag{6}$$

and by (3)

$$M_1(0) = O(p_1h+(K+1)c_3/\sqrt{n}) . \tag{7}$$

Next,

$$\int_0^1 M_2(t)\,dM_1(t) = \int_0^{K/(K+1)} M_2(t)\,dM_1(t) + \int_{K/(K+1)}^1 M_2(t)\,dM_1(t); \tag{8}$$

$$\int_0^{K/(K+1)} M_2(t)\,dM_1(t) = O\left(\int_0^{K/(K+1)} dM_1(t)\right), \tag{9}$$

since $M_2(1)=0$, $\operatorname{Var} M_2=O(1)$,

$$\int_0^{K/(K+1)} dM_1(t) = M_1(K/(K+1)) - M_1(0); \tag{10}$$

$$M_1(K/(K+1)) = O((K+1)(p_1h+c_3/\sqrt{n})). \tag{11}$$

Taking into account (4), we find

$$\int_{K/(K+1)}^1 M_2(t)\,dM_1(t) = O\left(\int_{K/(K+1)}^1 (K+1)\left(1-t+qc_3/\sqrt{n}\right)dM_1(t)\right). \tag{12}$$

Since either $q=0$ or $\operatorname{Var} M_1=O(1)$, then

$$\int\limits_{K/(K+1)}^{1} (K+1)\left(1-t+qc_3/\sqrt{n}\right) dM_1(t) =$$
$$= \int\limits_{K/(K+1)}^{1} (K+1)(1-t)\, dM_1(t) + O\left((K+1)\, c_3/\sqrt{n}\right). \quad (13)$$

We now integrate by parts:

$$\int\limits_{K/(K+1)}^{1} (K+1)(1-t)\, dM_1(t) = -M_1(K/(K+1) + \int\limits_{K/(K+1)}^{1} (K+1)\, M_1(t)\, dt. \quad (14)$$

By (3)

$$\int\limits_{K/(K+1)}^{1} (K+1)\, M_1(t)\, dt = O\left((K+1)\left(p_1 h\, |\ln h| + c_3/\sqrt{n}\right)\right). \quad (15)$$

Combining (5)-(15), we obtain

$$\int\limits_{0}^{1} M_1(t)\, dM_2(t) = O\left((K+1)\left(p_1 h\, |\ln h| + c_3/\sqrt{n}\right)\right),$$

which was to be proved.

LEMMA 2. Let

$$D = \{(t, x) : 0 \leqslant t \leqslant b,\ d_1(t) > x > d_2(t)\};\ d_1(t) > d_2(t);$$
$$|d_i(t+h) - d_i(t)| \leqslant Kh (0 \leqslant t \leqslant t+h \leqslant b);\ d_1(0) > 0 > d_2(0).$$

Then (see (2))

$$P_n(0, 0, (t, b], D) = O((K\sqrt{b}+1)((b-t)/b + c_3/\sqrt{nb})).$$

Proof of Lemma 2. For b=1 the proof follows easily if we note (2.16) from [1]. The general case can be proved in the same way as in [1, see (2.88), (2.33)].

LEMMA 3. For $m<n=O(m)$

$$\mathbf{P}(S_m < x,\ S_n < y) = \mathbf{P}(\xi(m/n) < x/\sqrt{n},\ \xi(1) < y/\sqrt{n}) + O(c_3/\sqrt{n}).$$

Proof of Lemma 3. Let F_k be the distribution function of S_k $(k=1,2,\ldots)$. It is easily seen that

$$\mathbf{P}(S_m < x, S_n < y) = \int_{-\infty}^{x} F_{n-m}(y-u)\, dF_m(u);$$

$$\mathbf{P}\left(\xi\left(\frac{m}{n}\right) < \frac{x}{\sqrt{n}},\ \xi(1) < \frac{y}{\sqrt{n}}\right) = \int_{-\infty}^{x} \Phi\left(\frac{y-u}{\sqrt{n-m}}\right) d_u\Phi\left(\frac{u}{\sqrt{m}}\right),$$

where

$$\Phi(u) = \int_{-\infty}^{u} \frac{1}{\sqrt{2\pi}} e^{-v^2/2} dv.$$

Then

$$\mathbf{P}(S_m < x,\ S_n < y) = \mathbf{P}(\xi(m/n) < x/\sqrt{n},\ \xi(1) < y/\sqrt{n}) + I_1 + I_2\,, \quad (16)$$

where

$$I_1 = \int_{-\infty}^{x} \left[F_{n-m}(y-u) - \Phi\left(\frac{y-u}{\sqrt{n-m}}\right)\right] d_u\Phi\left(\frac{u}{\sqrt{m}}\right);$$

$$I_2 = \int_{-\infty}^{x} F_{n-m}(u)\, d_u\left[F_m(u) - \Phi\left(\frac{u}{\sqrt{m}}\right)\right].$$

The following inequality due S.V. Nagaev is well known [4]:

$$|F_k(x\sqrt{k}) - \Phi(x)| = O\left(\frac{c_3}{\sqrt{k}(1+|x|)^3}\right).$$

Using this estimate, we find

$$I_1 = O\left(c_3 \int_{-\infty}^{x} \frac{1}{\sqrt{m}} e^{-u^2/2m}\left(1 + \left|\frac{y-u}{\sqrt{n-m}}\right|^3\right)^{-1} \frac{du}{\sqrt{n-m}}\right) =$$

$$= O\left(\frac{c_3}{\sqrt{m}} \int_{-\infty}^{\infty} \frac{du}{1+|u|^3}\right) = O(c_3/\sqrt{n}). \quad (17)$$

Integrating by parts, we obtain

$$I_2 = (F_m(x) - \Phi(x/\sqrt{m}))F_{n-m}(y-x) - \int_{-\infty}^{x} [F_m(u) - \Phi(u/\sqrt{m})] d_u F_{n-m}(y-u) ,$$

whence, using the Berry-Esseen estimate, we derive

$$I_2 = O(c_3/\sqrt{n}) . \tag{18}$$

The proof of Lemma 3 follows from (16)-(18).

LEMMA 4. For $m<n=O(m)$

$$\mathbf{P}(S_m < x,\ S_n \geqslant y) = P(S_m \geqslant -x,\ S_n < -y) + O(c_3/\sqrt{n}).$$

Lemma 4 follows from Lemma 4 and the symmetry of the distribution of the process $\xi(t)$.

The following lemma is, possibly, of independent interest.

LEMMA 5. Let $\tau_x = \inf\{m : S_m \geq x\}$, $\chi_x = S_{\tau_x} - x$ $(x>0)$ be respectively the time of the first exit and the magnitude of the first jump beyond the level x. Then

$$\mathbf{P}(S_n - 2\chi_x \geqslant x) = \mathbf{P}(S_n \geqslant x) + O(c_3/\sqrt{n}).$$

Proof of Lemma 5. Let $\bar{S}_n = \max_{1 \leq k \leq n} S_k$. Obviously,

$$\mathbf{P}(\bar{S}_n \geqslant x,\ S_n < x) = \sum_{k=1}^{n} \int_{-\infty}^{\infty} \mathbf{P}(\tau_x = k,\ S_k \in du)\,\mathbf{P}(S_{n-k} < x-u).$$

From Lemma 4 (for $y=-\infty$) we obtain that

$$\mathbf{P}(S_{n-k} < x-u) = \mathbf{P}(S_{n-k} \geqslant u-x) + O(c_3/\sqrt{n-k}),$$

Hence

$$\mathbf{P}(\bar{S}_n \geqslant x,\ S_n < x) = \sum_{k=1}^{n} \int_{-\infty}^{\infty} \mathbf{P}(\tau_x = k,\ S_k \in du)\, \mathbf{P}(S_{n-k} \geqslant u - x) +$$

$$+ R = \sum_{k=1}^{n} \int_{-\infty}^{\infty} \mathbf{P}(\tau_x = k,\ S_k \in du,\ S_n \geqslant 2u - x) + R =$$

$$= \mathbf{P}(S_n \geqslant 2\chi_x + x) + R, \quad (19)$$

where

$$R = O\left(\sum_{k=1}^{n} \mathbf{P}(\tau_x = k) \min(c_3/\sqrt{n-k},\ 1) \right) =$$

$$= O\left(\int_0^1 \min(c_3/\sqrt{n(1-t)},\ 1)\, P_n(0, 0, dt, G_x) \right); \quad (20)$$

$G_x = \{(t,u):\ 0 \leq t \leq 1,\ u < x\}$. Let

$$M_1(t) = \min(c_3/\sqrt{n(1-t)},\ 1); \quad (21)$$

$$M_2(t) = -P_n(0,\ 0,\ (1-t,\ 1],\ G_x). \quad (22)$$

By Lemma 2

$$M_2(t) = O(1-t+c_3/\sqrt{n}) .$$

Hence the conditions of Lemma 1 are satisfied with $b=1$, $K=0$, $p_1=p_3=p_4=0$, $p_2=q=1$, $\mathrm{Var}\ M_1=O(1)$. Therefore,

$$\int_0^1 M_1(t)\, dM_2(t) = O(c_3/\sqrt{n}). \quad (23)$$

From (20)-(23) we obtain $R=O(c_3/\sqrt{n})$, which together with (19) yield

$$\mathbf{P}(\bar{S}_n \geqslant x,\ S_n < x) = \mathbf{P}(S_n - 2\chi_x \geqslant x) + O(c_3/\sqrt{n}). \quad (24)$$

On the other hand,

$$\mathbf{P}(\bar{S}_n \geqslant x,\ S_n < x) = \mathbf{P}(\bar{S}_n \geqslant x) - \mathbf{P}(S_n \geqslant x). \quad (25)$$

According to the basic result of [2]:

$$\mathbf{P}(\bar{S}_n \geqslant x) = 2(1 - \Phi(x/\sqrt{n})) + O(c_3/\sqrt{n}). \quad (26)$$

The Berry-Esseen estimate plus (25), (26) yield

$$\mathbf{P}(\bar{S}_n \geqslant x,\ S_n < x) = \mathbf{P}(S_n \geqslant x) + O(c_3/\sqrt{n}).$$

The comparison of this estimate with (24) proves Lemma 5.

LEMMA 6. Let $x>0$, $0<r<p=O(r)$. Then

$$\mathbf{P}(\bar{S}_m \geqslant x,\ \ S_{m+r} < x,\ \ S_{m+p} \geqslant x) =$$
$$= \mathbf{P}(\bar{S}_m \geqslant x,\ S_{m+r} \geqslant x,\ S_{m+p} < x) + O(c_3/\sqrt{m})$$
$$\left(\bar{S}_m = \max_{1 \leqslant k \leqslant m} S_k \ \ (m = 1, 2, \ldots)\right).$$

Proof of Lemma 6. Let

$$R_1 = \mathbf{P}(\bar{S}_m \geqslant x,\ S_{m+r} < x,\ S_{m+p} \geqslant x). \tag{27}$$

We have

$$R_1 = \sum_{k=1}^{m} \int_{-\infty}^{\infty} \mathbf{P}(\tau_x = k,\ S_k \in du)\, \mathbf{P}(S_{m+r-k} < x - u,\ S_{m+p-k} \geqslant x - u). \tag{28}$$

Since $(m+p-k)/(m+r-k) \leq p/r = O(1)$ for $k \leq m$, then by Lemma 4

$$\mathbf{P}(S_{m+r+k} < x - u,\ \ S_{m+p-k} \geqslant x - u) = \mathbf{P}(S_{m+r-k} \geqslant$$
$$\geqslant u - x,\ \ S_{m+p-k} < u - x) + R_2, \tag{29}$$

where

$$R_2 = O(c_3/\sqrt{m+p-k}) = O(c_3/\sqrt{m-k}). \tag{30}$$

It follows from (28)-(30) that

$$R_1 = \sum_{k=1}^{m} \int_{-\infty}^{\infty} \mathbf{P}(\tau_x = k,\ S_k \in du)\, \mathbf{P}(S_{m+r-k} \geqslant u - x,\ S_{m+p-k} < u - x) + R_3 =$$
$$= \sum_{k=1}^{m} \int_{-\infty}^{\infty} \mathbf{P}(\tau_x = k,\ S_k \in du,\ S_{m+r} \geqslant 2u - x,\ S_{m+p} < 2u - x) + R_3 =$$
$$= \mathbf{P}(\bar{S}_m \geqslant x,\ S_{m+r} - 2\chi_x \geqslant x,\ S_{m+p} - 2\chi_x < x) + R_3, \tag{31}$$

where

$$R_3 = O\left(\sum_{k=1}^{m} \mathbf{P}(\tau_x = k) \min(c_3/\sqrt{m-k},\ 1)\right). \tag{32}$$

For m=n the sum on the right-hand side of (32) coincides with the sum on the right-hand side of (20). Hence (compare with (23))

$$R_3 = O(c_3/\sqrt{m}) . \tag{33}$$

Next,

$$\mathbf{P}(\bar{S}_m \geqslant x,\ S_{m+r} - 2\chi_x \geqslant x,\ S_{m+p} - 2\chi_x < x) =$$
$$= \mathbf{P}(\bar{S}_m \geqslant x,\ S_{m+r} \geqslant x,\ S_{m+p} < x) + O(\mathbf{P}(S_{m+r} \geqslant x) - \mathbf{P}(S_{m+r} - 2\chi_x \geqslant x)) +$$
$$+ O(\mathbf{P}(S_{m+p} \geqslant x) - \mathbf{P}(S_{m+p} - 2\chi_x \geqslant x)). \tag{34}$$

Using (27), (31), (33), (34), and Lemma 5, we complete proving Lemma 6.

Let

$$\tilde{G} = \{(t,\ x):\ 0 \leqslant t \leqslant 1,\ g_2(t) < x < g_1(t)\};$$
$$G = G_{y,n} = \{(t,\ x):\ 0 \leqslant t \leqslant 1,\ g_2(t) < x < g_{1,n}(t)\},$$

where

$$g_{1,n}(t) = \begin{cases} g_1(t) & \text{for } 0 \leqslant t \leqslant 1 - 1/n, \\ y & \text{for } 1 - 1/n < t \leqslant 1. \end{cases}$$

We put $P_n(t,x) = P_n(t,x,G)$.

Unfortunately, the function $g_{1,n}$ is discontinuous and, much less, satisfies the Lipschitz condition. However, the Lipschitz condition was actually used in [1] only for proving Lemmas 8 and 9. Lemmas 7-10 are respectively analogs of Lemmas 8, 9, 16, 17 in [1].

<u>LEMMA 1</u>. For $0 \leq t < 1$

$$P_n(t,\ x + l) - P_n(t,\ x) = O((K + 1/\sqrt{1-t})(l + c_3/\sqrt{n})).$$

The statements of Lemmas 8 in [1] and our Lemma 7 look alike, but in Lemma 7 $P_n(t,x)$ has a different meaning.

<u>Proof of Lemma 7</u>. Let

$$D = \{(t, x)\colon 0 \leqslant t \leqslant 1,\ d_1(t) > x > d_2(t)\}.$$

Let $A_x(D)=\{\tau_n(0,x,D)>n\}$, if $(0,x)\in D$, $A_x(D)=\varnothing$, otherwise $B(y)=\{\xi_n^{(2)}(0,x,n)<y\}$, $D-\ell=\{(t,x)\colon (t,x+\ell)\in D\}$. Obviously,

$$\mathbf{P}(A_x(D)) = P_n(0,\ x,\ D),\ \mathbf{P}(A_x(D-l)) = P_n(0,\ x+l,\ D),\ A_x(G) = A_x(\tilde{G})B(y).$$

Let the symbol + between the events designate here and in the sequel the symmetric difference. Then

$$\begin{aligned} A_x(G-l) + A_x(G) &= A_x(\tilde{G}-l)B(y-l) + A_x(\tilde{G})B(y) \subseteq \\ &\subseteq (B(y-l)+B(y)) \cup (A_x(\tilde{G}-l) + A_x(\tilde{G})). \end{aligned} \tag{35}$$

By the Berry-Esseen theorem

$$\mathbf{P}(B(y-l)+B(y)) = O(l + c_3/\sqrt{n}). \tag{36}$$

In [1, see (2.22) and further] it has actually been proved that

$$\mathbf{P}(A_x(\tilde{G}-p)) - \mathbf{P}(A_x(G_l)) = O((K+1)(l+c_3/\sqrt{n}))$$

for $p=0,\ell$, where $G_\ell=\tilde{G}\cap(\tilde{G}-\ell)$. Hence

$$\mathbf{P}(A_x(\tilde{G}-l) + A_x(\tilde{G})) = O((K+1)(l+c_3/\sqrt{n})). \tag{37}$$

From (35)-(37) it follows that

$$\begin{aligned} P_n(0,\ x+l,\ G) - P_n(0,\ x,\ G) &= O(\mathbf{P}(A_x(G-l) + A_x(G))) = \\ &= O((K+1)(l+c_3/\sqrt{n})), \end{aligned}$$

which proves the assertion of Lemma 7 for $t=0$. The move to the general case is the same as in [1, see (2.28)]. //

<u>LEMMA 8</u>. For $0\le t<t+h\le 1$:

A) $P_n(t+h,\ x) - P_n(t,\ x) = O((K^2 + 1/(1-t))h +$
$+ (K + 1/\sqrt{(1-t)}) \times c_3/\sqrt{n} + \sqrt{h/(1-t)});$

B) $P_n(t+h,\ x) - P_n(t,\ x) = O((K^2 + 1/(1-t))h +$
$+ (K + 1/\sqrt{1-t}) \times (c_3/\sqrt{n} + h|\ln h|/\sqrt{1-t})).$

<u>Proof of Lemma 8</u>. As in [1], we put

$$G_{h,1}=\{(t,x):\ 0\leqslant t\leqslant 1-h,\ g_2(t)<x<g_1(t)\};$$
$$G_{h,2}=\{(t,x):\ 0\leqslant t\leqslant 1-h,\ g_2(t+h)<x<g_1(t+h)\};$$
$$G_{h,3}=G_{h,1}\cap G_{h,2}.$$

From the proof of Lemma 9 in [1] it is seen that it is possible to assume without loss of generality: $x=0$, $t=0$. We consider the events:

$$A_{h,i}=\{\tau_n(0,0,G_{h,i})>n_h\}\quad (i=1,2,3);$$
$$B_h=\{\xi_n^{(2)}(0,0,n_h)<y\},\quad A=A_{0,1},\quad B=B_0.$$

Here and further $n_h=[n(1-h)]$. Obviously,

$$P_n(0,0)=\mathbf{P}(AB),\ P_n(h,0)=\mathbf{P}(A_{h,2}B_h).$$

Hence

$$P_n(h,0)-P_n(0,0)=O(\mathbf{P}(AB+A_{h,2}B_h))=O(\mathbf{P}(A+A_{h,2})+ + \mathbf{P}(B+B_h)).\quad (38)$$

Next,

$$\mathbf{P}(A+A_{h,2})\leqslant \mathbf{P}(A+A_{h,1})+\mathbf{P}(A_{h,1}+A_{h,3})+\mathbf{P}(A_{h,3}+A_{h,2});\quad (39)$$
$$\mathbf{P}(A+A_{h,1})=\mathbf{P}(A_{h,1}\setminus A)=P_n(0,0,(1-h,1],\bar{G})= =O((h+c_3/\sqrt{n})(K+1))\quad (40)$$

by Lemma 2. In [1] it has actually been proved (see the proof of Lemma 9 in [1]) that for $i=1,2$

$$\mathbf{P}(A_{h,i}+A_{h,3})=\mathbf{P}(A_{h,i})-\mathbf{P}(A_{h,3})=O((K+1)(Kh+c_3/\sqrt{n})).\quad (41)$$

From (39)-(41) it follows that

$$\mathbf{P}(A+A_{h,2})=O((K^2+1)h+(K+1)c_3/\sqrt{n}).\quad (42)$$

Next, by Lemma 4

$$\mathbf{P}(B+B_h)=\mathbf{P}(B\setminus B_h)+\mathbf{P}(B_h\setminus B)=\mathbf{P}(\xi(1-h)\geqslant y,\ \xi(1)<y)+ +\mathbf{P}(\xi(1-h)<y,\ \xi(1)\geqslant y)+O(c_3/\sqrt{n}),\quad (43)$$

since, without loss of generality, we can assume that $h<\frac{1}{4}$, which we shall in fact do.

We have

$$\mathbf{P}(\xi(1-h)\geqslant y, \xi(1)<y)=\int_y^\infty \frac{1}{\sqrt{2\pi(1-h)}}e^{-u^2/2(1-h)}\Phi\left(\frac{y-u}{\sqrt{h}}\right)du =$$

$$= O\left(\sqrt{h}\int_{-\infty}^{0}\Phi(u)\,du\right) = O(\sqrt{h}). \quad (44)$$

Similarly,

$$\mathbf{P}(\xi(1-h)<y,\ \xi(1)\geqslant y)=O(\sqrt{h}). \quad (45)$$

It follows from (43)-(45) that

$$\mathbf{P}(B+B_h)=O(c_3/\sqrt{n}+\sqrt{h}). \quad (46)$$

From (38), (42), (46) we prove Lemma 8-A for $x=t=0$. From (41) it follows that

$$\mathbf{P}(A_{h,1}+A_{h,2})=O((K+1)(Kh+c_3/\sqrt{n}))$$

and

$$P_n(h,\ 0)=\mathbf{P}(A_{h,2}B_h)=\mathbf{P}(A_{h,1}B_h)+O((K+1)(Kh+c_3/\sqrt{n})).$$

Hence

$$P_n(h,\ 0)-P_n(0,\ 0)=\mathbf{P}(A_{h,1}B_h)-\mathbf{P}(AB)+O((K+1)(Kh+c_3/\sqrt{n})). \quad (47)$$

Next,

$$|\mathbf{P}(A_{h,1}B_h)-\mathbf{P}(AB)|\leqslant|\mathbf{P}(B)-\mathbf{P}(B_h)|+|\mathbf{P}(\bar{A}_{h,1}B_h)-\mathbf{P}(\bar{A}B)|. \quad (48)$$

By the Berry-Esseen theorem

$$|\mathbf{P}(B)-\mathbf{P}(B_h)|=|\Phi(y)-\Phi(y/\sqrt{1-h})|+ + O(c_3/\sqrt{n(1-h)})=O(h+c_3/\sqrt{n}), \quad (49)$$

since for $0.5<a<1$, $x>0$

$$\Phi(x)-\Phi(ax)=(1-a)x\varphi(\theta x)=O(1-a),$$

where $a<\theta<1$, $\phi(u)=\frac{d}{du}\Phi(u)$. It is seen that

$$|\mathbf{P}(\overline{A}_{h,1}B_h)-\mathbf{P}(\overline{A}B)|\leqslant|\mathbf{P}(\overline{A}_{h,1}B_h)-\mathbf{P}(\overline{A}_{h,1}B)|+|\mathbf{P}(\overline{A}_{h,1}B)-\\ -\mathbf{P}(\overline{A}B)|; \qquad (50)$$

since $\overline{A}_{h,1}\subseteq\overline{A}$, then by (40)

$$\mathbf{P}(\overline{A}_{h,1}B)-\mathbf{P}(\overline{A}B)=O(\mathbf{P}(\overline{A}\setminus\overline{A}_{h,1}))=O((h+c_3/\sqrt{n})(K+1)). \qquad (51)$$

It is not hard to see that

$$\mathbf{P}(\overline{A}_{h,1}B_h)-\mathbf{P}(\overline{A}_{h,1}B)=\mathbf{P}(\overline{A}_{h,1}B_h\overline{B})-\mathbf{P}(\overline{A}_{h,1}B\overline{B}_h). \qquad (52)$$

We introduce two events: the event C_h that the trajectory (of a random walk starting at the point $(0,0)$) at an instant of time not exceeding n_h, leaves the domain $\tilde{G}$ (i.e., the event $\overline{A}_{h,1}$ occurs) and next at an instant of time not exceeding n, traverses the horizontal level y; and the event $\tilde{C}_h$ that the trajectory at an instant of time not exceeding n_{2h} leaves the domain $\tilde{G}$ (i.e., the event $\overline{A}_{2h,1}$ occurs) and next at an instant of time not exceeding n_{2h} crosses the level y. By "crossing" the level y after the instant of time m_1 and before the instant of time m_2, we mean the following:

$$(\xi_n^{(2)}(0,0,m_1)-y)(\xi_n^{(2)}(0,0,m_2)-y)\leqslant 0.$$

It is easy to see that $C_h\supseteq\overline{A}_{h,1}B_h\overline{B}\cup\overline{A}_{h,1}B\overline{B}_h$, with $C_h\subseteq\overline{A}_{h,1}$. Hence $\overline{A}_{h,1}B_h\overline{B}=C_hB_h\overline{B}$, $\overline{A}_{h,1}B\overline{B}_h=C_hB\overline{B}_h$, whence, since $\tilde{C}_h\subseteq C_h$, we obtain that

$$\mathbf{P}(\overline{A}_{h,1}B_h\overline{B})-\mathbf{P}(\tilde{C}_hB_n\overline{B})|+|\mathbf{P}(\overline{A}_{h,1}B\overline{B}_h)-\mathbf{P}(\tilde{C}_hB\overline{B}_h)|\leqslant\\ \leqslant 2\mathbf{P}(C_h\setminus\tilde{C}_h). \qquad (53)$$

It is not hard to see that

$$\mathbf{P}(C_h \setminus \widetilde{C}_h) \leqslant \mathbf{P}(\overline{A}_{h,1} \setminus \overline{A}_{2h,1}) +$$
$$+ \int_0^{1-2h} \int_{-\infty}^{\infty} P_n(0, 0, dt, dx, \widetilde{G}) P_n(t, x, (1-2h, 1], G_{y,x}). \quad (54)$$

Here $G_{y,x}=\{(t,u): 0\le t\le 1, u<y)\}$, if $x<y$; $G_{y,x}=\{(t,u): 0\le t\le 1, u>y\}$ otherwise.

By Lemma 2

$$\mathbf{P}(\overline{A}_{h,1} \setminus \overline{A}_{2h,1}) = O((K+1)(h + c_3/\sqrt{n}));$$

$$P_n(t, x, (1-2h, 1], G_{y,x}) = O(M_1(t)),$$

where

$$M_1(t) = \min\left(1, \frac{h}{1-t} + c_3/\sqrt{n(1-t)}\right).$$

Hence, as a consequence of (54),

$$\mathbf{P}(C_h \setminus \widetilde{C}_h) = O\left((K+1)(h + c_3/\sqrt{n})\right) + \int_0^{1-2h} M_1(t)\, dM_2(t), \quad (55)$$

where

$$M_2(t) = -P_n(0, 0, (t, 1-2h], \widetilde{G}).$$

Applying Lemma 2 again, we obtain that

$$M_2(t) = O((K+1)((1-2h-t) + c_3/\sqrt{n})).$$

Now, applying Lemma 1 with $b=1-2h$, $p_1=p_2=1$, $p_3=p_4=0$, $q=1$, $\mathrm{Var}\, M_1 \le 1$, we obtain

$$\int_0^{1-2h} M_1(t)\, dM_2(t) = O\left((K+1)\left(h\,|\ln h| + c_3/\sqrt{n}\right)\right),$$

which plus (55) yield

$$\mathbf{P}(C_h \setminus \widetilde{C}_h) = O((K+1)(h|\ln h| + c_3/\sqrt{n})). \quad (56)$$

Next, it is not hard to see that

$$\mathbf{P}(\widetilde{C}_h B_h \overline{B}) = \int_0^{1-2h} \int_{-\infty}^{\infty} P_n(0, 0, dt, dx, \widetilde{G})\, \mathbf{P}(A^{(1)}(t, x)); \tag{57}$$

$$\mathbf{P}(\widetilde{C}_h B \overline{B}_h) = \int_0^{1-2h} \int_{-\infty}^{\infty} P_n(0, 0, dt, dx, \widetilde{G})\, \mathbf{P}(A^{(2)}(t, x)), \tag{58}$$

where $A^{(1)}(t,x)$={the trajectory which began at a point (t,x) at the instant of time n_{1-t}, crosses the level y before the time n_{2h}, passing below the level y at the instant of time n_h and above the level y at the instant of time n} and the event $A^{(2)}(t,x)$ differs from the event $A^{(1)}(t,x)$ only in that the trajectory is above the level y at time n_h and below the level y at time n. According to Lemma 6 for $0\leq t<1-2h$

$$\mathbf{P}(A^{(1)}(t, x)) = \mathbf{P}(A^{(2)}(t, x)) + O(c_3/\sqrt{n(1-2h-t)}) \tag{59}$$

(one needs to take $m=[n(1-2h-t)]$, $r=n_h-n_{2h}$, $p=n-n_h$).

It follows from (57)-(59) that

$$\mathbf{P}(\widetilde{C}_h B_h \overline{B}) - \mathbf{P}(\widetilde{C}_h B \overline{B}_h) = O\left(\int_0^{1-2h} M_1(t)\, dM_2(t)\right), \tag{60}$$

where

$$M_1(t) = \min(1, c_3/\sqrt{n(1-2h-t)});$$

$$M_2(t) = -P_n(0, 0, (t, 1-2h], \widetilde{G}),$$

and

$$M_2(t) = O((K+1)(1-2h-t+c_3/\sqrt{n})).$$

Applying Lemma 1 again with $b=1-2h$, $p_1=p_3=p_4=0$, $p_2=q=1$, $\operatorname{Var} M_1(t)\leq 1$, we obtain

$$\int_0^{1-2h} M_1(t)\, dM_2(t) = O((K+1)\, c_3/\sqrt{n}). \tag{61}$$

From (60), (61) it follows that

$$\mathbf{P}(\widetilde{C}_h B_h \overline{B}) - \mathbf{P}(\widetilde{C}_h B \overline{B}_h) = O((K+1)c_3/\sqrt{n}),$$

which plus (52), (53), (56) yield

$$\mathbf{P}(\bar{A}_{h,1}B_h) - \mathbf{P}(\bar{A}_{h,1}B) = O((K+1)(h|\ln h| + c_3/\sqrt{n})).$$

Then, taking into account (47)-(51), we obtain

$$P_n(h, 0) - P_n(0, 0) = O((K^2+1)h + (K+1)(c_3/\sqrt{n} + h|\ln h|)),$$

thus proving Lemma 8-B for $t=x=0$. //

LEMMA 9. For $0\leq t<1$

$$\bar{P}_n(t, x) - P_n(t, x) = O\left(\left(K + \frac{1}{\sqrt{1-t}} \ln \frac{K+2}{1-t}\right) \frac{c_3}{\sqrt{n}}\right)$$

(here, as well as in [1], $\bar{P}_n(t,x)=S^{-1}(f_{12}^{(n)}S(P_n(t,x)))$, where S is the Fourier transform).

Proof of Lemma 9. The proof of Lemma 9 is basically the same as the proof of Lemma 16 in [1]. The difference lies in the following:

1. To obtain an estimate of the form (2.66) in [1], one needs instead of Lemma 9 of [1] to use Lemma 8-A, which leads to the estimate

$$P_n(\alpha + 1/n + t, x - y) = P_n(t, x - y) + O((K + 1/\sqrt{1-t})c_3/\sqrt{n});$$

2. instead of Lemma 8 in [1], one needs to use everywhere Lemma 7;

3. estimating the integral

$$I_1 = \iint\limits_{\{-(K+1/\sqrt{1-t})^{-2} < s < -\alpha\}} (P_n(t-s, x-y) - P_n(t+\alpha, x-y))\, q_{n2}(s, y)\, ds dy,$$

on the left-hand side of the formula (2.79) in [1], using Lemma 8-B, we obtain $I_1=O(I_2+I_3)$, where

$$I_2 = (K^2 + 1/(1-t))\, e^3 \int_{\alpha}^{(K+1/\sqrt{1-t})^{-2}} s^{-1/2} ds +$$

$$+ \frac{c_3^2}{\sqrt{n}} (K + 1/\sqrt{1-t}) \int_{\alpha}^{(K+1/\sqrt{1-t})^{-2}} s^{-3/2} ds;$$

$$I_3 = c_3 (K/\sqrt{1-t} + 1/(1-t)) \int_{\alpha}^{(K+1/\sqrt{1-t})^{-2}} \frac{\ln s}{s^{1/2}} ds.$$

It has been shown in [1] that (for $\alpha = c_3^2/n$)

$$I_2 = O(c_3(K + 1/\sqrt{1-t})).$$

Next

$$I_3 = O\left(c_3 \frac{\ln\left(K + 1 + \frac{1}{\sqrt{1-t}}\right)}{\sqrt{1-t}}\right) = O\left(\frac{c_3}{\sqrt{1-t}} \ln \frac{K+2}{1-t}\right).$$

Hence

$$I_1 = O\left(c_3\left(K + \frac{1}{\sqrt{1-t}} \ln \frac{K+2}{1-t}\right)\right).$$

This, as well as in [1], yields the estimate (compare with (2.84) in [1])

$$P_{n3}\left(t, x, \frac{c_3^2}{n}\right) = O\left(\left(K + \frac{1}{\sqrt{1-t}} \ln \frac{K+2}{1-t}\right) \frac{c_3}{\sqrt{n}}\right),$$

which plus the relations (2.69), (2.74), (2.86), (2.88) from [1] prove Lemma 9.

<u>LEMMA 10</u>. If $y_0 < x < y$, where $y_0 = g_2(1-1/n) + K/n$, then

$$\bar{P}_n(1 - 1/n, x) = 1 + O\left(n\left(\int_0^{1/n} \frac{s\,ds}{(x-y)^2 + s} + \int_0^{1/n} \frac{s\,ds}{(x-y_0)^2 + s}\right)\right);$$

if $x > y$, then

$$\overline{P}_n(1-1/n, x) = O\left(n \int_0^{1/n} \frac{s\,ds}{(x-y)^2+s} \right).$$

Proof of Lemma 10. We have (compare with (2.89) in [1])

$$\overline{P}_n(1-1/n, x) = \frac{n}{\sqrt{2\pi}} \int_0^{1/n} ds \int_{x-y}^{x-g_2(1-1/n+s)} e^{-u^2/2s} s^{-1/2} du.$$

If $x>y$, then, as in [1], we have the estimate

$$\overline{P}_n(1-1/n, x) = O\left(n \int_0^{1/n} ds \int_{x-y}^{\infty} e^{-u^2/2s} s^{-1/2} du \right) = O\left(n \int_0^{1/n} \frac{s\,ds}{(x-y)^2+s} \right).$$

One considers the case $y_0<x<y$ in the same way.

We complete now proving the Theorem. To do so, we need to repeat the final part of the proof of the main result obtained in [1], beginning right after the proof of Lemma 17. We need to make, however, the appropriate changes:

1. Instead of $\Pi_n = \Pi_n(R^1)$ we need to take $\Pi_n = \Pi_n((-\infty, y))$.

2. To estimate the integral

$$V_n = \int_{g_2(1-1/n)}^{g_1(1-1/n)} \overline{P}_n(1-1/n, x)\, \Pi_n(dx)$$

one needs to start from the representation

$$V_n = \Pi_n + \int_{g_2(1-1/n)}^{y} (\overline{P}_n(1-1/n, x) - 1)\, \Pi_n(dx) + \int_y^{\infty} \overline{P}_n(1-1/n, x)\, \Pi_n(dx)$$

and use Lemma 10 instead of Lemma 17 in [1]. This results in the estimate

$$V_n = \Pi_n + O(n^{-\frac{1}{2}} + Kc_3 n^{-1}) .$$

3. Instead of (2.100) in [1], by Lemma 9 we have

$$\overline{P}_n(t, g_1(t)) = O\left(\left(K + \frac{1}{\sqrt{1-t}} \ln \frac{K+2}{1-t}\right)\frac{c_3}{\sqrt{n}}\right).$$

Then, to estimate the integral

$$\int_0^{1-1/n} \overline{P}_n(t, g_1(t))\, \Pi^+(dt) = O\left(\int_0^{1-1/n} M_1(t)\, dM_2(t)\right),$$

where

$$M_1(t) = \left(K + \frac{\ln(K+2)}{\sqrt{1-t}} + \frac{1}{\sqrt{1-t}} \ln \frac{1}{1-t}\right)\frac{c_3}{\sqrt{n}};$$

$$M_2(t) = -\Pi^+\left(\left(t, \frac{n-1}{n}\right)\right),$$

we use Lemma 1, setting $b=1-1/n$, $p_1=q=0$, $p_2=p_3=p_4=1$. As a result, instead of (2.103) from [1] we have $V_n^+=O((K+1)c_3/\sqrt{n})$ and similarly $V_n^-=O((K+1)c_3/\sqrt{n})$.

4. An estimate of the form $\Pi_n(((n-1/n, 1))=O((K+1/n)$ does not need to hold here. However, using (2.106) from [1] and (44), (45), we have instead of (2.106) from [1]:

$$\Pi_n = P(0,0,D_y,1) + O(1/\sqrt{n}+K/n) .$$

Theorem 1 is proved under the assumption (2').

To prove Theorem 1 in the general case, we shall need the following auxiliary lemma (to be used later on as well).

LEMMA 11. Let

$$p_m(\varepsilon) = \sup_{\alpha,\beta} \mathbf{P}\left(S_m/\sqrt{m} \in (\alpha, \beta),\ S_{m-1}/\sqrt{m} \notin (\alpha+\varepsilon, \beta-\varepsilon)\right) + \\ + \sup_{\alpha,\beta} \mathbf{P}\left(S_m/\sqrt{m} \notin (\alpha+\varepsilon, \beta-\varepsilon),\ S_{m-1}/\sqrt{m} \in (\alpha, \beta)\right).$$

Then $p_m(\varepsilon)=O(c_3/\sqrt{m}+\varepsilon)$.

Proof of Lemma 11. The proof of this Lemma follows from (46) and from $\Phi(x+\varepsilon)-\Phi(x)=O(\varepsilon)$.

Thus, let the condition (2') be violated, for example, $y \geq g_1(1)-K/n$. By Lemma 11

$$P_n(0, 0, D_y, 1) - P_n(0, 0, D_y, 1-1/n) = \\ = O(p_n(2K/n)) = O(c_3/\sqrt{n} + K/n). \quad (62)$$

By (1)

$$P_n(0, 0, D_y, 1-1/n) - P(0, 0, D_y, 1-1/n) = \\ = O((K+1)c_3/\sqrt{n}). \quad (63)$$

Similarly to the formula (106) from [1] we have

$$P(0, 0, D_y, 1-1/n) - P(0, 0, D_{g1(1)}, 1) = O((K+1)/n).$$

Hence

$$P(0, 0, D_y, 1-1/n) = P(0, 0, D_y, 1) + O((K+1)/n).$$

This plus (62), (63) prove Theorem 1 for $y \geq g_1(1)-K/n$. The quantities $y \leq g_2(1)+K/n$ are considered in the same way. Theorem 1 is completely proved.

3. ESTIMATES OF STABILITY (WITH RESPECT TO TIME) OF SOLUTIONS OF BOUNDARY VALUE PROBLEMS

Let

$$G = \{(t, x) : 0 \leqslant t \leqslant 1,\ g_2(t) < x < g_1(t)\},$$
$$g_2(0) < 0 < g_1(0);\ g_2(t) < g_1(t);$$
$$|g_i(t+h) - g_i(t)| \leqslant Kh (0 \leqslant t \leqslant t+h \leqslant 1,\ i = 1, 2);$$
$$G_h = G_{h,y} \equiv \{(t, x) \in G : x < y\ (t = 1-h)\};$$
$$h_n =]nh[/n.$$

THEOREM 3. For $0 \leq h < 1$

$$P_n(0, 0, G_{h_n}, 1-h) - P_n(0, 0, G_0, 1) = O((K^2+1)h + (K+1)c_3/\sqrt{n}).$$

THEOREM 4. For $0\le h<1$

$$P(0,0,G_h,1-h)-P(0,0,G_0,1)=O((K^2+1)h).$$

REMARK 3. Using Theorem 3, it is possible to improve the estimate given in Lemma 8-B. Indeed, using the notation used in Section 2, we have

$$P_n\left(0,0,G_{h_n},1-h\right)=\mathbf{P}(A_{h,1}B_h);\ P_n(0,0,G_0,1)=\mathbf{P}(AB).$$

Hence, by Theorem 1 and the formula (47) we obtain

$$P_n(h,0)-P_n(0,0)=O\left((K^2+1)h+(K+1)\frac{c_3}{\sqrt{n}}\right).$$

This proves in the case t=x=0 the following corollary.

COROLLARY 3. For $0\le t<t+h\le 1$

$$P_n(t+h,x)-P_n(t,x)=O\left(\left(K^2+\frac{1}{1-t}\right)h+\left(K+\frac{1}{\sqrt{1-t}}\right)\frac{c_3}{\sqrt{n}}\right).$$

The move to the case $t\neq 0$ or $x\neq 0$ is the same as in proving Lemma 9 in [1].

Proof of Theorem 4. First we give three auxiliary assertions.

LEMMA 12. The inequalities

$$x-Kt\le\sqrt{t}\le x+Kt\,(K,t,x>0)$$

hold iff:

1. for $4Kx<1$ either $t\ge t_1$ or $t_3\le t\le t_2$;
2. for $4Kx\ge 1$ $t\le t_3$; here

$$t_{1,2}=\frac{1-2Kx\pm\sqrt{1-4Kx}}{2K^2},\quad t_3=\frac{1+2Kx-\sqrt{1+4Kx}}{2K^2}.$$

Proof of Lemma 12. We leave it to the reader to prove Lemma 12. We only note that $t_{1,2}$ are the roots of the equation $\sqrt{t}=x+Kt$, and t_3 is the (unique) root of the equation

$\sqrt{t}=x-Kt$, with $t_3 \le t_2 \le t_1$ (for $4Kx<1$).

LEMMA 13. Let $M^+(K)$ be the class of nonnegative (Lebesgue measurable) functions g satisfying the condition $|g(t)-g(0)| \le Kt \, (0 \le t \le 1)$. For $g \in M^+(K)$ let

$$I(g) = \int_0^1 \exp\{-g^2(t)/2t\} \, g(t) \, t^{-3/2} dt.$$

Then

$$M_K \equiv \sup\{I(g) : g \in \mathfrak{M}^+(K)\} = O(K+1).$$

Proof of Lemma 13. Define $\psi(g)=ge^{-g^2/2t}$ $(g>0)$. This function increases for $\sqrt{t}>g>0$ and decreases for $g>\sqrt{t}$. Next, for $g \in M^+(K)$ the inequalities $x-Kt \le g(t) \le x+Kt$ are satisfied, where $x=g(0)$. We assume that $x>0$, since otherwise $g(t) \le Kt$ and $I(g) \le K\int_0^1 t^{-\frac{1}{2}} dt=2K$. Then

$$I(g) \le \sum_{k=1}^{3} I_k , \tag{64}$$

where

$$I_k = \int_{E_k} \psi\left(g_k^0(t)\right) t^{-3/2} dt;$$

$$g_1^0(t) = \sqrt{t}, \; g_2^0(t) = x + Kt, \; g_3^0(t) = x - Kt;$$

$$E_1 = \{t : 0 < t \le 1, \; x - Kt \le \sqrt{t} \le x + Kt\};$$

$$E_2 = \{t : 0 < t \le 1, \; \sqrt{t} > x + Kt\};$$

$$E_3 = \{t : 0 < t \le 1, \; \sqrt{t} < x - Kt\}.$$

We estimate first I_1. By Lemma 12, $E_1 \subseteq [t_1,1] \cup [t_3,t_2]$ for $4Kx<1$ and $E_1 \subseteq [t_3,1]$ for $4Kx \ge 1$. Next, $t_1 \ge 1/4K^2$, $t_2/t_3=O(1)$ for $4Kx<1$ and $t_3 \ge 1/4K^2$ for $4Kx \ge 1$. Hence for $4Kx<1$

$$\int_{t_1}^{1} \psi(\sqrt{t})\, t^{-3/2}\, dt = O\left(\int_{1/4K^2}^{1} t^{-1} dt \right) = O(\ln(K+1));$$

$$\int_{t_3}^{t_2} \psi(\sqrt{t})\, t^{-3/2} dt = O\left(\int_{t_3}^{t_2} t^{-1} dt \right) = O\left(\ln \frac{t_2}{t_3} \right) = O(1),$$

and for $4Kx \geq 1$

$$\int_{t_3}^{1} \psi(\sqrt{t})\, t^{-3/2} dt = O\left(\int_{1/4K^2}^{1} t^{-1} dt \right) = O(\ln(K+1)).$$

Therefore,

$$I_1 = O(\ln(K+1)) . \tag{65}$$

Now we estimate I_2. We have

$$I_2 = O\left(\int_0^1 \psi(x+Kt)\, t^{-3/2} dt \right) =$$
$$= O\left(K \int_0^1 t^{-1/2} dt \right) + O\left(\int_0^\infty e^{-u^2/2} du \right) = O(K+1). \tag{66}$$

At last, we estimate I_3. Obviously, $E_3 \subseteq [0,t^*]$, where $t^* = \min(1, x/K)$. Hence $I_3 \leq f(K)$, where

$$f(K) = \int_0^{t^*} \psi(x-Kt)\, t^{-3/2} dt.$$

Since t^* does not increase as K increases, then for $K \neq x$

$$\frac{df(K)}{dK} \leqslant \int_0^{t^*} e^{-(x-Kt)^2/2t} [(x-Kt)^2 - t]\, t^{-3/2} dt \leqslant$$
$$\leqslant \int_0^1 \exp\{-(x-Kt)^2/2t\} \frac{(x-Kt)^2}{t} t^{-1/2} dt = O(1).$$

Moreover,

$$f(0) \leqslant \int_0^\infty e^{-u^2/2} du.$$

Hence

$$I_3 \leqslant f(0) + K \max_{\substack{0 \leqslant L \leqslant K \\ L \neq x}} \frac{df(L)}{dL} = O(K+1). \tag{67}$$

The proof of Lemma 13 follows from (64)-(67).

<u>LEMMA 14</u>. Let $W(h)$ $(0 \leq h < 1)$ be the event that the trajectory $\xi(t)$ exits a domain G before time $1-h$, whereupon it crosses for the first time the horizontal level y within the interval of time $(1-h,1]$. Then $P(W(h))-O((K^2+1)h)$.

<u>Proof of Lemma 14</u>. It is not hard to see that

$$P(W(h)) = I_1 + I_2 , \tag{68}$$

where

$$I_k = \int_0^{1-h} P_k(dt, G) P((1-h-t, 1-t], y - g_k(t));$$

$$P_k(E, G) = \mathbf{P}\{(\exists \tau \in E)(\forall t < \tau) \xi(t) \in G, \ \xi(\tau) = g_k(\tau)\};$$
$$P(E, u) = \mathbf{P}\{(\exists \tau \in E)(\forall t < \tau) \xi(t) \neq u, \ \xi(\tau) = u\};$$

E is a Borel set in R^1, $k=1,2$. We can assume without loss of generality that $h<1/4$.

From Lemma 2, letting n to ∞, we obtain due to the invariance principle

$$P_i((t_1,t_2],G) = O((K+1)(t_2-t_1)/t_2) \quad (i=1,2) , \tag{69}$$

where $0 \leq t_1 < t_2 \leq 1$. This implies, in particular, that

$$I_1 = I_3 + O((K+1)h) , \tag{70}$$

where

$$I_3 = \int_0^{1-2h} P_1(dt, G)\, P((1-h-t, 1-t], y - g_1(t)).$$

We put

$$I_{31} = \int_0^{1/2} P_1(dt, G)\, P((1-h-t, 1-t], y - g_1(t));$$
$$I_{32} = I_3 - I_{31}. \tag{71}$$

We have for $0 \le t \le 1-2h$

$$P((1-h-t, 1-t], y - g_1(t)) = 2\left| \Phi\left(\frac{g_1(t)-y}{\sqrt{1-t}}\right) - \Phi\left(\frac{g_1(t)-y}{\sqrt{1-t-h}}\right)\right| =$$
$$= 2h\cdot\varphi\left(\frac{g_1(t)-y}{\sqrt{1-t-\theta h}}\right)(1-t-\theta h)^{-3/2}|g_1(t)-y| \leqslant$$
$$\leqslant 2h\cdot\varphi\left(\frac{g_1(t)-y}{\sqrt{1-t}}\right)\cdot 2^{3/2}(1-t)^{-3/2}|g_1(t)-y|,$$

where $0<\theta<1$. Hence

$$I_{31} = O\left(h\cdot\max_x\{x\varphi(x)\}\int_0^{1/2} P_1(dt, G)\right) = O(h),$$

and taking into account (69)

$$I_{32} = O((K+1)hM_K) = O((K+1)^2 h),$$

by Lemma 13. The last two estimates yield, by Eq. (71), $I_3=O((K^2+1)h)$. This plus (70) imply that $I_1=O((K^2+1)h)$. Similarly, $I_2=O((K^2+1)h)$. Hence, noting (68), we see that Lemma 14 is proved.

We shall complete proving Theorem 4. Let

$$A_h = \{\xi(t) \in G(t)(0 \leqslant t \leqslant 1-h)\};$$
$$B_h = \{\xi(1-h) < y\},\ A = A_0,\ B = B_0.$$

The assertion of Theorem 4 is equivalent to

$$P(A_h B_h) - P(AB) = O((K^2+1)/h).$$

Arguing in the same way as in proving Lemma 8 (all the calculations become essentially simpler due to the continuity and symmetry of the process $\xi(t)$), we obtain the estimate $P(A_h B_h)-P(AB)=O((K+1)h)+O(P(W(h)))$, whereupon it remains only to use Lemma 14. //

Proof of Theorem 3. By Theorem 4 it suffices to show that for $0\le h\le 1/2$

$$\Delta_n(h) \equiv P_n\left(0, 0, G_{h_n}, 1-h\right) - P(0, 0, G_h, 1-h) = \\ = O\left((K+1)c_3/\sqrt{n} + (K^2+1)h\right), \quad (72)$$

which follows from Theorem 1 if nh is an integer. Next, by Theorem 4

$$P\left(0, 0, G_{\widetilde{h}_1}, 1-\widetilde{h}_1\right) - P\left(0, 0, G_{\widetilde{h}_2}, 1-\widetilde{h}_2\right) = O\left((K^2+1)\left(\widetilde{h}_1+\widetilde{h}_2\right)\right),$$

hence $\Delta_n(h)=\Delta_n(h_n)+O((K^2+1)(h+1/n))$, which proves (72) in the general case. //

4. PROOFS

We assume without loss of generality that F is continuous.

Proof of Theorem 1'. One can assume that $c=0$. Using Lemma 11, it is not hard to see that we can consider only the case where $(c-d)n$ is an integer. Through the strain (contraction) in t we arrive at $d=1$.

Now let

$$\tilde{D}_n = \{(t, x) : g_2(t) + K/n < x < g_1(t) - K/n;\ 0 \leqslant t \leqslant 1\}.$$

It is not hard to see that

$$0 \leqslant P_n(0, x, D, 1) - \widehat{P}_n(0, x, D, 1) \leqslant$$
$$\leqslant P_n(0, x, D, 1 - 1/n) - P_n(0, x, \bar{D}_n, 1). \quad (73)$$

By Theorem 3

$$P_n(0, x, \bar{D}_n, 1) = P_n(0, x, \bar{D}_n, 1 - 1/n) + O((K+1)c_3/\sqrt{n}). \quad (74)$$

Next (see (2.22) and further in [1]),

$$P_n(0, x, D, 1 - 1/n) - P_n(0, x, \bar{D}_n, 1 - 1/n) =$$
$$= O((K+1)(K/n + c_3/\sqrt{n})) = O((K+1)c_3/\sqrt{n}). \quad (75)$$

From (73)-(75) the assertion of Theorem 1' follows for $\hat{P}_n(c,x,D,d)$. An estimate for $\bar{P}_n(c,x,D,d)$ obtains similarly.//

Proof of Theorems 2 and 2'. Theorems 2 and 2' can be proved by induction in s. Let $s=2$. Then Theorems 2 and 2' follow from Theorems 1 and 1'. Let $s>2$ and let Theorems 2 and 2' be proved for domains having no more than $s-1$ discontinuity points.

Let $Q_n = Q_n(a,x,D,b)$ be a general designation for the functions P_n^r, $\bar{P}_n^r$, P_n, P (we recall that the function $P_n^r(a,x,D,b)$ is defined only for domains D such that nt_j are integers). We can assume without loss of generality that $|\Delta_j|>2/n$.

For brevity let $\tau =]nt_{s-2}[/n$. Then

$$Q_n(a, x, D, b) = \int_{D(\tau+0)} Q_n(\tau, u, D, b)\, d_u Q_n(a, x, D, \tau, u),$$

where

$$Q_n(a, x, D, t, u) = Q_n(a, x, D_u, t);$$
$$D_u = D \setminus (\{t\} \times [y, \infty)).$$

Hence

$$(Q_n - P)(a,x,D,b) = I_1 + I_2\,, \quad (76)$$

where

$$I_1 = \int_{D(\tau+0)} (Q_n - P)(\tau, u, D, b)\, d_u Q_n(a, x, D, \tau, u);$$

$$I_2 = \int_{D(\tau+0)} P(\tau, u, D, b)\, d_u((Q_n - P)(a, x, D, \tau, u)).$$

We put

$$D^{(i)} = \{(t, x) : x \in \delta_i(t+0)(\tau \leqslant t < b),$$
$$x \in \delta_i(b-0) \cap D(b)(t=b)\} \quad (i = 1, \ldots, m_{s-1}).$$

Then for $u \in \delta_i(\tau+0)$

$$Q_n(\tau, u, D, b) = Q_n(\tau, u, D^{(i)}, b) \qquad (i=1, \ldots, m_{s-1}) .$$

Next, by Theorems 1 and 1'

$$(Q_n - P)(\tau, u, D^{(i)}, b) = O(m(b)(K_{s-1} + 1/\sqrt{|\Delta_{s-1}|})c_3/\sqrt{n}).$$

We then conclude that

$$I_1 = O(m(b)(K_{s-1} + 1/\sqrt{|\Delta_{s-1}|})c_3/\sqrt{n}). \tag{77}$$

Next, integrating by parts, we have

$$I_2 = -\Sigma , \tag{78}$$

where

$$\Sigma = \sum_{i=1}^{m_{s-1}} \int_{\delta_i(\tau+0)} (Q_n - P)(a, x, D, \tau, u)\, d_u P\left(\tau, u, D^{(i)}, b\right).$$

By Lemma 11 for $u \in D(\tau+0)$

$$Q_n(a, x, D, \tau, u) - Q_n(a, x, D, t_{s-2}, u) =$$
$$= O\left(P_{n(\tau-a)}(K_{s-1}/n)\right) = O\left(\frac{c_3}{\sqrt{n\left(t_{s-2} - a\right)}} + \frac{K_{s-1}}{n}\right). \tag{79}$$

By induction,

$$|(Q_n - P)(a, x, D, t_{s-2}, u)| \leqslant \Psi^0_{s-1} \frac{c_3}{\sqrt{n}}, \tag{80}$$

where $\psi^0_{s-1}=\psi_{s-1}$ for $Q_n=P^r_n$, $\psi^0_{s-1}=\hat{\psi}_{s-1}$ for $\hat{P}_n$, $\bar{P}^r_n$, P.
From (79), (80) it follows that

$$|\Sigma| \leqslant (\Psi^0_{s-1} + O(\Delta\hat{\lambda}_{s-1})) \frac{c_3}{\sqrt{n}} \cdot \sum_{i=1}^{m_{s-1}} V_i, \tag{81}$$

where

$$V_i = \operatorname*{Var}_{u\in\delta_i(\tau+0)} P(\tau, u, D^{(i)}, b) \quad (i = 1, \ldots, m_{s-1}).$$

To estimate V_i we need the following lemma.

LEMMA 15. Let $D=\{(t,x):d^-(t)<x<d^+(t)\ (c\leq t\leq d)\}$, $d^\pm$ be continuous functions and let $\xi_x(t)=\xi_0(t)+x$, where $\xi_0(t)$ is a continuous random process on $[c,d]$, $\xi_0(c)=0$ a.s.,

$$\pi_{x,D}(E_m) = \mathbf{P}(\xi_x(t)\in D(t)(c\leqslant t<d);\ \xi_x(d)\in E_m),$$

where E_m is the union of no more than m intervals. Then

$$\operatorname*{Var}_{-\infty<x<\infty} \pi_{x,D}(E_m) \leqslant 2m.$$

Proof of Lemma 15. It suffices, as is easily seen, to show that for $-\infty<\alpha_-<\alpha_+<\infty$

$$\operatorname*{Var}_{x\in D(c)} \pi_{x,D}((\alpha_-, \alpha_+)) \leqslant 2. \tag{82}$$

Suppose for $x\in D(c)$, $\pi^{\pm}_{x,D}(\alpha_\pm)$ are the probabilities of the events that either:

1. $\xi_x(t)$ exits D on $[c,d)$, $\tau_x \xi_x(\tau_x)=d^\pm(\tau_x)$, at the time of the first exit; or

2. $\xi_x(t)\in d(t)$ $(c\leq t<d)$, but $\pm\xi_x(d)\geq\pm\alpha_\pm$. Then

$$\pi_{x,D}((\alpha_-, \alpha_+)) = 1 - \pi^+_{x,D}(\alpha_+) - \pi^-_{x,D}(\alpha_-), \tag{83}$$

$\pi^+_{x,D}(\alpha_+)$ not decreasing in x, $\pi^-_{x,D}(\alpha_-)$ not increasing in x. Hence (83) follows from (82). Lemma 15 is proved.

Now we continue proving Theorems 2, 2'. From (81) by Lemma 15 it follows that

$$|\Sigma| \leqslant 2m(t_{s-1})\left(\Psi_{s-1}^{0} + O\left(\Delta\hat{\lambda}_{s-1}\right)\right) c_3/\sqrt{n}. \tag{84}$$

Summarizing (76)-(78), (84), we see that

$$|(Q_n - P)(a, x, D, b)| \leqslant 2m(t_{s-1})\left[\Psi_{s-1}^{0} + O\left(\lambda_{s-1} + \Delta\hat{\lambda}_{s-1}\right)\right] c_3/\sqrt{n};$$

in other words,

$$|(Q_n - P)(a, x, D, b)| \leqslant \Psi_s^0 c_3/\sqrt{n},$$

where

$$\Psi_s^0 = \mu_{s-1}\left[\Psi_{s-1}^0 + L\hat{\lambda}_{s-1}\right](s > 2);\ \Psi_2^0 = L\mu_1\hat{\lambda}_1;$$

$L>0$ is an absolute constant. //

To prove Corollaries 2, 2', it suffices to note that $\mu_j \equiv 2$ $(j=1,\ldots,s-1)$; $\Delta\hat{\lambda}_j = O(\lambda_{j-1}+\lambda_j)$ $(j=2,\ldots,s-1)$.

REFERENCES

[1] S.V. Nagaev. "On the Speed of Convergence in a Boundary Problem, I, II." *Theory Prob. Applications*, 2 (1970): 163-186; 3 (1970): 403-429.

[2] A.I. Sakhanenko. "On the Speed of Convergence in a Boundary Problem." *Theory Prob. Applications*, 2 (1974): 399-403.

[3] A.A. Borovkov. *Probability Processes in the Queueing Theory*. Moscow: Nauka, 1972. (In Russian.)

[4] S.V. Nagaev. "Some Limit Theorems for Large Deviations." *Theory Prob. Applications*, 2 (1965): 214-235.

ON ESTIMATES OF THE RATE OF CONVERGENCE IN THE INVARIANCE PRINCIPLE

A.I. Sakhanenko

1. INTRODUCTION

Let $\xi_1,\ldots,\xi_n$ be a sequence of independent random variables such that $M\xi_i=0$, $\sum D\xi_i=1$. (Here and below, if not stated otherwise, the sums, products and maxima are taken over i running through the values from 1 to n.) Let

$$t_i=\sum_{j=1}^{i} \mathbf{D}\xi_j, \quad t_0=0,\ L_s=\Sigma\,\mathbf{M}\,|\,\xi_i\,|^s, \quad s>2.$$

We construct on the interval [0,1] a continuous broken line ξ with vertices at points $\left(t_i,\xi(t_i)=\sum_{j=1}^{i}\xi_j\right)$. The objective of this paper is to develop estimates of the proximity between the distributions in C(0,1), generated by the process ξ and the standard Wiener process w. For estimates in terms of the Lyapunov relation L_s we prove and strengthen the results announced in [1].

We assume that the space C(0,1) of continuous functions on [0,1] has the norm $\|x\|=\max_{0\le t\le 1}|x(t)|$, which generates the σ-algebra of Borel sets. Also, we denote by $B^{(\varepsilon)}$ the ε-neighborhood of the set B. We define the Lévy-Prokhorov distance $\lambda(\xi,w)$ between the distributions of the processes ξ and w in the space C(0,1), assuming $\lambda(\xi,w)\le\varepsilon$, if for all measurable B

$$\mathbf{P}(\xi\in B)\leqslant\mathbf{P}(w\in B^{(\varepsilon)})+\varepsilon,\ \ \mathbf{P}(w\in B)\leqslant\mathbf{P}(\xi\in B^{(\varepsilon)})+\varepsilon.$$

The estimate of proximity between the distributions of ξ and w is most frequently accepted as the estimate of the distance $\lambda(\xi,w)$. The first result of this kind belongs to Yu.V. Prokhorov [2] who obtained

$$\lambda(\xi,w) = o(L_3^{1/4}|\ln L_3|^2) \qquad \text{as} \quad L_3 \to 0 .$$

A.A. Borovkov improved this estimate [3]:

$$\lambda(\xi,w) \leq cL_s^{1/(s+1)} , \quad 2<s\leq 3 , \tag{1}$$

where c is an absolute constant. Most of the other results on this problem (see References in [3]) were obtained under the assumption that $\xi_1,\dots,\xi_n$ are identically distributed. The best result among them is, apparently, the well-known result obtained by Komlos, Major, and Tusnady [4], which, in particular, implies (1) for all $s>2$, but with the constant depending on s and the distribution $\sqrt{n}\xi_n$.

If we have an estimate of the Lévy-Prokhorov distance $\lambda=\lambda(\xi,w)$, we obtain

$$\Delta(B) \equiv |\mathbf{P}(\xi \in B) - \mathbf{P}(w \in B)| \leqslant \mathbf{P}(w \in (B')^{(\lambda)}) + \lambda$$

for any measurable set B with boundary B'. In particular, for Lipschitz sets, i.e., for sets whose boundary B' satisfies the condition

$$\mathbf{P}(w \in (B')^{(\varepsilon)}) \leqslant C_B\varepsilon \quad \forall \varepsilon > 0, \tag{2}$$

we have

$$\Delta(B) \leqslant (C_B + 1)\lambda(\xi, w). \tag{3}$$

In [3] one can find an example of a set B_0 such that

$$\Delta(B_0) \geq cL_s^{1/s} . \tag{4}$$

In this paper, we bridge the gap between the estimates (1) and (4). In Theorem 1 we give a lower bound for the distance between the distributions, which shows that (1) is non-improvable; in Theorem 2 we obtain an estimate for $\Delta(B)$ implying the existence of an absolute constant c for which, if $L_s \leq 1$

$$\Delta(B) \leqslant cC_B\left(1+\ln\left(1+C_B^{-1}\right)\right)L_s^{1/s}\left(1+\ln\left(L_s^{-1/s}\right)\right)^{3(s-1)/2s}, \tag{5}$$

which is essentially more precise than the estimate given by (1) and (3). We prove thereby, in particular, that for $\Delta(B)$ it is possible to obtain a sharper result than that which follows from (3).

In proving Theorem 2 we use a slightly modified method of a common probability space, suggested by A.V. Skorokhod.

2. THE ESTIMATE FROM BELOW FOR THE LÉVY-PROKHOROV DISTANCE

Let

$$\varepsilon = \max\{y:\ y \leqslant \sum \mathbf{P}(|\xi_i| \geqslant y)\};\quad \sigma_i^2 = \mathbf{D}\xi_i;$$
$$\sigma^2 = \max \sigma_i^2;\quad \delta = \sqrt{\sigma^2\left(1+3\ln(\sigma^{-2})\right)}/8.$$

THEOREM 1. $\lambda \equiv \lambda(\xi, w) \geq \max\{\varepsilon/2 - 4\delta, \delta\}$. By the Chebyshev inequality

$$\begin{gathered}\varepsilon \leqslant \max\{y:\ y \leqslant L_s y^{-s}\} = L_s^{1/(s+1)},\\ \sigma \leqslant \max\left(\mathbf{M}|\xi_i|^s\right)^{1/s} \leqslant L_s^{1/s}.\end{gathered} \tag{6}$$

If a sequence of a series is such that

$$\Sigma\mathbf{P}\left(|\xi_i| > c_1 L_s^{1/(s+1)}\right) \geqslant c_2 L_s^{1/(s+1)},$$

then Theorem 1 implies the estimate $\lambda \geq cL_s^{1/(s+1)}$ for sufficiently small L_s.

A concrete example of the distribution of ξ_i can be found in [5].

Proof of Theorem 1. We put

$$A(y)=\{x\in C(0,\ 1): \max|x(t_i)-x(t_{i-1})|\geqslant y\}$$

and note that

$$(A(\varepsilon))^{(\varepsilon/2-4\delta)}\subset A(\varepsilon-2(\varepsilon/2-4\delta))=A(8\delta).$$

Therefore, to prove the inequality

$$\lambda \geq \varepsilon/2 - 4\delta \tag{7}$$

it suffices to show that

$$\mathbf{P}(\xi\in A(\varepsilon))\geqslant\varepsilon/2,\quad \mathbf{P}(w\in A(8\delta))\leqslant 4\delta. \tag{8}$$

We have

$$\mathbf{P}(\xi\in A(\varepsilon))=1-\Pi(1-\mathbf{P}(|\xi_i|\geqslant\varepsilon))\geqslant 1-\exp(-\Sigma\mathbf{P}(|\xi_i|\geqslant\varepsilon)\geqslant 1-e^{-\varepsilon}.$$

To derive from this inequality the first inequality in (8), it suffices to note that $\varepsilon \leq 1$ by (6) for $s=2$.

Next

$$\mathbf{P}(w\in A(8\delta))\leqslant\Sigma\mathbf{P}(|w(t_i)-w(t_{i-1})|>8\delta)\leqslant\Sigma\Phi_0(8\delta\sigma_i^{-1}), \tag{9}$$

where

$$\Phi_0(x)\equiv(2/\sqrt{2\pi})\int_x^\infty e^{-v^2/2}dy\leqslant(2/\sqrt{2\pi})\,x^{-1}e^{-x^2/2}.$$

Using the inequalities

$$(8\delta)^2\geqslant\sigma^2 \quad\text{and}\quad (8\delta)^2\geqslant\sigma_i^2+\sigma_i^2\ln\sigma_i^{-2}+2\sigma_i^2\ln\sigma^{-2},$$

we have

$$\Sigma\Phi_0(8\delta\sigma_i^{-1}) \leqslant \Sigma(2/\sqrt{2\pi})(\sigma_i/\sigma)\exp(-(1/2)-(1/2)\ln\sigma_i^{-2}-\ln\sigma^{-2}) \leqslant$$
$$\leqslant \Sigma(2/\sqrt{2\pi})(\sigma_i^2/\sigma)e^{-1/2}\sigma^2 \leqslant (1/2)\sigma \leqslant 4\delta.$$

This and (9) imply the second inequality in (8), which fully proves (7).

Now we prove that

$$\lambda \geq \delta \; . \tag{10}$$

Let $\sigma=\sigma_j$. We denote by $B(y)$ the set of all trajectories $x(t)$ whose maximal deviation on the interval (t_{j-1},t_j) from the line segment connecting $(t_{j-1},x(t_{j-1}))$ and $(t_j,x(t_j))$, is greater than y. We note that $(B(2\delta))^{(\delta)}\subset B(0)$; $P(\xi\in B(0))=0$. Thus, to prove (10) it suffices to verify that

$$P(w \in B(2\delta)) \geq \delta \; . \tag{11}$$

Because the process w is homogeneous,

$$\mathbf{P}(w\in B(2\delta)) = \mathbf{P}\Big(\max_{0<t<\sigma^2}|w(t)-t\sigma^{-2}w(\sigma^2)| > 2\delta\Big) =$$
$$= \mathbf{P}\Big(\max_{0\leqslant t\leqslant 1}|w(t)-tw(1)| > 2\delta\sigma^{-1}\Big).$$

Since the distribution of the process $w(t)-tw(1)$ <u>coincides</u> with the distribution of the Brownian bridge $w_0(t)$ (the correlation functions of these processes <u>coincide</u>), we have (see [6, p. 640])

$$\mathbf{P}(w\in B(2\delta)) = \mathbf{P}\Big(\max_{0<t<1}|w_0(t)| > 2\delta\sigma^{-1}\Big) = 2\sum_{k=1}^{\infty}(-1)^k\exp(-2(2k\delta\sigma^{-1})^2)$$

The relation (11) follows now from the inequalities

$$\mathbf{P}(w\in B(2\delta)) \geqslant 2\sum_{k=1}^{2}(-1)^k\exp(-2(2k\delta\sigma^{-1})^2) = 2e^{-1/8}\sigma^{3/4} - 2e^{-1/2}\sigma^3;$$
$$8\delta = \sqrt{\sigma^2(1+3\ln\sigma^{-2})} \leqslant \sqrt{12}\,\sigma^{3/4},\; \sigma \leqslant 1.$$

The assertion of the Theorem follows from (7) and (11).

3. ESTIMATES FOR LIPSCHITZ SETS

For $a>0$ we put

$$L_4(a) = \Sigma \mathbf{M}\{|\xi_i|^4;\ |\xi_i| < a\};\ H_s(a) = \Sigma \mathbf{M}\{|\xi_i|^s;\ |\xi_i| \geqslant a\};$$
$$g(a) = a(1 + \ln(1 + a^{-1}))$$

and note that

$$C_B \le g(C_B) \le C_B+1\ ; \qquad g(C_B)\to 0 \quad \text{as} \quad C_B\to 0\ .$$

THEOREM 2. For $0<a\le e^{-1}$ for any set B satisfying (2)

$$\Delta(B) \leqslant g(C_B)[84(L_4(a))^{1/4}|\ln a|^{3/4} + 41a|\ln a|^{3/2} +$$
$$+ 26(H_2(a))^{1/2}|\ln a|^{1/2} + 2H_1(a)].$$

COROLLARY. For $L_s\le 1$ we have (5), more precisely,

$$\Delta(B) \leqslant 285 g(C_B) L_s^{1/s}(1 + \ln(L_s^{-1/s}))^{3(s-1)/2s}.$$

To prove Corollary, it suffices to put in Theorem 2 $a=e^{-1}L_s^{1/s}(1+\ln(L_s^{-1/s}))^{-3/2s}$ and make use, for $0<r\le s\le 4$, of the inequalities

$$\mathbf{M}\{|\xi_i|^4;\ |\xi_i| < a\} \leqslant a^{4-s}\mathbf{M}|\xi_i|^s;\ \mathbf{M}\{|\xi_i|^r;\ |\xi_i| \geqslant a\} \leqslant a^{r-s}\mathbf{M}|\xi_i|^s,$$

implying

$$L_4(a) \leqslant a^{4-s}L_s;\ H_1(a) \leqslant a^{1-s}L_s;\ H_2(a) \leqslant a^{2-s}L_s.$$

Proof of Theorem 2. Let $F_1(x),\dots,F_n(x)$ be the distribution functions of the random variables $\xi_1,\dots,\xi_n$, respectively. We construct the random variables $\xi_1,\dots,\xi_n$ on a common probability space with a Wiener process $w(t)$, $0\le t<\infty$, and an independent of w sequence $\nu=(\nu_1,\dots,\nu_n)$ of independent random variables uniformly distributed on $[0,1]$. (More precisely, we construct a new sequence of random variables, distributed identically to the initial sequence $\xi_1,\ldots,\xi_n$.)

Let $u_i=F_i^{-1}(\nu_i)$. Since $Mu_i=0$, we put $v_i=F_i^{-1}(G_i(\nu_i))$, where $G_i(u)$ is uniquely defined as the solution of the equation

$$\int_u^{G_i(u)} F_i^{-1}(t)\,dt=0 \tag{12}$$

and $u_iv_i<0$ for $u_i\neq 0$. We now fix $a>0$ and denote by $I=I(\nu,a)$ the set of those i for which $\max\{|u_i|,|v_i|\}\geq a$. We define the sequence of independent vectors $(\xi_i,\tau_i=T_i-T_{i-1})$, $\tau_i\geq 0$, $T_0=0$, as follows: let $\xi_i=u_i$ and $\tau_i=0$ for $i\in I$; otherwise we define T_i as the first instant after T_{i-1} when the process w reaches either the level $w(T_{i-1})+u_i$ or the level $w(T_{i-1})+v_i$, and we put $\xi_i=w(T_i)-w(T_{i-1})$. Since u_i has the distribution function F_i, and ξ_i for $i\in I$ is defined the same as in the Skorokhod representation, the $\xi_i,\ldots,\xi_n$ possesses the required properties.

We note that by the properties of stopping times τ_i (see [7, 8]) for $d(\nu)\equiv\Sigma\mathbf{M}(\tau_i^2\,|\,\nu)\geqslant\mathbf{D}(T_n\,|\,\nu)$

$$\begin{aligned}&\mathbf{M}d(\nu)\leqslant 8\Sigma\mathbf{M}\{|\xi_i|^4;\ i\notin I\}\leqslant 8L_4(a);\\ &\mathbf{M}\{\tau_i;\ i\notin I\,|\,\nu\}=-u_iv_i;\quad \mathbf{M}T_n=\Sigma\mathbf{M}\{\xi_i^2;\ i\notin I\}\leqslant 1.\end{aligned} \tag{13}$$

We denote by ξ' the broken line with vertices at points $(t_i,\xi'(t_i)=w(T_i))$. In this case

$$\|\xi-w\|\leqslant\|\xi-\xi'\|+\|\xi'-w\|;\quad \|\xi-\xi'\|\leqslant\zeta\equiv\sum_{i\in I}|\xi_i|.$$

It follows from Lemmas 3 and 4 that $P(\|\xi-w\|>\alpha\,|\,\nu)\leq\beta$, for

$$\alpha=6\sqrt{6(1+b)|\ln a|(\delta+\gamma)}+a+\zeta;\quad \beta=(3+\mathbf{M}(T_n|\nu))ae^{-b},$$

where a and b are fixed, $b>0$, $0<a\leq e^{-1}$, and δ and γ are defined in Lemma 3.

We emphasize the point that the variables α, β, γ, δ and ζ depend on ν, but do not depend on w. Hence for the set B satisfying (2), we have the relations

$$\mathbf{P}(\xi \in B|\nu) \leqslant \mathbf{P}(w \in B^{(\alpha)}|\nu) + \beta \leqslant \mathbf{P}(w \in B) + C_B\alpha + \beta;$$
$$\mathbf{P}(\xi \in B) \leqslant \mathbf{P}(w \in B) + C_B\mathbf{M}\alpha + \mathbf{M}\beta.$$

Considering instead of the set B its complement, we obtain similarly inequalities in the opposite direction. Therefore,

$$\Delta(B) \leq C_B M\alpha + M\beta \ .$$

Putting now $b=\ln(1+C_N^{-1})$ and estimating $M\alpha$ and $M\beta$ using (13) and Lemmas 5, 6, we obtain the estimate required in the Theorem.

4. AUXILIARY ASSERTIONS

LEMMA 1.

$$\mathbf{P}(\tau_i > t|\nu) \leqslant 4t^{-1}\mathbf{M}(\tau_i|\nu)\exp(-2t|u_i - v_i|^{-2}).$$

Proof. Omitting temporarily the subscript i, we put

$$A = \pi \min\{|u|, |v|\}/|u-v|;\ K = \pi^2 t/(4|u-v|^2) \geqslant 2t/|u-v|^2.$$

For $\max\{|u|,|v|\}\leq a$ from [8, pp. 133-134] we have

$$\mathbf{P}(\tau > t|\nu) = \frac{4}{\pi}\sum_{m=0}^{\infty}\frac{\sin((2m+1)A)}{2m+1}\exp(-(2m+1)^2 2K).$$

Using the inequalities $|\sin x|\leq|x|$ and $e^{-|x|}\leq|x|^{-1}$, we obtain

$$\mathbf{P}(\tau > t|\nu) \leqslant \frac{4}{\pi}\sum_{m=0}^{\infty} AA(m)e^{-K}/K,$$

where

$$A(m) = 1/[2(2m+1)^2 - 1]; \quad \sum_{m=0}^{\infty} A(m) \leqslant A(0) + \int_0^{\infty} A(x)\,dx \leqslant 1.$$

We have finally

$$\mathbf{P}(\tau > t|\nu) \leqslant 4\pi^{-1}AK^{-1}e^{-K} \leqslant 16\pi^2 \min\{|u|, |v|\}|u-v|e^{-K}t^{-1} \leqslant$$
$$\leqslant 4|uv|e^{-K}t^{-1} = 4M(\tau|\nu)e^{-K}t^{-1}.$$

LEMMA 2. For $x \geq y \geq a^2$

$$\mathbf{P}(\psi \equiv \max|T_i - \mathbf{M}(T_i|\nu)| > x|\nu) \leqslant 2(ed(\nu)/(xy))^{x/y} +$$
$$+ 4\mathbf{M}(T_n|\nu)y^{-1}e^{-y/2a^2}.$$

Proof. Since $\tau_i \geq 0$ and $M(\tau_i|\nu) \leq a^2$, then from [9] (see Theorem 4) and from [10] we have

$$\mathbf{P}(\psi > x|\nu) = \mathbf{P}\left(\max\left|\sum_{j=1}^{i}(\tau_j - \mathbf{M}(\tau_j|\nu))\right| > x|\nu\right) \leqslant$$
$$\leqslant 2\exp(x/y - (d(\nu)/y^2 + x/y)\ln(xy/d(\nu)+1)) + \Sigma\mathbf{P}(|\tau_i -$$
$$- \mathbf{M}(\tau_i|\nu)| > y|\nu) \leqslant 2(ed(\nu)/(xy))^{x/y} + \Sigma\mathbf{P}(\tau_i > y|\nu).$$

Applying next Lemma 1, we obtain the required assertion.

LEMMA 3. For $b>0$, $0<a\leq e^{-1}$,

$$\gamma = \max|\mathbf{M}(T_i|\nu) - t_i|, \quad \delta = (1+b)[\sqrt{8d(\nu)|\ln a|} + 6a^2|\ln a|^2]$$

we have

$$\mathbf{P}(\mu \equiv \max|T_i - t_i| > \delta + \gamma|\nu) \leqslant (2 + \mathbf{M}(T_n|\nu))ae^{-b}.$$

Proof. The assertion follows from the relation $\mu \leq \psi + \gamma$ and from Lemma 2 for $y = \delta/((1+b)|\ln a|)$, $x = \delta + \gamma$, if we note that

$$xy \geqslant e^2 d(\nu); \quad x/y \geqslant (1+b)|\ln a|;$$
$$y/2a^2 \geqslant (1+b)|\ln a|; \quad |\ln a| = -\ln a \geqslant 1.$$

LEMMA 4. For $1 \geq x \geq a^2 \geq \sigma^2$

$$\mathbf{P}(\|w - \xi'\| > 6\sqrt{6(1+b)|\ln a|x} + a|\nu) \leqslant ae^{-b} + \mathbf{P}(\mu > x|\nu).$$

Proof. Since $\xi'(t_i)=w(T_i)$, for $|T_i-t_i|\le x$, $x\ge\sigma^2$ and $t=(t_{i-1},t_{i+1})$ we have

$$|w(t)-\xi'(t)|\leqslant|w(t)-w(t_i)|+|w(t_i)-w(T_i)|+ \\ +|\xi'(t_i)-\xi'(t)|\leqslant 2\eta_i'+a,$$

where

$$\eta_i' = \max_{|y-t_i|<x} |w(y)-w(t_i)| .$$

Let

$$\eta_j(x) = \max_{jx<y<(j+1)x} |w(y)-w(jx)|$$

and note that

$$\max_i \eta_i' \leqslant 3 \max_{0<j<1/x} \eta_j(x) \equiv 3\eta(x).$$

Therefore

$$\mathbf{P}(\|w-\xi'\|>6z+a|\nu)\leqslant\mathbf{P}(\eta(x)>z)+\mathbf{P}(\mu>x|\nu).$$

By the well-known properties of the maximum of a Wiener process (see, for instance, [6, p. 371])

$$\mathbf{P}(\eta(x)>z)\leqslant(1+1/x)\mathbf{P}(\eta_1(x)>z)\leqslant(1+1/x)2\Phi_0(z/\sqrt{x})\leqslant \\ \leqslant(1+1/x)4(2\pi)^{-1/2}z^{-1}\sqrt{x}\exp(-z^2/2x).$$

The last inequality for $z=\sqrt{6(1+b)|\ln a|x}$ together with $a^2(1+1/x)\le 3/2$ implies the required assertion.

LEMMA 5. $M\zeta\le 2H_i(a)$.

Proof. Two cases are possible:

$$0 \le F_i(a) \le G_i(F_i(a)) \quad \text{or} \quad G_i(F_i(a)) < F_i(-a+0) < 0 .$$

We consider the first case and use (12). We have

$$\mathbf{M}\{|\xi_i|;\ i\in I\}=2\mathbf{M}\{\xi_i;\ \xi_i\geqslant a\}\leqslant 2\mathbf{M}\{|\xi_i|;\ |\xi_i|\geqslant a\}. \tag{14}$$

Summarizing, we obtain the required result.

LEMMA 6. $M\gamma \leq 8(8L_4(a))^{1/2} + 3H_2(a)$.

Proof. Let

$$\tau_i' = M(\tau_i \mid \nu) - M\tau_i, \quad \gamma_1 = \max |M(T_i \mid \nu) - MT_i| = \max \left| \sum_{j=1}^{i} \tau_i' \right|,$$
$$\gamma_2 = \max |MT_i - t_i|.$$

We note that

$$\gamma_2 \leqslant \Sigma \left| \sigma_i^2 - M\tau_i \right| = \Sigma M \{\xi_i^2;\ i \in I\} \leqslant 3H_2(a),$$

since by (14)

$$M\{\xi_i^2;\ i \in I\} \leqslant M\{\xi_i^2;\ |\xi_i| \geqslant a\} + aM\{|\xi_i|;\ i \in I\} \leqslant 3M\{\xi_i^2;\ |\xi_i| \geqslant a\}.$$

To estimate $M\gamma_1$, we use the fact that the random variables $\tau_1', \ldots, \tau_n'$ are independent, since τ_i' depends only on ν_i, and $M\tau_i'=0$. From [11, p. 303] and (13) we have

$$M\gamma_1 \leqslant 8M \left| \Sigma\tau_i' \right| \leqslant 8 \left(D\left(\Sigma\tau_i'\right) \right)^{1/2} \leqslant 8\left(Md(\nu)\right)^{1/2} \leqslant 8\left(8L_4(a)\right)^{1/2}.$$

The inequality $\gamma \leq \gamma_2 + \gamma_2$ immediately gives us the required result.

REFERENCES

[1] A.I. Sakhanenko. "Estimates of the Rate of Convergence in the Invariance Principle." *Soviet Math. Dokl.*, 15, 6 (1974): 1752-1755.

[2] Yu.V. Prokhorov. "Convergence of Random Processes and Limit Theorems in Probability Theory." *Theory Prob. Applications*, 2 (1956): 157-214.

[3] A.A. Borovkov. "On the Rate of Convergence for the Invariance Principle." *Theory Prob. Applications*, 2 (1973): 207-225.

[4] J. Komlos, P. Major, and G. Tusnady. "An Approximation of Partial Sums of Independent R.V.'-s and simple D.F." *Z. Wahrscheinlichkeitstheorie verw. Gebiete*, 34, 1 (1976): 33-58.

[5] T.V. Arak. "On an Estimate of A.A. Borovkov." *Theory Prob. Applications*, 2 (1975): 372-373.

[6] I.I. Gikhman and A.V. Skorokhod. *Introduction to the Theory of Random Processes*. Philadelphia: Scripta Technica, W.B. Saunders Co., 1969.

[7] A.V. Skorokhod. *Studies in the Theory of Random Processes*. Reading, Massachusetts: Scripta Technica, Addison-Wesley Pub. Co., 1965.

[8] S. Sawyer. "A Remark on the Skorokhod Representation." *Z. Wahrscheinlichkeitstheorie verw. Gebiete*, 23, 1 (1972): 67-74.

[9] D.Kh. Fuk and S.V. Nagaev. "Probability Inequalities for Sums of Independent Random Variables." *Theory Prob. Applications*, 4 (1971): 643-660.

[10] A.A. Borovkov. "Notes on Inequalities for Sums of Independent Variables." *Theory Prob. Applications*, 3 (1972): 556-557.

[11] J.L. Doob. *Stochastic Processes*. New York: John Wiley, 1953.

Part 2

LIMIT THEOREMS FOR RANDOM PROCESSES AND THEIR APPLICATIONS

ON ASYMPTOTICALLY OPTIMAL TESTS FOR TESTING COMPLEX CLOSE HYPOTHESES

A.A. Borovkov and A.I. Sakhanenko

This article consists of two parts. In the first part we present a general principle which can be referred to as the "statistical invariance principle," enabling us to design asymptotically powerful criteria for testing close hypotheses. This principle is closely connected with ideas suggested by Le Cam, in particular, with the concept of continguality, and is a generalization of the results obtained by Chibisov, for a one-dimensional case in 1969. The second half of the paper is devoted to the construction of asymptotically optimal minimax criteria for some special classes of hypotheses.

Everywhere, if not stipulated otherwise, limits are taken as $n \to \infty$, and the region of integration is the entire space.

1. THE GENERAL FORM OF THE STATISTICAL INVARIANCE PRINCIPLE

To elucidate statement of the problem, we begin with a finite-dimensional parametric case. Let $X=(x_1,\ldots,x_n)$,

$x_i \in R$, be the sample of the distribution P_θ on the line R with given density $p(x,\theta)$, where $\theta=(\theta_1,\ldots,\theta_k)$ is an unknown parameter.

We consider close alternatives for the hypothesis $H^0=\{\theta=\theta^0\}$, i.e., alternatives corresponding to the values of the parameter

$$\theta = \theta^0 + h/b(n), \qquad h \in R^k , \tag{1}$$

where $b(n)\to\infty$ and $b(n)\le\sqrt{n}$. We assume without loss of generality that $\theta^0=0$. Then, if the density $p(x,\theta)$ is a smooth function of θ at the point $\theta=0$, the density of distribution p_θ with respect to P_0 is representable as

$$\frac{dP_\theta}{dP_0} = 1 + \frac{\gamma_n(x)}{b(n)}; \tag{2}$$

$$\gamma_n(x) \to \gamma(x) = \sum_{j=1}^{h} h_j f_j(x), \tag{3}$$

where $f_j(x)=\left.\frac{\partial p(x,\theta)}{\partial\theta_j}\right|_{\theta=0}$. In this representation, using a linear transformation of the parameter h it is possible to arrange so that the new functions f_j in (3) are orthonormal with respect to measure P_0.

Now we consider the hypothesis H that the parameter θ has some a priori distribution in R^k. More precisely, we assume that h in (1) is random and its distribution Q^n converges weakly to a distribution Q. This means that elements of the sample x_i for the hypothesis H have the random density (2) and that this density is a random process, and $\gamma_n(x)$ converges in some sense to a limit process $\gamma(x)$ of the form

$\sum_{j=1}^{k}\gamma_j f_j(x)$. The distribution of γ is in some sense degenerate, since it is defined by the distribution of the random variable $(\gamma_1,\ldots,\gamma_k)$ in R^k.

The foregoing makes the following statement of the problem meaningful. We assume that the "parameter" θ is given in an arbitrary space. This allows us to consider a broad class of statistical problems, including those which are regarded as non-parametric in the usual classification (i.e., such that the parameter is absent). As regards the observable quantities x_i, no assumption will be made. We denote by $(X,\sigma(X))$ the measurable space in which x_i assume their values.

Thus, let $X_n=(x_{n1},\ldots,x_{nn})$, $x_{ni}\in X$, be a sample of the distribution P_{γ_n} on $(X,\sigma(X))$ with density

$$\frac{dP_{\gamma_n}}{dP_0}=1+\frac{\gamma_n(x)}{b(n)},\quad b(n)\to\infty,$$

with respect to the given distribution P_0. We shall test the hypothesis $H^0=\{\gamma_n=0\}$ against the complex hypothesis $H=H_n=\{\gamma_n\neq 0\}$.

If the set of admissible trajectories γ_n is sufficiently rich, many approaches and results of the finite-dimensional case lose their force. For example, the method of maximal likelihood becomes meaningless.

We assume that on the set of alternatives $\{\gamma_n\}$ there is given an a priori distribution Q^n. This means that γ_n can be regarded as X_n-independent random process (or field) of an arbitrary, in contrast to the finite-dimensional case, nature. In this section we assume that $b(n)=\sqrt{n}$ and the distribution Q^n of the processes γ_n approximates the distribution Q of a process γ. The problem is to find the asymptotic form of the most powerful test.

We now introduce the Hilbert space $L_2=L_2(X,\sigma(X),P_0)$ of measurable functions on X with scalar product $(f,g)=\int f(x)g(x)dP_0(x)$ and the norm $||f||=(f,f)^{\frac{1}{2}}$. Furthermore, we denote by L_2^0 the subspace of L_2 consisting of functions f such that $(f,1)=0$. We assume that there exists a process γ in $L_2^0(P(||\gamma||<\infty)=1)$, which is the limit of γ_n in the following sense (the symbol => between the random variables designates the weak convergence of their distributions):

A) $(f,\gamma_n)\Rightarrow(f,\gamma)$ for all bounded and measurable functions. Since $(\gamma_n,1)=0$, we have $\gamma\subset L_2^0$;

$$\text{B}_1)\quad \rho_n \equiv \sqrt{n} \int_{\gamma_n(x)\geqslant\sqrt{n}} \gamma_n(x)\,dP_0(x) \Rightarrow 0;$$

$$\text{B}_2)\quad \int_{\gamma_n(x)<\sqrt{n}} \gamma_n^2(x)\,dP_0(x) \Rightarrow \|\gamma\|^2.$$

Here B_1 and B_2 are conditions of compactness of Q^n, and A is one of the weak convergence of Q^n for a special class of sets formed by the subspaces $(f,\gamma)>t$. These conditions are close to minimal ones in view of the theorem formulated later. We can note also that the condition $P(||\gamma||<\infty)=1$ does not necessarily imply loss of generality, since otherwise limit measures corresponding to the hypotheses H_0 and H are singular. We can also note that conditions A, B_1, B_2 do not require that $\gamma_n\in L_2$.

For every sequence $a=(a_1,a_2,\dots)$, let $a^{(k)}$ denote the "truncated" sequence, i.e., $a^{(k)}=(a_1,\dots,a_k,0,0,\dots)$. For the square summable sequences a and b we shall use the notation $(a,b)=\sum_{j=1}^{\infty} a_j b_j$, $\|a\|=(a,a)^{\frac{1}{2}}$. This should not cause confusion since it is always clear from the context to which of the spaces, L_2 or ℓ_2, the elements considered belong.

Further, each test is by definition a critical function ϕ (i.e., $\phi(X)$ is the probability for a given sample X to reject the hypothesis H_0). Let $\alpha_n(\phi)$ and $\beta_n(\phi)$ denote the significance level and the power of ϕ, respectively, when the sample is of size n.

Fix α, $0<\alpha<1$. We call the sequence of tests $\{\phi_n^*\}$ asymptotically most powerful if $\alpha_n(\phi_n^*)\to\alpha$ and for any other sequence ϕ_n, with the property $\alpha_n(\phi_n)\to\alpha$, we have

$$\liminf\,(\beta_n(\phi_n^*) - \beta_n(\phi_n)) \geq 0 .$$

We formulate now the "statistical invariance principle" mentioned above.

<u>THEOREM 1</u>. The following three assertions are true.

I. For any orthonormal system $\{g_j\}$ in L_2^0 finite-dimensional distributions of the sequence $G_n=(G_{n1},G_{n2},\dots)$, where $G_{nj}=n^{-\frac{1}{2}}\sum_{i=1}^{n} g_j(x_{ni})$, converge weakly, under the hypothesis H_0, to distributions of the sequence $\xi=(\xi_1,\xi_2,\dots)$ of independent random variables with standard normal distribution.

II. Under the hypothesis $H_n=H_n(Q^n)$ that the distribution of x_{ni} is P_{γ_n}, and if the conditions A, B_1, B_2 are satisfied, the finite-dimensional distributions G_n converge weakly to the distributions $\xi+\zeta=(\xi_1+\zeta_1, \xi_2+\zeta_2, \ldots)$, where the sequences $\zeta=(\zeta_1, \zeta_2, \ldots)=((g_1, \gamma), (g_2, \gamma), \ldots)$ and ξ are independent.

To formulate the third basic assertion of the principle, we denote by $Y=(Y_1, Y_2, \ldots)$ the observation (sample of size unity) over a random sequence $\eta=(\eta_1, \eta_2, \ldots)$. Let either simple hypothesis hold with respect to η:

1) the hypothesis h^0: the distribution η coincides with the distribution ξ;

2) the hypothesis h: the distribution η coincides with the distribution $\xi+\zeta$.

By limit tests we mean those for testing the hypothesis h^0 against h; and we designate them by the letter ψ in contrast to prelimit tests which we designate by the letter ϕ. From the Neymann-Pearson lemma one can easily obtain the explicit form (see Remark 2 below) of the most powerful limit test ψ^* at the level α.

III. If the subspace generated by the orthonormal system $\{g_j\}$ is the support of the distribution of the process γ, i.e.,

$$\lim_{k\to\infty} Q\left(\|\gamma\|^2 - \sum_{j=1}^{k} (\gamma, g_j)^2 > \delta\right) = 0 \quad \forall\delta>0, \tag{4}$$

then there exists a sequence $k_0(n)$ such that the sequence of tests $\phi_n^*(X_n)=\psi^*(G_n^{(k(n))})$ for all $k(n)\to\infty$, $k(n)\le k_0(n)$, will be asymptotically most powerful.

The term "invariance principle" in our formulation is justified because the nature of limit distributions therein depends neither on the system of functions $\{g_j\}$ chosen nor on the main distribution P_0.

<u>REMARK 1</u>. Considering the finite-dimensional case and passing then to the limit, it is not hard to see that the density $q(y)$ of the distribution $\xi+\zeta$ with respect to the distribution ξ at a point $y=(y_1,y_2,\ldots)$ is

$$q(y)=\mathbf{M}\exp[(y,\ \zeta)-\|\zeta\|^2/2]. \tag{5}$$

<u>REMARK 2</u>. The following alternative holds: either $P(\|\gamma\|=0)=1$ and $q(y)\equiv 1$, or

$$\mathbf{P}(\|\gamma\|=0)<1, \tag{6}$$

and then $P(q(\xi)=c)=0$ for all c. To prove this alternative, it is enough to note that from (6) and the equality $\|\gamma\|^2=\|\xi\|^2$ there follows the existence of k such that $P(|\zeta_k|=0)<1$, and therefore

$$\frac{d^2q(y)}{dy_k^2}=\mathbf{M}\zeta_k^2\exp[(y,\zeta)-\|\zeta\|^2/2]>0 \quad \forall y.$$

The last relation immediately implies that $P(q(\xi)=c)=0$, since for any cut

$$\mathbf{P}(q(\xi)=c/\xi_1=y_1,\ \ldots,\ \xi_{k-1}=y_{k-1},\ \xi_{k+1}=y_{k+1},\ \ldots)=0.$$

Hence we assume throughout that $\psi^*\equiv\alpha$, for $P(\|\gamma\|=0)=1$, and when (6) is satisfied ψ^* coincides with the indicator of the set $\{q(y)>c_\alpha\}$, where c_α is the unique solution of the equation $P(q(\xi)>c_\alpha)=\alpha$.

<u>REMARK 3</u>. Part III of Theorem 1 shows that using a finite number of statistics $G_{n1},\ldots,G_{nk}$, it is possible to construct

the test $\psi^*(G_n^{(k)})$ with parameters as close as possible to the parameters of the most powerful test.

REMARK 4. The condition (4) will, apparently, be satisfied if the system $\{g_j\}$ is complete.

REMARK 5. The assertion in part III of Theorem 1 is not, generally speaking, valid if $k(n)\equiv\infty$. This is evidenced by Example 2 given at the end of this section.

We formulate now an assertion from which Theorem 1 will follow as a particular case. We agree to call the function $\psi(y)$ defined on the space of sequences, almost continuous, if for each k the function $\psi(y^{(k)})$ is continuous almost everywhere with respect to Lebesgue measure in R^k and

$$\lim_{k\to\infty} \mathbf{P}(|\psi(\xi)-\psi(\xi^{(k)})|>\delta)=0 \quad \forall\delta>0. \tag{7}$$

Below, in Lemma 2, we shall show that the function q defined in (5) and therefore ψ^* are examples of almost continuous functions. By $\alpha(\psi)$ and $\beta(\psi)$ we shall denote the level of significance and the power of the limit test ψ.

THEOREM 2. Let conditions A, B_1, B_2 be satisfied and let the sequence of tests $\phi_n(X_n)$ be given. Then

I. If the test ψ satisfies the following conditions:

$$\psi \text{ is almost continuous ;} \tag{8}$$

$$\lim_{k\to\infty}\overline{\lim_{n\to\infty}} P_0(|\varphi_n(X_n)-\psi(G_n^{(k)})|>\delta)=0 \quad \forall\delta>0, \tag{9}$$

then $\alpha_n(\phi_n)\to\alpha(\psi)$ and $\beta_n(\phi_n)\to\beta(\psi)$.

II. If the orthonormal system $\{g_j\}$ satisfies the condition (4) and (9) is satisfied for $\psi=\psi^*$, the sequence of tests ϕ_n is asymptotically most powerful.

Let us compare this assertion with Chibisov's results [1]. It is not hard to see that if $X=R$ and all the functions $\{g_j\}$ are step functions, then the fact that the conditions (8) and (9) are satisfied guarantees, in the terminology used in [1], that the sequence belongs to the class ω. The assertion of Theorem 2 will follow in this case from Chibisov's results, if there is a probability space $(\Theta,\sigma(\Theta),\pi)$ and measurable mappings $\Delta(\cdot,\cdot)$ and $\Delta_n(\cdot,\cdot)$ from $X\times\Theta$ in X, such that $\gamma_n(x)=\Delta_n(x,\theta)$ and $\gamma(x)=\Delta(x,\theta)$ and the difference $\Delta_n-\Delta$ satisfies the condition

$$\text{C)}\quad P_0\left(\left|\sum_{i=1}^{n}[\Delta_n(x_{ni},\theta)-\Delta(x_{ni},\theta)]\right|>\sqrt{n}\delta\right)\to 0\quad \forall\delta>0$$

uniformly in $\theta\in\Theta_1$ for any compact $\Theta_1\subset\Theta$ (here and below by a compact we mean any $\Theta_1\subset\Theta$ such that the set $\{\Delta(\cdot,\theta):\ \theta\in\Theta_1\}$ is relatively compact in L_2^0).

It is necessary, therefore, to construct measures Q and Q^n on a common probability space, such that condition C is satisfied, which is, although possible, hard to do. We note that while condition C is used actually in the proofs in [1], the assumptions that $X=R$ and that g_j are steplike are needed therein only to describe the limit problem, which is different from our description. Hence, we shall assume in the sequel that the assertions of Theorem 2 are proved in [1] for any measure $\bar{Q}$ and the sequence of measures $\bar{Q}_n$ satisfying condition C instead of conditions A, B_1, B_2.

We also note that since $\Delta_n(x,\theta)\geq-\sqrt{n}$, $\int\Delta^2(x,\theta)dP_0(x)<\infty$ and $\int\Delta_n(x,\theta)dP_0(x)=\int\Delta(x,\theta)dP_0(x)=0$, it is not hard to obtain from the criterion of degenerate convergence (see [2, p. 331]) that C is equivalent to the following two conditions, more convenient than (3.3)-(3.5) in [1].

$$C_1)\ \rho_n(\theta)\equiv\sqrt{n}\int_{\Delta_n(x,\theta)\geqslant\sqrt{n}}\Delta_n(x,\theta)\,dP_0(x)\to 0;$$

$$C_2)\ \int_{\Delta_n(x,\theta)<\sqrt{n}}(\Delta_n(x,\theta)-\Delta(x,\theta))^2dP_0(x)\to 0,$$

where the convergence holds uniformly in θ on any compact of Θ.

Proof of Theorem 2. We fix the orthonormal sequence $\{g_j\}$ in L_2^0, satisfying the condition (4). Furthermore, let

$$\gamma_n^*(x)=\min\{\gamma_n(x),\sqrt{n}\};\quad \zeta_{nj}=(\gamma_n^*,g_j);\quad \zeta_n^{(k)}=(\zeta_{n1},\ldots,\zeta_{nk},0,0,\ldots);$$

$$\gamma^{(k)}(x)=\gamma(x)-\sum_{j=1}^{k}\zeta_j g_j(x);\quad \gamma_n^{(k)}=\gamma_n^*(x)-\sum_{j=1}^{k}\zeta_{nj}g_j(x).$$

We can rewrite the condition (4) in this case as:

$$\lim_{k\to\infty}\mathbf{P}(\|\gamma^{(k)}\|>\delta)=0\quad\forall\delta>0. \tag{10}$$

Let $\mu_n^{(k)}$ denote the Lévy-Prokhorov distance between the distributions in R^k, corresponding to the random vectors $\zeta^{(k)}$ and $\zeta_n^{(k)}$. By Lemma 1, below, $\mu_n^{(k)}\to 0$ for all k. Hence it is possible to find a sequence $k(n)\to\infty$ such that $\mu_n^{(k(n))}\to 0$.

As π we take the measure in $\Theta=[0,1]\times\ell_2$, generated by the distribution of the pair (ζ_0,ζ), where ζ_0 and ζ are independent and ζ_0 is distributed uniformly on $[0,1]$. Also, let $\Delta(x,\theta)=\gamma(x)$. By the Strassen theorem (see [3]) there exists a function $\Delta_n(x,\theta)$ such that $\Delta_n(x,\theta)$ is identically distributed with $\gamma_n(x)$ and for $\gamma_n(x)=\Delta_n(x,\theta)$

$$\pi\left(\theta: \left\|\zeta_n^{(k(n))} - \zeta^{(k(n))}\right\| > \mu_n^{(k(n))}\right) \leqslant \mu_n^{(k(n))}.$$

Letting next $\gamma_n(x)=\Delta_n(x,\theta)$, from the last relation and (10) we have

$$\left\|\zeta_n^{(k(n))} - \zeta\right\| \leqslant \left\|\zeta_n^{(k(n))} - \zeta^{(k(n))}\right\| + \left\|\gamma^{(k(n))}\right\| \Rightarrow 0.$$

Using this convergence and rewriting B_2 in the form

$$\|\gamma_n^*\|^2 = \|\zeta_n^{(k(n))}\|^2 + \|\gamma_n^{(k(n))}\|^2 \Rightarrow \|\gamma\|^2,$$

we obtain $\|\gamma_n^{(k(n))}\| \Rightarrow 0$, which plus B_1 plus (10) imply that there will be a sequence $\delta(n)\to 0$ such that for the event

$$\bar{\Theta}_n = \{\|\zeta_n^{(k(n))} - \zeta^{(k(n))}\| \leqslant \delta(n),\ \|\gamma_n^{(k(n))}\| \leqslant \delta(n), \|\gamma^{(k(n))}\| \leqslant \delta(n), \rho_n(\theta) \leqslant \delta(n)\}$$

we have $\pi\{\theta: \theta \notin \bar{\Theta}_n\} \leqslant \delta(n)$.

Now we define the functions $\bar{\Delta}_n$, letting $\bar{\Delta}_n(x,\theta)=\Delta_n(x,\theta)$ for $\theta\in\bar{\Theta}_n$, and $\bar{\Delta}_n(x,\theta)=0$ otherwise. The sequence $\bar{\Delta}_n$ thus constructed satisfies conditions C_1 and C_2, since for $\theta\in\bar{\Theta}_n$

$$\sqrt{n} \int\limits_{\bar{\Delta}_n(x,\theta)\geqslant\sqrt{n}} \bar{\Delta}_n(x,\theta)\, dP_0(x) \leqslant \delta(n);$$

$$\|\bar{\Delta}_n(x,\theta) - \Delta(x,\theta)\| \leqslant \|\zeta_n^{(k(n))} - \zeta^{(k(n))}\| + \|\gamma_n^{(k(n))}\| + \|\gamma^{(k(n))}\| \leqslant 3\delta(n).$$

Therefore, the assertion of Theorem 2 holds for the sequence of distributions $\bar{Q}^n$ corresponding to the processes $\gamma_n=\Delta_n(\cdot,\theta)$. Denoting by $\bar{\beta}_n(\phi)$ the power of the test ϕ under the hypothesis $H_n(\bar{Q}^n)$, it is not hard to notice that

$$|\beta_n(\varphi) - \beta_n(\varphi)| \leqslant \pi(\theta: \Delta_n(\cdot,\theta) \neq \Delta_n(\cdot,\theta)) \leqslant \delta(n) \to 0.$$

From this it follows that for any sequence of the tests ϕ_n

the limits of their power under the alternatives $H_n(Q^n)$ and $H_n(\bar{Q}^n)$ coincide. Theorem 2 is proved.

LEMMA 1. $\zeta_n^{(k)} \Rightarrow \zeta^{(k)}\ \forall k$.

Proof. Let $g \in L_2$. We denote by g^c the function g truncated at the level c, i.e., the product of g and the indicator function of the set $\{x: |g(x)| \le c\}$. Since g^c is a bounded function, by condition A $(g^c, \gamma_n) \Rightarrow (g^c, \gamma)$, and therefore it is possible to choose $c(n) \to \infty$, $c(n) \le \sqrt{n}$, such that $(g^{c(n)}, \gamma_n) \Rightarrow (g, \gamma)$. This plus B_1, B_2 plus the inequality

$$|(g, \gamma_n^*) - (g^{c(n)}, \gamma_n)| \leqslant \|g - g^{c(n)}\| \|\gamma_n^*\| + c(n) \int |\gamma_n^* - \gamma_n| \, dP_0(x)$$

imply $(g, \gamma_n^*) \Rightarrow (g, \gamma)\ \forall g \in L_2$.

Since the last relation is satisfied for all g of the form $\sum_{i=1}^{k} \alpha_i g_i$, then (see, for instance, [4, p. 36]) Lemma 1 is proved.

LEMMA 2. The function $q(y)$ is almost continuous.

Proof. The function $q(y^{(k)})$ is continuous in $y^{(k)}$, since

$$\exp(y_j z_j - z_j^2/2) \leqslant \exp(y_j^2/2), \ |z_j| \exp(y_j z_j - z_j^2/2) \leqslant \exp(y_j^2),$$

and hence

$$\left| \frac{q}{q y_j} q(y^{(k)}) \right| \leqslant \exp(\|y^k\|^2/2 + y_j^2/2) < \infty.$$

We prove now that the function $q(y)$ satisfies the condition (7). We note that for $z \in \ell_2$ the random variable (ξ, z) has a normal distribution, and therefore

$$\mathbf{M} \exp[(\xi, z) - \|z\|^2/2] = 1;$$

$$\mathbf{M} | \exp[(\xi, z) - \|z\|^2/2] - 1 \| = (2\pi)^{-1/2} \int \| \exp[-(t - \|z\|)^2/2] -$$

$$- \exp[-t^2/2] | \, dt = 2(2\pi)^{-1/2} \int_{-\|z\|/2}^{\|z\|/2} \exp(-t^2/2) \, dt \leqslant 2 \min\{\|z\|, 1\}.$$

Using the relations obtained plus the independence of ζ, $\xi^{(k)}$ and $\xi - \xi^{(k)}$, we have

$$\mathbf{P}(|q(\xi) - q(\xi^{(k)})| > \delta) \leqslant \delta^{-1}\mathbf{M}|q(\xi) - q(\xi^{(k)})| =$$

$$= \delta^{-1}\mathbf{M}\{\mathbf{M}_\zeta \exp[(\xi^{(k)}, \zeta^{(k)}) - \|\zeta^{(k)}\|^2/2]\mathbf{M}_\zeta |\exp[(\xi - \xi^{(k)}, \zeta - \zeta^{(k)}) -$$

$$- \|\zeta - \zeta^{(k)}\|/2] - 1|\} \leqslant 2\delta^{-1}\mathbf{M} \min\{\|\zeta - \zeta^{(k)}\|, \ 1\} \leqslant$$

$$\leqslant 2\delta^{-1}(\varepsilon + \mathbf{P}(\|\zeta - \zeta^{(k)}\| > \varepsilon)).$$

Since $\|\zeta - \zeta^{(k)}\| = \|\gamma^{(k)}\|$, passing to the limit first as $k \to \infty$, and then as $\varepsilon \to 0$, and using (10), we obtain Lemma 2.

EXAMPLE 1. Let γ be a Gaussian process. For simplicity, we assume that the normally distributed random variables $\zeta_1, \zeta_2, \ldots$ are independent. Also, we denote $a_j = \mathrm{M}\zeta_j$, $\lambda_j = \mathrm{D}\zeta_j$. It is not hard to calculate that

$$q(y) = \left(\prod_{j=1}^{\infty} (1 + \lambda_j)^{-1/2} \right) \exp\left[\frac{1}{2} \sum_{j=1}^{\infty} \frac{\lambda_j y_j^2 + 2a_j y_j - a_j^2}{1 + \lambda_j} \right].$$

Therefore, when (6) is satisfied (i.e., for $0 < \sum_{j=1}^{\infty} (\lambda_j + a_j^2) < \infty$) the most powerful limit test ψ^* coincides with the indicator of a set of the form

$$S(y) \equiv \sum_{j=1}^{\infty} \frac{\lambda_j y_j^2 + 2a_j y_i}{1 + \lambda_j} > c_\alpha.$$

We note that for all $k(n) \to \infty$ the sequence of tests $\psi^*(G_n^{(k(n))})$ satisfies the condition (9), since for $k(n) > k$

$$\mathbf{M}\left|S\left(G_n^{(k(n))}\right)-S\left(G_n^{(k)}\right)\right|\leqslant\mathbf{M}\left(\sum_{j=k+1}^{k(n)}\lambda_j G_{nj}^2\right)+$$
$$+2\left(\mathbf{M}\left|\sum_{j=k+1}^{k(n)}a_j(1+\lambda_j)^{-1}G_{nj}\right|^2\right)^{1/2}\leqslant\sum_{j=k+1}^{\infty}\lambda_j+$$
$$+2\left(\sum_{j=k+1}^{\infty}a_j^2\right)^{1/2}\to 0 \text{ for } k\to\infty.$$

Therefore, the sequence of tests corresponding to the indicator of the set $S(G_n^{(k(n))})>c_\alpha$, is most powerful for all $k(n)\to\infty$, in particular for $k(n)\equiv\infty$.

EXAMPLE 2. Let P_0 be the uniform distribution on $[0,1]$ and let $\{g_{mj}\}$ be triangular array of functions on $[0,1]$ of the form

$$g_{mj}(x)=\begin{cases}2^{m/2} & \text{for } (j-1)2^{-m}\leq x<(2j-1)2^{-m-1},\\ -2^{m/2} & \text{for } (2j-1)2^{-m-1}\leq x<j2^{-m},\\ 0 & \text{otherwise},\end{cases}$$

where $1\leq j\leq 2^m$, $m=0,1,2,\ldots$ In this case the system of functions $\{g_{mj}\}$ is orthonormal and orthogonal to $g_0\equiv 1$. We introduce the independent triangular arrays $G_n=\{G_{nmj}\}$ and $\zeta=\{\zeta_{mj}\}$, where

$$G_{nmj}=n^{-1/2}\sum_{i=1}^{n}g_{mj}(x_{ni});\ \mathbf{P}(\zeta_{mj}=\pm 1,\ \zeta_{rs}=0$$
$$\forall(r,s)\neq(m,j))=2^{-2m-1}.$$

Also let

$$\gamma(x)=\sum_{m,j}\zeta_{mj}g_{mj}(x),\ \ \gamma_n(x)=\sum_{m=0}^{m(n)}\sum_j\zeta_{mj}g_{mj}(x),$$

where $n/2<2^{m(n)}\leq n$. The most powerful limit test ψ^* coincides with the indicator of a set of the form

$$q(y) \equiv e^{1/2} \sum_{m,j} 2^{-2m-2} [\exp(y_{mj}) + \exp(-y_{mj})] > c_\alpha.$$

We note that in this case the test $\psi^*(G_n)$ will not be asymptotically optimal. Indeed, for each n, with probability 1 we can find an interval of the form $[(j_0-1)2^{-m_0}, j_0 2^{-m_0})$, into which only one element of the sample $x_{n1},\ldots,x_{nn}$ (we denote it in the sequel by x^*) falls. Then, for $m>m_0$ it is possible to find an interval $[(j_m-1)2^{-m}, j_m 2^{-m})$ imbedded into $[(j_0-1)2^{-m_0}, j_0 2^{-m_0})$, containing exactly one element of the sample (in fact, the same x^*). Next, we have

$$G_{nmj_m} = n^{-1/2} g_{mj_m}(x^*) = \pm 2^{m/2} n^{-1/2};$$

$$q(G_n) \geqslant e^{1/2} \sum_{m=m_0}^{\infty} 2^{-2m-2} \exp|G_{nmj_m}| = e^{1/2} \sum_{m=m_0}^{\infty} 2^{-2m-2} \exp(2^{m/2} n^{-1/2}) \equiv \infty.$$

REMARK 6. In the considerations above, the simple hypothesis H^0 was expressed as $\{\gamma_n=0\}$. However, the assertions similar to Theorems 1 and 2 will hold as well for a "symmetric" problem, when two hypotheses H_n^0 and H_n^1 are tested, where the hypothesis H_n^i states that the sample $X_n=(x_{n1},\ldots,x_{nn})$ is taken from the distribution with density $1+n^{-\frac{1}{2}}\gamma_n^i(x)$ with respect to the fixed probability measure P_0 on X, and $\gamma_n^i(x)$ is the trajectory of a random process with values in X.

Assume that for each $i=0,1$ there exist limit processes γ^i with values in L_2^0, such that $\gamma_n=\gamma_n^i$ and $\gamma=\gamma^i$ satisfy conditions A, B_1, B_2. In this case the assertion of Theorem 2.I holds. Also for $H_n=H_n^i$ and $\zeta_n=\zeta_n^i$ the assertion of Theorem 1.II is true. The limit theorem becomes thus: using the sample of size Y over the random sequence η, test the hypothesis h^0 against the hypothesis h^1, where h^i states that the distribution of η coincides with $\xi+\zeta^i$. The density of the distribution $\xi+\zeta^1$ with respect to the distribution $\xi+\zeta^0$ at the point y is $q^1(y)/q^0(y)$, where $q^i(y)$ coincides with $q(y)$ in (5) for $\zeta=\zeta^i$. According to Remark 2, the optimal limit test ψ^* of the level α will be defined uniquely as the indicator of a set of the form $\{q^1(y)/q^0(y)>c_\alpha\}$, if at least for one i the process $\gamma=\gamma^i$ satisfies (6), and $\psi^*=\alpha$ otherwise. The assertions of Theorem 1.III and Theorem 2.II hold for ψ^* thus constructed, if the orthonormal system $\{g_j\}$ satisfies (4) for $\gamma=\gamma^0$ and $\gamma=\gamma^1$.

2. ASYMPTOTICALLY MINIMAX TESTS

We consider first the finite-dimensional parametric case. Let the relations (1), (2) be satisfied and let the distributions Q^n of the processes γ_n converge weakly to the distribution Q (A^n and Q are distributions in R^k).

We have the following assertion which we can borrow from [5]. Let $q^n(y)$ and $q(y)$ denote respectively the densities of the distributions Q^n and Q with respect to Lebesgue measure at the point y.

THEOREM 3. Let $b(n)=o(\sqrt{n})$, and let the distributions Q^n and Q be such that there exist densities $q^n(0)$ and $q(0)>0$ at the point $y=0$; and let in addition to the convergence $Q^n \Rightarrow Q$ there be a convergence of densities $q^n(0)\to q(0)>0$. Next let c_α be the solution of the equation

$$\mathbf{P}(\|\xi\|^2 > c_\alpha) = \mathbf{P}(\xi_1^2 + \ldots + \xi_k^2 > c_\alpha) = \alpha, \tag{11}$$

where the vector $\xi=(\xi_1,\ldots,\xi_k)$ consists of independent normally distributed random variables. Then for any Q among all the tests ϕ_n at the asymptotic level α, the test ϕ_n^* coinciding with the indicator of the set

$$\left\{X_n : \sum_{j=1}^{k}\left(n^{-1/2}\sum_{i=1}^{n} f_j(x_i)\right)^2 > c_\alpha\right\} \tag{12}$$

will be asymptotically optimal for testing the hypotheses H^0 against the hypothesis $H(Q^n)$. This means that for probabilities of the errors of the second kind $\limsup (1-\beta_n(\phi_n^*))/(1-\beta_n(\phi_n)) \leq 1$ is satisfied for any test ϕ_n with $\alpha_n(\phi_n) \to \alpha$.

It is not hard to verify that the test ϕ_n^* coincides with maximum likelihood test. It yields, in some sense, a universal test for testing H^0 against a complex hypothesis $H=\{\theta=h/b(n),\ h\neq 0\}$, because it does not depend on Q.

Are such universal tests possible in the general (nonparametric) case for $b(n)=\sqrt{n}$? The answer to this question is, strictly speaking, negative; however, when the question is formulated somewhat differently, it is possible to construct tests close to (12) under some conditions.

In what follows we shall consider only the case $b(n)=\sqrt{n}$.

Let, for example, the limit process γ be Gaussian with zero mean, and let the sample space X be arbitrary. It is well known [4, p. 247] that in this case there exists an orthonormal system $\{g_j\}$ in X, such that $\gamma(x)=\sum_{j=1}^{\infty}\zeta_j g_j(x)$, where ζ_j are independent and normally distributed with zero means and dispersions λ_j. The functions g_j are defined as eigenfunctions of the correlation operator $Tg(x)=\int g(y)R(x,y)dP_0(y)$, where $R(x,y)$ is the correlation function of the process γ, and λ_j are the eigenvalues corresponding to g_j.

Example 1 in Section 1 implies the following.

<u>COROLLARY</u>. Under the conditions formulated the asymptotically optimal test ϕ_n^* coincides with the indicator of the set

$$\left\{X_n : \sum_{j=1}^{\infty} \lambda_j (1+\lambda_j)^{-1} G_{nj}^2 > C_\alpha\right\},$$

where C_α is the solution of the equation

$$\mathbf{P}\left(\sum_{j=1}^{\infty} \lambda_j (1+\lambda_j)^{-1} \xi_j^2 > C_\alpha\right) = \alpha$$

(G_{nj} and ξ_j are as in Theorem 1).

From this assertion it follows, in particular, that the test ϕ_n^* essentially depends on the limit measure Q. This is also true, obviously, in the non-parametric case.

The assumption concerning the Gaussianness of the limit process γ could be justified in many cases (for example, in verifying non-algorithmic generator of random integers). It is, however, unlikely that we will know the form of the correlation function $R(x,y)$ determining the form of the asymptotically optimal test (i.e., the system of functions g_j and integers λ_j). It is therefore desirable to obtain tests not depending, if possible, on the limit distribution Q (similar to Theorem 3).

Thus, let us consider the problem: given a finite system of orthonormal functions $\{g_1, \ldots, g_k\}$ (generally speaking, in no way connected with the correlation function $R(x,y)$), about which we know (hypothesis H) that

$$\mathbf{M} \sum_{j=1}^{k} (\gamma, g_j)^2 \geqslant \Lambda > 0. \tag{13}$$

This means that we know a priori that γ contains "harmonics" $g_1, \ldots, g_k$ and the sum "power" of the projection of γ onto a subspace spanned on $g_1, \ldots, g_k$, is not smaller than Λ. Otherwise, the distribution of the process γ is arbitrary.

We complete now $\{g_1, \ldots, g_k\}$ to the system satisfying (4) and denote by Ψ^α the class of sequences of the tests $\{\phi_n\}$ satisfying for some $\psi(y)=\psi(y,\{\phi_n\})$ the conditions (8), (9) and $\alpha(\psi)=\alpha$. From Theorem 2 it follows that for any $\{\phi_n\} \in \Psi^\alpha$ we have

$$\alpha_n(\phi_n) \to \alpha \quad \text{and} \quad \beta_n(\phi_n) \to \beta(\psi) = \beta(\psi(\cdot,\{\phi_n\}), Q)$$

(for the notation see Theorem 1).

Under the assumptions made above the following theorem holds.

THEOREM 4. In the class Ψ^α, the test $\psi^*(G_n^{(k)})$, where $\psi^*(y)$ is the indicator of the set $\{\|y^{(k)}\| > c_\alpha\}$ and c_α is the solution of Eq. (11), is the asymptotically minimax test for testing the hypothesis H_0 against H.

The asymptotically minimax property is understood here as minimaxness for limit probabilities of errors of the second kind:

$$\sup_Q [1-\beta(\psi^*, Q)] = \min_\psi \sup_Q [1-\beta(\psi, Q)],$$

where supremum is taken over all the distributions Q satisfying (13) and minimum is taken over all ψ with $\alpha(\psi)=\alpha$. We see that the test (12) is valid under the new conditions as well. The assertion of Theorem 4 will follow from Theorem 2 and Theorem 5 below.

As in Theorem 1, let $\xi=(\xi_1,\xi_2,\ldots)$ and $\zeta=(\zeta_1,\zeta_2,\ldots)$ be independent sequences, $\xi_1,\xi_2,\ldots$ being independent with standard normal distribution and let $\zeta^{(k)}=(\zeta_1,\ldots,\zeta_k,0,0,\ldots)$. By $\mathcal{T}$ we denote the set of all orthonormal transformations in R^k. Furthermore, let $T\zeta$ be a vector composed of the first k coordinates of the image under T of the first k coordimates of ζ, and all the succeeding coordinates replaced by zeros so that $T\zeta=(T\zeta)^{(k)}=T(\zeta^{(k)})$.

Assume that we are given a set $\mathcal{F}$ of distributions Q_ζ of the sequence ζ, having the property:

$$\text{if } Q_\zeta \in \mathcal{F} \text{ then } Q_{T\zeta} \in \mathcal{F} \ \forall\, T \in \mathcal{T} \ . \qquad (14)$$

Test the hypothesis $H^0=\{\zeta=0\}$ against $H=\{Q_\zeta\in\mathcal{F}\}$ using the sample Y of size one from the distribution corresponding to $\xi+\zeta$. As earlier, we designate by $\alpha(\psi)=M\psi(\xi)$ and $\beta(\psi,Q_\zeta)=M\psi(\xi+\zeta)$ the level of significance and, respectively, the power of the test ψ for the fixed distribution Q_ζ for the alternative H.

<u>THEOREM 5</u>. The test ψ^* defined in Theorem 4 is minimax (minimizing the maximal possible error of the second kind) among tests of the level α:

$$\inf_{\mathcal{F}} \beta(\psi^*, Q_\zeta) = \max \inf_{\mathcal{F}} \beta(\psi, Q_\zeta),$$

where the maximum is taken over all ψ such that $\alpha(\psi)=\alpha$.

From this assertion it follows that Theorem 4 will hold also when the hypothesis H states that (13) is satisfied not in the entire class of distributions of γ, but only in a subclass supporting the rotation operator (see (14)), for example, in the class of all Gaussian distributions.

Proof of Theorem 5. From the relation

$$\inf_{\mathcal{F}} \beta(\psi(y), Q_\zeta) \leqslant \inf_{\mathcal{F}} \beta\left(\psi(y), Q_{\zeta^{(k)}}\right) = \inf_{\mathcal{F}} \beta\left(\psi\left(y^{(k)}\right), Q_\zeta\right)$$

it follows that we can restrict ourselves to tests depending only on $y^{(k)}$. Further, from the results obtained in Section 3 of Chapter 8 in [6] it follows that the minimax test needs to be sought among the tests invariant relative to the group of transformations T, since the hypotheses H_0 and H are invariant relative T by (14) and the identical distributivity of $\xi^{(k)}$ and $T\xi^{(k)}$. However, among the invariant tests, the test ψ^* possesses an even stronger property than the minimaxness; in fact ψ^* is uniformly most powerful (see [6, p. 415] for $r=s=q=k$). //

Now we consider the case where the carrier of the distribution Q is known to be separated from zero by a hyperplane. We fix $\alpha\in(0,1)$, and for $\theta\in\ell_2$, $\theta\neq 0$, we denote by ψ_0 the test coinciding with the indicator of the set $(y,\theta)>c_\alpha\|\theta\|$, where c_α is the solution of the equation

$$\Phi_0(c_\alpha) = (2\pi)^{-1/2} \int_{c_\alpha}^{\infty} \exp(-t^2/2)\, dt = \alpha. \tag{15}$$

By δ_ζ we denote the probability measure concentrated at the point $\zeta \in \ell_2$. In this case from Theorem 1.II it follows that

$$\beta(\psi_\theta, \delta_\zeta) = \Phi_0(c_\alpha - (\zeta, \theta)/\|\theta\|). \tag{16}$$

Let Γ be a closed convex set in ℓ_2, not containing zero, and let F be the set of all distributions on Γ. In this case there exists a very simple minimax test on the basis of sample Y from the distribution of $\xi+\zeta$, for testing the hypotheses $H^0=\{\zeta=0\}$ and $H=\{Q_\zeta \in F\}$.

<u>THEOREM 6</u>. Let θ^* be the point of Γ nearest to zero. Then $\psi_{\theta*}$ is a minimax test of the level α, i.e.,

$$\inf_{\mathcal{F}} \beta(\psi_{\theta*}, Q_\zeta) = \max \inf_{\mathcal{F}} \beta(\psi, Q_\zeta), \tag{17}$$

where the maximum is taken over all ψ such that $\alpha(\psi)=\alpha$.

<u>Proof</u>. We note that the test $\psi_{\theta*}$ possesses two properties: it is optimal in testing the hypothesis $H^0=\{\zeta=\theta\}$ against the alternative $H^*=\{\zeta=\theta^*\}$ and $\min_{\zeta\in\Gamma} \beta(\psi_{\theta*}, \delta_\zeta)=\beta(\psi_{\theta*}, \delta_{\theta*})$. Therefore, by Theorem 1 in [6, p. 446], the distribution on F, assigning the unit mass to the point $\delta_{\theta*}$ is the least favorable and the test $\psi_{\theta*}$ is minimax among tests of the level α.

<u>REMARK 1</u>. Let $\Gamma_0=\Gamma\cup\{\zeta:-\zeta\in\Gamma\}$, and let F_0 be the set of distributions on Γ_0. Then the test coinciding with the indicator of the set $|(y,\theta^*)|>c_{\alpha/2}\|\theta^*\|$ is minimax among all the tests at the level α. To prove this it suffices to note that the distribution assigning the equal masses to the points $\delta_{\theta*}$ and $\delta_{-\theta*}$ will be the least favorable.

REMARK 2. If Γ is an arbitrary set, we define θ^* as the point nearest to zero in the closure of the convex hull $\bar{\Gamma}$ of the set Γ and assume that $\bar{\Gamma}$ does not contain zero. In this case $\psi_{\theta*}$ is minimax among linear tests of the level α, i.e., (17) is satisfied when the maximum is taken over all ψ of the form ψ_θ. By (16) it is enough to prove that

$$\inf_{\zeta\in\Gamma} (\zeta, \theta^*)/\|\theta^*\| = \|\theta^*\| = \max_\theta \inf_{\zeta\in\Gamma} (\zeta, \theta)/\|\theta\| ,$$

which follows from the fact that the hyperplane $(y,\theta^*)=\|\theta^*\|^2$ is tangent to the set Γ and separates it from zero; and for any other hyperplane of the form $(y,\theta)= \|\theta^2\|$ with these properties we have $\|\theta\| \leq \|\theta^*\|$.

REMARK 3. If the point θ^* defined in Remark 2 does not belong to Γ, then the test $\psi_{\theta*}$ cannot be minimax in the class of all tests of the level α. Indeed, let Γ consist of two points: $a=(1,1,0,0,0,\ldots)$ and $b=(1,-1,0,0,0,\ldots)$. Then $\theta^*=(1,0,0,0,\ldots)\notin\Gamma$. It is easy to see that in this case the least favorable distribution on F assigns equal probabilities to the points δ_a and δ_b, and therefore by the Neymann-Pearson lemma, the optimal test at the level α coincides with the indicator of a set of the form

$$c_\alpha < \frac{(2\pi)^{-1/2}\exp\left[-(y_1-1)^2/2-(y_2-1)^2/2\right]+(2\pi)^{-1/2}\exp\left[-(y_1-1)^2/2-(y_2+2)^2/2\right]}{2(2\pi)^{-1/2}\exp\left[-y_1^2/2-y_2^2/2\right]} = 2^{-1}e^{-1}e^{y_1}\left(e^{y_2}+e^{-y_2}\right).$$

Similarly, we can see now that $\psi_{\theta*}(G_n)$ is asymptotically minimax in the class of tests Ψ^α, where the asymptotic minimaxness is understood in the same sense as in Theorem 4. Let-

$g^*(x)=\sum_{j=1}^{\infty}\theta_j^* g_j(x)$, we can represent the test $\psi_{\theta*}(G_n)$ as the indicator of the set

$$\left\{X_n : n^{-1/2}\sum_{i=1}^{n} g^*(x_{ni}) > c_\alpha \| g^* \|\right\},$$

where c_α is defined in (15).

To conclude the discussion, we consider an application of the Theorem to pattern recognition problems. Assume that we need to test the hypothesis H^0 against the hypothesis H_{γ_0}, where the density $1+n^{-\frac{1}{2}}\gamma_0(x)$ is unknown and estimated through a separate sample of size m. Using the known estimates for the density, it is possible to obtain the estimate $\tilde{\gamma}_m(x)$ for $\gamma_0(x)$ and construct for γ_0 a "confidence sphere" Γ_m in L_2 of radius ε centered at the point $\tilde{\gamma}_m$. Letting $\|\tilde{\gamma}_m\|>\varepsilon$ and γ_0 Γ_m, and applying the Theorem, we obtain $g^*=\tilde{\gamma}_m(1-\varepsilon/\|\tilde{\gamma}_m\|)$, so that the asymptotically optimal, in the sense mentioned, test is given by

$$\left\{n^{-1/2}\sum_{i=1}^{n} \tilde{\gamma}_m(x_{ni}) > c\right\}.$$

REFERENCES

[1] D.M. Chibisov. "Transition to the Limiting Process for Deriving Asymptotically Optimal Tests." *Sankhya*, A31, 3 (1969): 241-258.

[2] M. Loève. *Probability Theory*. Princeton, New Jersey: Van Nostrand, 1963.

[3] G. Schay. "Nearest Random Variables With Given Distributions." *Ann. Probability*, 2, 1 (1974): 163-166.

[4] I.I. Gikhman and A.V. Skorokhod. *Introduction to the Theory of Random Processes.* Philadelphia: W.B. Sauders Co., 1969.

[5] A.A. Borovkov. "Asymptotically Optimal Tests for Compound Hypotheses." *Theory Prob. Applications*, 3 (1975): 447-469.

[6] E.L. Lehmann, *Testing Statistical Hypotheses.* New York: John Wiley, 1959.

ON MODERATELY LARGE DEVIATIONS FROM THE INVARIANT MEASURE

A.A. Mogul'skij

1. INTRODUCTION

Let R be the linear space of finite measures μ given on the Borel σ-algebra B of some metric space X, and let $P \subset R$ be the class of probability measures on B. We consider the random process $\xi(t)$, $0 \leq t$, with values in X, whose trajectories are Borel measurable with probability 1. Let

$$\hat{P}_t(A) = \int_0^t \chi_A(\xi(y))\,dy$$

be the sojourn time t of the process $\xi(y)$ in the set $A \in B$. Then under some mild assumptions, the random probability measures $t^{-1}\hat{P}_t$ and $t\to\infty$ converge weakly with probability 1 to a nonrandom probability measure P.

In [1-5], coarse theorems on probabilities of large (of order O(1)) deviations of the random measure $t^{-1}\hat{P}_t$ from the measure P are obtained for various concrete processes $\xi(t)$. Under some additional conditions the following relations hold true:

I) $\lim_{t\to\infty} t^{-1}\ln \mathbf{P}(t^{-1}\hat{P}_t \in G) = -\inf_{\mu\in G} I(\mu)$ for some class of the sets $G\subseteq \mathcal{P}$;

II) $\lim_{t\to\infty} t^{-1}\ln \mathbf{M}\exp\{\langle f, \hat{P}_t\rangle\} = S(f)$, where $f=f(x)$ runs through the space D of measurable bounded real functions on X, $\langle f,\mu\rangle=\int f(x)\mu(dx)$ for $\mu\in\mathcal{R}$. The functions $I(\mu)$, $\mu\in\mathcal{P}$ and $S(t)$, $f\in D$, are related by

$$I(\mu) = \sup_{f\in D}\{\langle f,\mu\rangle - S(f)\};\ R(f) = \sup_{\mu\in\mathcal{P}}\{\langle f,\mu\rangle - I(\mu)\}. \tag{1}$$

In this paper moderately large deviations (of order o(1)) of the random measure $t^{-1}\hat{P}_t$ from P are considered. The following nonrigorous arguments will lead us to the suitable results. As a rule, the minimum of the function $I(\mu)$ is zero for $\mu=P$; hence it is appropriate to expect that for any measure $\mu\in\mathcal{P}_0=\{\mu\in\mathcal{R}:\langle 1,\mu\rangle=\mu(X)=0\}$

$$I(P+\alpha\mu) = \alpha^2 I_0(\mu) + o(\alpha^2) \quad \text{as} \quad \alpha\to 0 . \tag{2}$$

Noting (1), we can derive from (2) that for any function $f\in D_0=\{f\in D:\langle f,P\rangle=0\}$ as $\alpha\to 0$, $S(\alpha f)=\alpha^2 S_0(f)+o(\alpha^2)$, and the functions $I_0(\mu)$, $\mu\in\mathcal{R}_0$; $S_0(f)$, $f\in D_0$, satisfy the relations (1). The following relations ($x(t)/\sqrt{t}\to\infty$, $x(t)=0(t)$ as $t\to\infty$)

are analogs of relations I and II:

$$\mathrm{I}_0)\ \lim_{t\to\infty} \frac{t}{x^2(t)} \ln \mathbf{P}\left(x^{-1}(t)\left(\hat{P}_t - tP\right) \in G\right) = -\inf_{\mu\in G} I_0(\mu)$$

for some class of sets $G \subseteq R_0$;

$$\mathrm{II}_0)\ \lim_{t\to\infty} \frac{t}{x^2(t)} \ln \mathbf{M} \exp\left\{\frac{x(t)}{t} \langle f, \hat{P}_t \rangle\right\} = S_0(f) \quad \text{for} \quad f \in D_0.$$

In this paper we obtain coarse limit theorems on probabilities of moderately large deviations for homogeneous Markov chains and homogeneous Markov processes under the assumptions close to those considered earlier in [3]. We have basically used the methods developed earlier in [1-5]. Note that [2] is the first work in which moderately large deviations as well as large deviations are considered for random measures.

2. STATEMENTS OF RESULTS IN THE DISCRETE CASE

Let X be a metric compact, and let $\xi_0, \xi_1, \ldots$ be a homogeneous Markov chain with stationary distribution P. Let the transition probabilities $\Pi(x,dy)$ of the chain be absolutely continuous with respect to P (briefly $\Pi(x,dy) \ll P(dy)$), and let the transition probabilities $\pi(x,y) = \Pi(x,dy)/P(dy)$ for all $x, y \in X$ satisfy the inequalities

$$0 < a \leq \pi(x,y) \leq A < \infty \ . \tag{3}$$

We note that if we assume that the transition probabilities are absolutely continuous with respect to an arbitrary probability measure μ, then from the condition (3) for appropriate densities there follows the existence of a stationary distribution P such that the densities of the transition probabilities chosen relative to P, will satisfy (3) [6]. The choice of $\pi(x,y)$ relative P is convenient for further considerations since in this case the function $\pi^*(x,y)=\pi(y,x)$ can be interpreted as the density of the transition probabilities relative P of a chain (ξ_n^*).

Next we define the operator Π mapping the space D of the bounded measurable functions $f(x)$, $x\in X$, letting

$$\Pi f(x) = \int \pi(x, y) f(y) P(dy) = M_x f(\xi_1).$$

Next, let

$$\Pi^{(1)}(x, dy) = \Pi(x, dy), \ldots, \Pi^{(k+1)}(x, dy) = \int \Pi^{(k)}(x, dz)\, \Pi(z, dy);$$

under the assumptions made [6, Section 7 of Chapter V] the sequence of measures $\sum_{k=1}^{n} \left(\Pi^{(k)}(x, dy) - P(dy)\right)$ converges as $n\to\infty$ in variation uniformly in $x\in X$ to a measure, say, $r(x,dy)$. The operator $Rf(x)=f(x)+\int r(x,dy)f(y)$ is the one-to-one mapping of the hyperplane $D_0=\{f\in D: \langle f,P\rangle=0\}$ into D, and for any function $f\in D$, $R(I-\Pi)f=f$, where I is the identity operator. Let

$$(f, g) = \int f(x)\, g(x)\, P(dx), \quad \|f\| = (f, f)^{1/2}.$$

for f, $g\in D$. For the function $f\in D$ we set $S_0(f)=\frac{1}{2}(\|Rf\|^2-\|\Pi Rf\|^2)$, and for the measure $\mu\in \mathcal{R}_0=\{\mu\in\mathcal{R}: \langle 1,\mu\rangle=\mu(X)=0\}$ we set

$$I_0(\mu) = \sup_{f\in D_0} \{\langle f, \mu\rangle - S_0(f)\}.$$

It is more convenient to write

$$I_0(\mu) = \begin{cases} \frac{1}{2}(m, (I-\Pi) R_C (I-\Pi^*) m), & \text{if } \mu \ll P,\ m=\frac{\partial\mu}{\partial P}, \\ \infty & \text{otherwise}, \end{cases} \quad (4)$$

where the operator R_C is the corresponding operator of R for the chain (ξ_n^C) with the densities of transition probabilities $\pi^C(x,y)=\int\pi(x,z)\pi(y,z)P(dz)$; the operator Π^* is the corresponding operator of Π for the chain (ξ_n^*) with the densities $\pi^*(x,y)=\pi(y,x)$.

To formulate the theorem, we need also to specify the topology in the space $\mathcal{R}$ of all finite measures on $(X,\mathcal{B})$. Let $\alpha=(\alpha)$ denote the family of all finite subalgebras of the σ-algebra $\mathcal{B}$. We define the topology τ in $\mathcal{R}$ using the family of the semi-norms

$$\rho_\alpha(\mu) = \max_{A\in\alpha} |\mu(A)|, \quad \alpha \in \mathrm{A}. \tag{5}$$

Thus, the convergence of measures in the topology τ is the convergence of values of measures on any measurable set.

We recall also the definition of the random measure

$$\widehat{P}_n(dy) = \sum_{i=0}^{n-1} \delta_{\xi_i}(dy),$$

where $\delta_x(dy)$ denotes the probability measure singular at the point $x\in X$.

<u>THEOREM 1</u>. Let $x(n)/\sqrt{n}\to\infty$, $x(n)/n\to 0$ as $n\to\infty$. Then for any $x\in X$:

a) for any τ-open set $U\subseteq\mathcal{R}_0$

$$\liminf_{n\to\infty} \frac{n}{x^2(n)} \ln \mathbf{P}_x\left(x^{-1}(n)\left(\widehat{P}_n - nP\right) \in U\right) \geqslant -\inf_{\mu\in U} I_0(\mu); \tag{6}$$

b) for any τ-closed set $U\subseteq\mathcal{R}_0$

$$\limsup_{n\to\infty} \frac{n}{x^2(n)} \ln \mathbf{P}_x\left(x^{-1}(n)\left(\widehat{P}_n - nP\right) \in U\right) \leqslant -\inf_{\mu\in U} I_0(\mu); \tag{7}$$

c) for any function $f\in D_0$

$$\lim_{n\to\infty} \frac{n}{x^2(n)} \ln \mathbf{M}_x \exp\left\{\frac{x(n)}{n}\langle f, \widehat{P}_n\rangle\right\} = S_0(f). \tag{8}$$

Note that the topology τ has been considered in [5], where the theorems on large deviations (of order O(1)) are obtained for empirical measures on a separable complete metric space. The topology τ is, in essence, contained (in proofs) in [4]; mostly, however, it is the topology $\|\mu\|_{B_1} = \sup_{A\in B_1} |\mu(A)|$, where $B_1 \subseteq B$ is the class of measurable sets, satisfying these two conditions:

a) all the measures $\mu\in R$ are defined uniquely by their values on B_1;

b) for any $\delta>0$ there are finite algebras α_1, α_2 of measurable sets with P-measure of the zero boundary, such that for any set $A\in B_1$ we can find sets $A_1\in\alpha_1$ and $A_2\in\alpha_2$ such that $(A\backslash A_1)\cup(A_1\backslash A)\subseteq A_2$, $P(A_2)\leq\delta$.

To assert <u>a priori</u> that there is at least one norm $\|\cdot\|_{B_1}$ is, apparently, not possible. In the cases where such a norm exists the assertions of Theorem 1 hold also for the topology generated by this norm.

3. STATEMENTS OF RESULTS IN THE CONTINUOUS CASE

Let X be a metric compact and let $\xi(t)$; $0\leq t$, be a homogeneous Markov process with the transition probabilities P(t,x,dy) and the stationary distribution P(dy). Let $P(t,x,dy)=p(t,x,y)P(dy)$ and for each $\delta>0$ let $a=a(\delta)$, $A=A(\delta)$ such that for $t\geq\delta$, $x,y\in X$

$$0 < a \leq p(t,x,y) \leq A < \infty . \tag{9}$$

Furthermore, let the semi-group of operators $T_t f(x)=\int f(y)p(t,x,y)P(dy)$ map onto itself the space C of the functions $f(x)$ continuous on X. Also, let the semi-group be strongly continuous in t ((T_t) is a semi-group of the class C_0 [7]). Under these assumptions (see [8, p. 135]) we can see that the sample trajectories of the process $\xi(t)$ are right continuous and have left limits for all $t\geq 0$; the domain of definition $\mathcal{D}$ of the infinitesimal operator L of the semi-group (T_t) is dense in C [7]. Next, under the assumptions made (see [6]), the sequence of measures $\int_0^t (p(u,x,y)-1)P(dy)du$ converges as $t\to\infty$ in variation uniformly in $x\in X$ to a measure, say, $r(x,dy)$. The operator $Rf(x)=\int f(y)r(x,dy)$ maps the hyperplane $C_0=\{f\in C: \langle f,P\rangle=(f,1)=0\}$ into $\mathcal{D}$, and $LRf=f$ for any function $f\in C_0$.

For the function f of C_0 let

$$S_0(f) = -(f,Rf) , \tag{10}$$

and for any measure $\mu\in\mathcal{R}_0=\{\mu\in\mathcal{R}: \langle 1,\mu\rangle=0\}$ let

$$I_0(\mu)=\sup_{f\in C_0}\{\langle f,\mu\rangle - S_0(f)\}.$$

The formula (4) becomes

$$I_0(\mu)=\begin{cases} -\frac{1}{2}(m, LR_C L^* m), & \text{if} \quad \mu<<P \text{ и } \frac{d\mu}{dP}=m, \\ \infty, & \text{if} \quad \mu</<P, \end{cases} \tag{11}$$

where L^* denotes the infinitesimal operator of the process $\xi^*(t)$ with density $p^*(t,x,y)=p(t,y,x)$, the operator R_C is the operator of R for the process $\xi_C(t)$ to which the infinitesimal operator $L_C=L+L^*$ corresponds.

Recall the definition of the random measure

$$\widehat{P}_t(dy) = \int_0^t \delta_{\xi(u)}(dy)\,du.$$

In the space R we consider the topology τ from Section 2.

THEOREM 2. Let $x(t)/\sqrt{t}\to\infty$, $x(t)/t\to 0$ as $t\to\infty$. Then for any $x\in X$:

a) for any τ-open set $u\in R_0$

$$\liminf_{t\to\infty} \frac{t}{x^2(t)} \ln \mathbf{P}_x\left(x^{-1}(t)\left(\widehat{P}_t - tP\right) \in U\right) \geqslant - \inf_{\mu\in U} I_0(\mu);$$

b) for any τ-closed set $U\in R_0$

$$\limsup_{t\to\infty} \frac{t}{x^2(t)} \ln \mathbf{P}_x\left(x^{-1}(t)\left(\widehat{P}_t - tP\right) \in U\right) \leqslant - \inf_{\mu\in U} I_0(\mu);$$

c) for any function $f\in C_0$

$$\lim_{t\to\infty} \frac{t}{x^2(t)} \ln \mathbf{M}_x \exp\left\{\frac{x(t)}{t}\langle f, \widehat{P}_t\rangle\right\} = S_0(f).$$

4. PROOF OF THEOREM 1

LEMMA 1. For any function $f\in D_0$ we can find $\varepsilon_0>0$, $A<\infty$ and a family of functions $f_\varepsilon\in D_0$, $0<\varepsilon\leq\varepsilon_0$, such that as $\varepsilon\to 0$

$$f_\varepsilon = \varepsilon f + \varepsilon^2 \tilde{f}_\varepsilon \tag{12}$$

where $\sup_{x\in X} |\tilde{f}_\varepsilon(x)|<A$, and as $\varepsilon(n)\to 0$, $\varepsilon(n)\sqrt{n}\to\infty$ as $n\to\infty$ for all $x\in X$

$$\lim_{n\to\infty} \frac{1}{n\varepsilon^2(n)} \ln \mathbf{M}_x \exp\{\langle f_{\varepsilon(n)}, \widehat{P}_n\rangle\} = S_0(f). \tag{13}$$

Proof. For $f\in D_0$ let $\phi=Rf$. Also, let

$$f_\varepsilon = \log(1-\varepsilon\varphi)/(1-\varepsilon\Pi\varphi) - (\log(1-\varepsilon\varphi)/(1-\varepsilon\Pi\varphi),\ 1). \tag{14}$$

It is seen that for sufficiently small $\varepsilon>0$ the function f_ε belongs to D_0; as $\varepsilon\to 0$ (12) is satisfied and (see (14) and the definition of $S_0(f)$ in Section 1)

$$-(\log(1-\varepsilon\varphi)/(1-\varepsilon\Pi\varphi),\ 1)=\varepsilon^2 S_0(f)+O(\varepsilon^2). \qquad (15)$$

By the formula (2.1) from [5], we can write

$$\mathbf{M}_x\left(e^{\langle w,\widehat{P}_n\rangle}\cdot v(\xi_{n-1})\right)=u(x), \qquad (16)$$

where $u(x)=1-\varepsilon\phi(x)$; $v(x)=1-\varepsilon\Pi\phi(x)$; $w(x)=f_\varepsilon(x)+(\log(u/v),1)$. From (16) we obtain

$$\mathbf{M}_x\left(e^{\langle f_\varepsilon,\widehat{P}_n\rangle}\cdot v(\xi_{n-1})\right)=e^{-n(\log(u|v),1)}\cdot u(x). \qquad (17)$$

Since for sufficiently small $\varepsilon>0$ we have for $x\in X$

$$0<a\le u(x)\le A<\infty, \qquad 0<a\le v(x)\le A<\infty,$$

then (17) implies (13) due to (15). //

Let α be an arbitrary finite subalgebra of $\mathcal{B}$, and let $D_{0,\alpha}$ be those functions D_0 which are measurable with respect to α,

$$I_{0,\alpha}(\mu)=\sup_{f\in D_{0,\alpha}}\{\langle f,\mu\rangle-S_0(f)\}.$$

Using the assertions of Lemma 1 and repeating the considerations used in proving Lemmas 1.1 and 1.2 in [4], it is possible to show that for any $\varepsilon>0$, $\mu\in\mathcal{R}_0$, $x\in X$

$$\liminf_{n\to\infty}\frac{n}{x^2(n)}\ln\mathbf{P}_x\left(x^{-1}(n)\,\rho_\alpha\left(\widehat{P}_n-nP-x(n)\,\mu\right)<\varepsilon\right)\geqslant I_{0,\alpha}(\mu), \qquad (18)$$

where the half-norm ρ_α is defined by (5), and

$$\limsup_{n\to\infty}\frac{n}{x^2(n)}\ln\mathbf{P}_x\left(x^{-1}(n)\left(\widehat{P}_n-nP\right)\notin\Phi_\alpha^\varepsilon(s)\right)\leqslant -s, \qquad (19)$$

where $\Phi_\alpha^\varepsilon(s) = \{\mu \in \mathcal{R}_0 : \rho_\alpha(\mu, \mu') < \varepsilon,\ I_{0,\alpha}(\mu') \leqslant s\}$.

Next, using (18) and (19) and repeating the considerations of Section 3 in [4] (see also [5]), we obtain (6) and (7). The relation (8) follows from (6), (7) (see [9]). //

To derive the formula (4) we use the scalar product in D_0

$$(f, g)_0 = (Rf, Rg) - (\Pi Rf, \Pi Rg);$$

the Cauchy inequality for $f \in D_0$ implies $\|f\|_0 = 0 \Leftrightarrow f = 0$. The operators Π^* and R^*, Π_C and R_C are defined for the chains (ξ_n^*), (ξ_n^C), from Section 2. We have the following:

$$(\Pi f, g) = (f, \Pi^* g); \quad (Rf, g) = (f, R^* g),$$

and the scalar products $(\cdot,\cdot)$ and $(\cdot,\cdot)_0$ are related by

$$(f, g) = (f, (I - \Pi) R_C (I - \Pi^*) g)_0.$$

Next, since $S_0(f) = \frac{1}{2}\|f\|_0^2$, then for $\mu \ll P$ for $m = \partial\mu/\partial P$

$$I_0(\mu) = \sup_{f \in D_0} \{(f, m) - S_0(f)\} = \sup_{f \in D_0} \{(f, (I - \Pi) R_C (I - \Pi^*) m)_0 - 1/2 \cdot \|f\|_0^2\}$$

from which the first part of (4) follows. If, however, $\mu <\!\!\backslash\!\!< P$, we use the inequality $\|f\|_0^2 \leq A^2 \|f\|^2$, where $A^2 < \infty$, which follows from Lemma 7.1 in [6, p. 203]. From this inequality we obtain

$$A^2 I_0(\mu) \geqslant \tilde{I}_0(\mu) \equiv \sup_{f \in D_0} \{\langle f, \mu\rangle - 1/2 \cdot \|f\|^2\}.$$

Since $\tilde{I}_0(\mu) = \infty$ for $\mu <\!\!\backslash\!\!< P$, the second part of (4) is obtained.

5. PROOF OF THEOREM 2

LEMMA 2. For any function $f \in C_0$ we can find $\varepsilon_0 > 0$ and $A < \infty$

and a family of functions $f_\varepsilon \in C_0$, $0<\varepsilon\leq\varepsilon_0$, such that

$$f_\varepsilon = \varepsilon f + \varepsilon^2 \tilde{f}_\varepsilon , \tag{20}$$

where $\sup_{x\in X} |\tilde{f}_\varepsilon(x)| \leq A$, and for $\varepsilon(t)\to 0$, $\varepsilon(t)\sqrt{t}\to\infty$ as $t\to\infty$ for all $x\in X$

$$\lim_{t\to\infty} \frac{1}{t\varepsilon^2(t)} \ln \mathbf{M}_x \exp\{\langle f_{\varepsilon(t)}, \widehat{P}_t\rangle\} = S_0(f), \tag{21}$$

where $S_0(f)$ is defined by (10).

Proof. In the continuous case instead of (16) we have

$$\mathbf{M}_x \exp\left\{-\left\langle \frac{Lu}{u}, \widehat{P}_t \right\rangle\right\} u(\xi(t)) = u(x), \tag{22}$$

which holds for any positive function $u\in\mathcal{D}$ [3]. For the specified function $f\in C_0$ let $\phi = Rf \in \mathcal{D}$ and

$$f_\varepsilon = -(L(1-\varepsilon\varphi))/(1-\varepsilon\varphi) + (L(1-\varepsilon\varphi)/(1-\varepsilon\varphi),\ 1).$$

It is seen that $f_\varepsilon \in C_0$ and (20) is satisfied. It is easy to see, next, that as $\varepsilon\to 0$

$$(L(1-\varepsilon\varphi)/(1-\varepsilon\varphi),\ 1) = -\varepsilon(L\varphi/(1-\varepsilon\varphi),\ 1) = \varepsilon^2 S_0(f) + o(\varepsilon^2). \tag{23}$$

By (22)

$$\mathbf{M}_x e^{\langle f_\varepsilon, \widehat{P}_t\rangle} (1-\varepsilon\varphi(\xi(t))) = (1-\varepsilon\varphi(x))\, e^{-(L\varepsilon\varphi/(1-\varepsilon\varphi),1)t},$$

hence by (23) we obtain

$$\lim_{t\to\infty} \frac{1}{t\varepsilon^2(t)} \ln \mathbf{M}_x e^{\langle f_\varepsilon, \widehat{P}_t\rangle} (1-\varepsilon\varphi(\xi(t))) = S_0(f). \tag{24}$$

Since for sufficiently small $\varepsilon>0$ ($\leq\varepsilon_0$) we have $0<a\leq 1-\varepsilon\phi(x)\leq A<\infty$, then (21) follows from (24). //

The rest of the proof of Theorem 2 coincides fully with that of Theorem 1.

The derivation of the formula (12) is analogous to that

of (4). To this end, we need the scalar product

$$(f,g)_0 = -[(f,Rg)+(g,Rf)] ,$$

so that (see (10)) $S_0(f)=\frac{1}{2}||f||_0^2$, and in addition, use the relations

$$-(f,g)_0 = (f, R^*(L+L^*)Rg) ; \quad -(f,g) = (f, LR_cL^*g)_0 .$$

REFERENCES

[1] I.N. Sanov. "On the Probability of Large Deviations for Random Variables." *Matem.Sb.*, 42 (84) (1957): 11-14. (In Russian.)

[2] A.A. Borovkov. "Boundary Value Problems for Random Walks and Large Deviations in Function Spaces." *Theory Prob. Applications*, 4 (1967): 575-595.

[3] M.D. Donsker and S.R.S. Varadhan. "Asymptotic Evaluation of Certain Markov Processes Expectations for Large Time," I. *Comm. Pure Appl. Math.*, 27 (1975): 1-47.

[4] Jürgen Gärtner. "On Large Deviations from the Invariant Measure." *Theory Prob. Applications*, 1 (1977): 24-39.

[5] P. Groenboom, J. Oosterhoff, and F.H. Ruymgart. *Large Deviation Theorems for Empirical Probability Measures*. Preprint. Amsterdam: Mathematical Center, 1976.

[6] J.L. Doob. *Stochastic Processes*. New York: John Wiley and Sons, 1953.

[7] K. Iosida, *Functional Analysis*. Moscow: Mir, 1963. (In Russian.)

[8] E.B. Dynkin. *Markov Processes*. New York: Academic Press, 1965.

[9] S.R.S. Varadhan. "Asymptotic Probabilities and Differential Equations." *Comm. Pure Appl. Math.*, 19, 3 (1966): 261-286.

A LOCAL THEOREM FOR BELLMAN-HARRIS CRITICAL PROCESSES WITH DISCRETE TIME

V.A. Topchij

1. INTRODUCTION

In this paper we consider age-dependent branching processes (the Bellman-Harris processes) with discrete time $\xi(n)$, $n=0,1,\ldots$ The random variables $\xi(n)$ assume integer values which we interpret to be the size of the population at time n. For the description of these processes see [1] or [2] (Model 3).

Let the process $\xi(n)$ be defined by integer random variables η (the lifetime of particles) with distribution function $F(k)=P\{\eta\le k\}$ (possibly $F(0)\neq 0$) and ζ (the number of descendants) with generating function $h(z)=\sum h_k z^k$.

Let

$$P_n(z) = Mz^{\xi(n)} \; ; \qquad p_{nk} = P\{\xi(n)=k\} \; ;$$

$$Q_n(z) = 1-P_n(z) \; ; \qquad P_n = P_n(0) \; ; \qquad Q_n = Q_n(0) \; .$$

Next we put $A=h'(1)$; $B=h''(1)$; $\mu=M\eta$; $\mu_k=M\eta^k$; $\alpha=B/2\mu$; $f_k=F(k)-F(k-0)$; $q_k=\sum_{j=k+1} f_j$; $f(u)=\sum_{k=0}^{\infty} f_k u^k$.

We consider the critical case, i.e., $A=1$. From the results obtained in Sevast'yanov [3] it follows that for the extinction probability of the process the equality

$$\lim_{n\to\infty} Q_n n = 1/\alpha \tag{1}$$

is satisfied (provided the third moments in η and ζ are finite). However, as Goldstein [4] has shown, for (1) to hold it suffices that the second moment in ζ exists and the conditions $k^2[1-F(k)]\to 0$ as $k\to\infty$ and $F(0)=0$ are satisfied (it is easy to get rid of the last condition). We note that the method employed in our article enables one to give a different proof of Goldstein's result (including the processes with an arbitrary distribution of lifetime of particles).

In 1966, Kestin, Ney, and Spitzer [5] proved the following theorem for the Galton-Watson processes (the latter coincide with the Bellman-Harris processes for $P\{\eta=1\}=1$).

<u>THEOREM 0</u>. Suppose $\xi(n)$ is a Galton-Watson process; $A=1$, $B>0$, $M\zeta^2 \ln \zeta<\infty$ and the greatest common divisor of s such that $h_s\neq 0$, equals d, then for every constant $0<c<\infty$

$$\lim_{\substack{k,n\to\infty \\ k/n<c}} n^2 e^{kd/n\alpha} p_{nkd} = d/\alpha^2. \tag{2}$$

Furthermore,

$$\sup_{k,n\geq 1} n^2 p_{nk} < \infty. \tag{3}$$

At the same time with [5], an article of S.V. Nagaev and

R. Mukhamedkhanova [6] was published, in which a similar result is proved, however, under more restrictive assumptions, which makes the proof much simpler.

To simplify the formulation of the main result proved in the sequel, we introduce the following condition:

a) $\xi(n)$ is not a Galton-Watson process, neither is it a process reducible to a Galton-Watson process by decreasing the time-scale with $d\neq 1$.

THEOREM 1. Let $A=1$, $B>0$, $M\zeta^2 \ln \zeta<\infty$, $\mu_3<\infty$ and let condition "a" be satisfied. Then for every $0<c<\infty$ we have the equality

$$\lim_{\substack{k,n\to\infty \\ k/n<c}} n^2 e^{k/n\alpha} p_{nk} = 1/\alpha^2 \tag{4}$$

and the estimate (3).

While proving the Theorem, in addition to condition "a" we assume that

b) the greatest common divisor of k such that $f_k\neq 0$, equals 1;

c) $F(0)=0$.

Note that the formulation of Theorem 1 is invariant with respect to d defined in Theorem 0. If condition "a" is not satisfied, it is possible to use the reduction of the case $d\neq 1$ to the case $d=1$, which is done in [5]; then from Theorem 1 we obtain Theorem 0. It is easy to get rid of condition "b" by redefining the time-scale.

For processes with $F(0)\neq 0$ one can construct a sequence of critical identically distributed processes with $F(0)=0$, the sum of a random number of which has the same finite-dimensional distributions as the initial process has. This is sufficient for proving Theorem 1 in the general case, if it is proved under condition "c".

To make the last assertion more concrete, we define the sequence of independent Bellman-Harris processes $\xi_i(t)$ by the distribution function of the lifetime of particles $F^*(k)=(F(k)-F(0))(1-F(0))^{-1}$ and by the generating function of the number of descendants of each particle $h^*(z)=h(P_0(z))$, where $P_0(z)$ is found from the equation $P_0(z)=h(P_0(z))F(0)+(1-F(0))z$.

Let ν be a random variable with the generating function $P_0(z)$. Then $\xi(n)$ is representable as $\xi(n)=\sum_{i=1}^{\nu}\xi_i(n)$.

We omit here the detailed proof of the case where conditions "b" and "c" are violated and restrict ourselves only to the arguments given above concerning condition "c". First we make a few remarks concerning the method of proof.

The proof of the local limit theorem for Galton-Watson processes in [5] rests on the investigation of the asymptotic behavior of $P_n(x)$ for real x. Unlike [5], we use the asymptotic analysis of the behavior of the sequence $P_n(z)$ on the unit circle in the complex plane (a similar approach for Galton-Watson processes is used in [6]). The analysis of $P_n(z)$ on the unit circle in the complex plane for Bellman-Harris processes has, apparently, never been made. Estimation of $|Q_n(z)|$ from below presents the main difficulty here.

As is well known [Chapter 6 in 1], for $|z|\leq 1$, $P_n(z)$ is the unique solution of the system of equations

$$P_n(z) = \sum_{k=0}^{n} h(P_{n-k}(z)) f_k + zq_n, \quad n = \overline{0, \infty}. \tag{5}$$

Relying on this system, Nagaev [7] obtained a representation for $Q_n(z)$, which in our case has the form

$$Q_n(z) = (1-z) + \sum_{k=0}^{n} (1 - h(P_{n-k}(z)) - Q_{n-k}(z))\, u_k \tag{6}$$

(we shall derive it in Section 2 and define u_k as well). This representation is the starting point of the proof.

We note that if the first four moments of η and ζ are finite, the proof can be made simpler and essentially shorter applying the results in [6].

We wish to focus the reader's attention on the facts to be used in the sequel without further mention.

By z we mean the complex-valued argument, $|z| \le 1$. All the functions of z are analytic in the domain $|z|<1$ and continuous up to the boundary. Hence if $\phi(z)=\sum_0^\infty c_k z^k$ for $|z|<1$ and is uniformly bounded, then $\phi(e^{it})=\sum_0^\infty c_k e^{itk}$ (i.e., it can be regarded as the Fourier series on $|z|=1$).

For any function $\phi(z)=\sum_{k=0}^{\infty} a_k z^k$ we define the linear functional $C_k(\phi(z))=a_k$ and the norm $\|\phi(z)\|=\sum_0^\infty |a_k|$. If $\|\phi(z)\|<\infty$, we write $\phi(z)\in \ell_1$.

In what follows we shall use the expressions $(n\alpha+1/(1-z))^{-1}$ and $Q_n(z)(n\alpha+1/(1-z))$, which for $z=1$ we define to be 0 and 1, respectively, which, since $Q_n'(1)=-1$ (see [2]), extends the definition of the functions by continuity.

2. AUXILIARY RESULTS

Let us derive (6). It follows from (5) that

$$Q_k(z) = (1-z)\,q_k + \sum_{j=1}^{k} (1 - h(P_{k-j}(z)))\,f_j. \tag{7}$$

Let

$$q(z,u) = \sum_{k=0}^{\infty} Q_k(z)\,u^k. \tag{8}$$

From (7) and (8) we obtain

$$q(z,u) = \frac{1-f(u)}{1-u}(1-z) + q(z,u)\,f(u) + \varphi(z,u)\,f(u), \tag{9}$$

where

$$\varphi(z,u) = \sum_{0}^{\infty} [1 - h(P_k(z)) - Q_k(z)]\,u^k.$$

From (9) we derive the identity

$$q(z,u) = \frac{1-z}{1-u} + \frac{\varphi(z,u)\,f(u)}{1-f(u)}. \tag{10}$$

We write the coefficients for u^k in (10) and put $u_k = C_k\left(\frac{f(z)}{1-f(z)}\right)$, and obtain

$$Q_n(z) = (1-z) + \sum_{k=0}^{n} (1 - h(P_{n-k}(z)) - Q_{n-k}(z))\,u_k. \tag{11}$$

Later on we shall need the estimates u_k. Since they are very crucial we write them as a theorem. But first we note that $\frac{f(z)}{1-f(z)} = \frac{1}{1-f(z)} - 1$.

<u>THEOREM 1</u>. Let $f^{(s)}(1)<\infty$ for $s\geq 2$ and let condition "b" of Section 1 be satisfied. Then

$$u_k - u_{k-1} = O(k^{-s}) \tag{12}$$

and

$$u_k - 1/\mu = O(k^{-s+1}) . \tag{13}$$

Furthermore, for s=3

$$\sum k^2 |u_k - u_{k+1}| < \infty . \tag{14}$$

In addition, let $b_k = u_k - 1/\mu$.

<u>COROLLARY</u>. For $f'''(1)<\infty$

$$\sum |b_k| < \infty \tag{15}$$

and

$$\sum k|b_k| < \infty . \tag{16}$$

The estimates (12) and (13) yield a corollary of Theorem 1 in the article of B.A. Rogozin [8] (this contains also earlier works from which these estimates can be obtained).

Note the equivalence of (8) and the condition

$$\left(\frac{1-z}{1-f(z)}\right)'' \in l_1. \tag{17}$$

It follows from (12) that

$$\left(\frac{1-z}{1-f(z)}\right)' = \sum_1^\infty (b_k - b_{k-1})\, kz^{k-1} \in l_1. \tag{18}$$

Direct computations yield

$$\left(\frac{1-z}{1-f(z)}\right)' = -\frac{1}{1-f(z)} + \frac{(1-z)f'(z)}{(1-f(z))^2} \tag{19}$$

and

$$\left(\frac{1-z}{1-f(z)}\right)'' = -\frac{2f'(z)}{(1-f(z))^2} + \frac{(1-z)f''(z)}{(1-f(z))^2} + \frac{2(1-z)(f'(z))^2}{(1-f(z))^3}. \tag{20}$$

From (20), (19), (18) and the relation $\frac{f''(z)(1-z)}{1-f(z)} \in \ell_1$, which holds by Theorem 2 and the estimate (12) it follows that (17) holds if

$$-\frac{2f'(z)}{(1-f(z))^2} + \frac{f''(z)f'(z)(1-z)^2}{(1-f(z))^3} + \frac{2(1-z)(f'(z))^2}{(1-f(z))^3} \in l_1. \tag{21}$$

It is easy to determine

$$\left\| \frac{-2(1-f(z))+2f'(z)(1-z)+f''(z)(1-z)^2}{(1-z)^3} \right\| =$$

$$= \left\| \frac{2\sum_{k=2}^{\infty} f_k \sum_{i=0}^{k-2} z^i \sum_{j=0}^{k-i-2} \left(z^{k-i-2}-z^j\right)}{1-z} \right\| = 0\,(\Sigma f_k k^3) < \infty.$$

By virtue of the relation $\frac{f'(z)(1-z)^3}{(1-f(z))^3} \in l_1$, which holds by Theorem 2 and the estimate (12), the last estimate implies (21) and therefore (17). //

We proceed to prove the Corollary. The relation (15) follows immediately from (13). The inequality (10) is proved if we establish that

$$\left(\frac{1}{1-f(z)} - \frac{1}{\mu(1-z)}\right)' = \left(\frac{\mu(1-z)-(1-f(z))}{\mu(1-z)(1-f(z))}\right)' \in l_1$$

which follows from the obvious relations

$$\left(\frac{1-z}{1-f(z)}\right)' \in l_1$$

and

$$\frac{\mu(1-z)-(1-f(z))}{(1-z)^2} = \sum_{k=2}^{\infty} f_k \sum_{i=1}^{k-1} \sum_{j=0}^{i-1} z^j,$$

and also from the condition $f'''(1)<\infty$. //

3. THE REPRESENTATION FOR $Q_n(z)$ AND $Q_n(z)-Q_n$

Throughout this section we assume that

$$M\eta^3 < \infty \qquad \text{and} \qquad M\zeta^2 \ln \zeta < \infty . \tag{22}$$

Let

$$N=[n/2],\quad \sum_a^b=\sum_{[a]}^{[b]},$$

if a or b are non-integers, and let

$$p_n=\frac{1}{n}\sum_{k=1}^{n}h_k k^3+\sum_{n+1}^{\infty}h_k k^2,$$

where $h_k=C_k(h(z))$, $p_0=1$.

In the sequel we write $\omega_n(z)=\tilde{O}(\phi_n)$, if $\exists\ c_1<\infty$ such that $\forall\ n\|\omega_n(z)/\phi_n\|\le c_1$. If for $\omega_n(z)$ there exists $\bar{c}_1$ such that $\forall\ n|\omega_n(z)/\phi_n|\le\bar{c}_1$ ($|\omega_n(z)/\phi_n|\to 0$ as $n\to\infty$) uniformly in $|z|<1$, we write $\omega_n(z)=O_n(\phi_n)=O(\phi_n)$ ($\omega_n(z)=o_n(\phi_n)=$ $=o_n(\phi_n)$). The function ϕ_n can depend on z as well.

LEMMA 1. We have the representation

$$Q_n(z) = Q_{n-1}(z) - \alpha Q_{n-1}^2(z) + \frac{\tilde{O}(p_{n-1})}{(\alpha(n-1)+1/(1-z))^2}\ . \tag{23}$$

Proof. First we investigate the properties of p_n. For any $0<\alpha_1\le\alpha_2<\infty$ there exists $c_{\alpha_1\alpha_2}$ such that for arbitrary n_1 and n satisfying the condition $\alpha_1 n\le n_1\le\alpha_2 n$, the inequality

$$p_n/p_{n_1}\le c_{\alpha_1\alpha_2} \tag{24}$$

is satisfied as well as the relations

$$p_n = o_n(1) \quad\text{and}\quad 1/(n+1) = O(p_n)\ . \tag{25}$$

Indeed, for $n\le n_1$

$$p_n=\frac{1}{n}\sum_{1}^{n}h_k k^3+\sum_{n+1}^{\infty}h_k k^2\leqslant\frac{1}{n}\sum_{1}^{n}h_k k^3+\sum_{n_1+1}^{\infty}h_k k^2\leqslant$$

$$\leqslant(\alpha_2+1)\left[\frac{1}{n_1}\sum_{1}^{n_1}h_k k^3+\sum_{n_1+1}^{\infty}h_k k^2\right]=(\alpha_2+1)\,p_{n_1},$$

and for $n>n_1$

$$p_n = \frac{1}{n}\sum_1^n h_k k^3 + \sum_{n+1}^{\infty} h_k k^2 \leqslant \frac{1}{n_1}\sum_1^n h_k k^3 + \sum_{n+1}^{\infty} h_k k^2 \leqslant$$

$$\leqslant \sum_1^{n_1} h_k k^3 + \sum_{n_1+1}^{n} h_k k^2 \frac{n}{n_1} + \sum_{n+1}^{\infty} h_k k^2 \leqslant (1+\alpha_1^{-1}) \times$$

$$\times \left(\frac{1}{n_1}\sum_1^{n_1} h_k k^3 + \sum_{n_1+1}^{\infty} h_k k^2 \right) = (1+\alpha_1^{-1})\, p_{n_1}.$$

The fact that $p_n \to 0$ as $n\to\infty$, follows from $\sum_1^{\infty} h_k k^2 < \infty$ and the estimate

$$\frac{1}{n}\sum_1^n h_k k^3 = \frac{1}{n}\sum_1^{\sqrt{n}-1} h_k k^3 + \frac{1}{n}\sum_{\sqrt{n}}^{n} h_k k^3 \leqslant \frac{1}{\sqrt{n}}\sum_1^{\infty} h_k k^2 + \sum_{\sqrt{n}}^{n} h_k k^2.$$

The estimate $1/(n+1)=O(p_n)$ is obvious by the definition of p_n.

By the definition of $h(z)$ and $P_n(z)$ we obtain

$$\frac{1-h(P_n(z)) - Q_n(z) + \frac{B}{2} Q_n^2(z)}{Q_n^2(z)} = \frac{\sum_{k=1}^{\infty} h_k \sum_0^{k-1} P_n^i(z) - 1 + \frac{B}{2} Q_n(z)}{Q_n(z)} =$$

$$= B/2 - \sum_{k=2}^{\infty} h_k \sum_1^{k-1}\sum_0^{i-1} P_n^j(z) = \frac{B}{2} - \sum_{k=2}^{\infty} h_k \sum_0^{k-2} (k-j-1) P_n^j(z) =$$

$$= \sum_{k=2}^{n} h_k \left(\frac{k(k-1)}{2} - \sum_{j=0}^{k-2} (k-j-1) P_n^j(z) \right) + \sum_{k=n+1}^{\infty} h_k \left(\frac{k(k-1)}{2} - \right.$$

$$\left. - \sum_{j=0}^{k-2} P_n^j(z)\cdot(k-j-1) \right) = \sum_{k=2}^{n} h_k \sum_{j=0}^{k-2} (k-j-1) \sum_{i=0}^{j-1} P_n^i(z)(1-P_n(z)) +$$

$$+ \sum_{k=n+1}^{\infty} h_k \left(\frac{k(k-1)}{2} - \sum_{j=0}^{k-2} (k-j-1) P_n^j(z) \right.$$

Because $P_n(z)=\sum_{k=0}^{\infty} p_{nk} z^k$, $p_{nk}\geq 0$, $P_n(1)=1$ and (1) is satisfied,

$$Q_n(z)=\tilde{O}\left(\frac{1}{n+1}\right). \tag{25a}$$

These facts together with the obvious equality $P_n(z)=\tilde{O}(1)$ imply

$$1-h(P_n(z))-Q_n(z)+\frac{B}{2}Q_n^2(z)=Q_n^2(z)\cdot\tilde{O}\left(\sum_{k=2}^{n}h_k\frac{k^3}{n}+\sum_{n+1}^{\infty}h_kk^2\right)=$$
$$=Q_n^2(z)\tilde{O}(p_n). \tag{26}$$

Since $Q_n(1)=Q_n(0)-\sum_1^\infty p_{nk}=0$ and $P_n(1)=1$ (see [7] or [8]), it follows from (1) that there exists $c_2<\infty$ such that

$$\|Q_n(z)(n\alpha+(1-z)^{-1})\|=\left\|\left(Q_n(0)-\sum_1^\infty p_{nk}z^k\right)n\alpha+\right.$$
$$\left.+(1-z)^{-1}\sum_1^\infty p_{nk}(1-z^k)\right\|\leqslant c_2. \tag{27}$$

Direct computations yield

$$\left\|\frac{n\alpha+(1-z)^{-1}}{k\alpha+(1-z)^{-1}}\right\|=\left\|1+\frac{(n-k)\alpha}{\alpha k+(1-z)^{-1}}\right\|=\left\|1+\right.$$
$$\left.+\frac{(n-k)\alpha(1-z)}{(\alpha k+1)(1-\alpha kz(\alpha k+1)^{-1})}\right\|=\left\|1+\frac{(n-k)\alpha}{\alpha k+1}\left(1-\sum_0^\infty\frac{(\alpha k)^i z^{i+1}}{(\alpha k+1)^{i+1}}\right)\right\|=$$
$$=\frac{\alpha n+1}{\alpha k+1}+\frac{|n-k|\alpha}{\alpha k+1}. \tag{28}$$

Taking the representation (11) for $Q_n(z)$ and $Q_{n-1}(z)$, it is easy to see that

$$Q_{n-1}(z)-Q_n(z)=-(1-h(P_n(z))-Q_n(z))/\mu+\sum_{k=0}^{N}[h(P_{n-k}(z))+$$
$$+Q_{n-k}(z)-h(P_{n-k-1}(z))-Q_{n-k-1}(z)]b_k-\sum_{k=N+2}^{n}[1-h(P_{n-k}(z))-$$
$$-Q_{n-k}(z)](b_k-b_{k-1})-b_{N+1}[1-h(P_{n-N-1}(z))-Q_{n-N-1}(z)]. \tag{29}$$

This plus (26) yield the estimate

$$Q_{n-1}(z) - Q_n(z) = \alpha Q_n^2(z) + \sum_{k=0}^{n} \frac{B}{2}\left(Q_{n-k}^2(z) - Q_{n-k-1}^2(z)\right) b_k +$$
$$+ \sum_{k=N+2}^{n} \frac{B}{2} Q_{n-k}^2(z)(b_k - b_{k-1}) + b_{N+1} Q_{n-N-1}^2(z) + Q_n^2(z)\widetilde{O}(p_n) +$$
$$+ \sum_{k=0}^{N} \left[Q_{n-k}^2(z)\widetilde{O}(p_{n-k}) + Q_{n-k-1}^2(z)\widetilde{O}(p_{n-k-1})\right] b_k +$$
$$+ \sum_{k=N+2}^{n} Q_{n-k}^2(z)\widetilde{O}(p_{n-k})(b_k - b_{k-1}) + b_{N+1} Q_{n-N-1}^2(z)\widetilde{O}(p_{n-N-1}). \quad (30)$$

The estimates (12), (27) and (28) allow us to establish that

$$\left\| \sum_{k=N+2}^{n} (n\alpha + (1-z)^{-1})^2 \frac{B}{2} Q_{n-k}^2(z)(b_k - b_{k-1}) \right\| =$$
$$= O\left(\sum_{k=N+2}^{n} \left\| \frac{n\alpha + (1-z)^{-1}}{(n-k)\alpha + (1-z)^{-1}} \right\|^2 \frac{1}{n^3} \right) = O\left(\sum_{k=N+2}^{n} \frac{n^{-1}}{(n-k+1)^2} \right) = O(1/n). \quad (31)$$

Similarly the estimates (15), (24), (27) and (28) imply

$$\sum_{k=0}^{N} \left[Q_{n-k}^2(z)\widetilde{O}(p_{n-k}) + Q_{n-k-1}^2(z)\widetilde{O}(p_{n-k-1})\right] b_k = \left(n\alpha + \frac{1}{1-z}\right)^{-2} \times$$
$$\times \sum_{k=0}^{N} \widetilde{O}(p_n) b_k = \frac{\widetilde{O}(p_n)}{(n\alpha + 1/(1-z))^2}. \quad (32)$$

From (13) and (27) we obtain

$$b_{N+1} Q_{n-N-1}^2 = \frac{\widetilde{O}(n^{-2})}{(n\alpha + 1/(1-z))^2}. \quad (33)$$

Using the estimates (25), (27), (31)-(33), we obtain

$$Q_{n-1}(z) - Q_n(z) = \alpha Q_n^2(z) + \sum_{k=0}^{N} \frac{B}{2}\left(Q_{n-k}^2(z) - \right.$$
$$\left. - Q_{n-k-1}^2(z)\right) b_k + \frac{\widetilde{O}(p_n)}{(\alpha n + 1/(1-z))^2}. \quad (34)$$

Multiplying (34) by $Q_{n-1}(z)+Q_n(z)$ and applying the estimates (15), (27) and (28) to the right-hand side of the obtained identity, we have

$$Q_{n-1}^2(z) - Q_n^2(z) = \frac{1}{(n\alpha + 1/(1-z))^3}\tilde{O}(1). \tag{35}$$

The equality (35), with (28) and (15) taken into account, enables us to have

$$\sum_{k=0}^{N} \frac{B}{2}\left(Q_{n-k}^2(z) - Q_{n-k-1}^2(z)\right) b_k = \sum_{k=0}^{N} \frac{\tilde{O}(1)}{((n-k)\alpha + (1-z)^{-1})^3} b_k$$
$$\frac{\tilde{O}(1)}{(n\alpha + (1-z)^{-1})^3} \sum |b_k| = \frac{\tilde{O}(1)}{(n\alpha + (1-z)^{-1})^3}. \tag{36}$$

We note that by analogy with (28):

$$\left\| \frac{1}{n\alpha + (1-z)^{-1}} \right\| = \frac{1}{n\alpha + 1} \left\| 1 - \sum_{i=0}^{\infty} \frac{(\alpha n)^i z^{i+1}}{(\alpha n + 1)^{i+1}} \right\| = \frac{2}{1 + n\alpha}. \tag{37}$$

From (34), (36), (37) and (25) we obtain

$$Q_{n-1}(z) - Q_n(z) = \alpha Q_n^2(z) + \frac{1}{(\alpha n + 1/(1-z))^2} \tilde{O}(p_n),$$

which proves the Lemma by virtue of (35), (37) and (25).

<u>LEMMA 2</u>. Let $x' \geq x \geq 0$, $D_n(x', x) = D_n = P_n(x') - P_n(x)$. Then there exists a constant $d > 0$ such that $\forall$ n

$$D_n \geqslant d \frac{x' - x}{(n+1)^2}. \tag{38}$$

For proving Lemma 2, we need several auxiliary assertions.

<u>LEMMA 3</u>. For $n \geq m_0$ let there exist a constant $c_3 > 0$ such that $|(Q_n(z)(n\alpha + (1-z)^{-1}))^{-1}| c_3$ is satisfied for $|z| \leq 1$. Then

$$Q_n(z) = \frac{1 + O(\bar{p}_n)}{\alpha n + 1/(1-z)}, \tag{39}$$

where

$$\bar{p}_n = \frac{1}{n} \sum_{k=1}^{n} p_k.$$

Proof. We show first that for any fixed k

$$Q_k^{-1}(z) = 1/1(1-z)+\tilde{O}(1) , \qquad (40)$$

if $P_k(z)\neq 1$ for $z\neq 1$ and $|z|\leq 1$ (this condition is satisfied for $k\geq m_0$).

Indeed,

$$\frac{1}{Q_k(z)} - \frac{1}{1-z} = \frac{1}{(1-z)\left(\sum_{n=1}^{\infty} p_{kn} \sum_0^{n-1} z^i\right)} - \frac{1}{1-z} = \frac{\sum_{n=2}^{\infty} p_{kn} \sum_{i=1}^{n-1}\sum_{j=0}^{i-1} z^j}{\sum_{n=1}^{\infty} p_{kn} \sum_0^{n-1} z^i}.$$

The numerator and denominator on the right-hand side of the last equality belong to ℓ_1 (since $P_n''(1)<\infty$, see [2, p. 295]).

Since $P_k(z)\neq 1$ (for $z\neq 1$), then

$$\sum_{n=1}^{\infty} p_{kn} \sum_0^{n-1} z^i = \frac{1-P_k(z)}{1-z} \neq 0,$$

therefore, by the Wiener theorem (see [9, p. 331. Russian translation: Moscow: Nauka, 1968]) $Q_n^{-1}(z)-1/(1-z)\in\ell_1$ for $|z|=1$.

This proves (40).

Let $L_n=Q_n^{-1}(z)$.

From Lemma 1 and the conditions of Lemma 3 for $n>m_0$ we obtain

$$L_n = L_{n-1}\left[1-\alpha Q_{n-1}(z) + \frac{O(p_{n-1})}{\alpha(n-1)+1/(1-z)}\right]^{-1},$$

which by (25), (27), (37) and by this Lemma becomes

$$L_n = L_{n-1}(1 + \alpha Q_{n-1}(z) + (n\alpha + (1-z)^{-1})^{-1} O(p_{n-1})) =$$
$$= L_{n-1} + \alpha + O(p_{n-1}) .$$

Summing the equalities obtained over n, we have

$$L_k = L_{m_0} + \alpha(k - m_0) + O\left(\sum_{i=m_0+1}^{k} O(p_{i-1})\right). \tag{41}$$

It is easy to compute that

$$\sum_{1}^{n} p_i \sim \sum_{i=1}^{n} (1 + \ln n - \ln i)\, i^3 h_i + n \sum_{n+1}^{\infty} i^2 h_i$$

and write

$$\bar{p}_n = \frac{1}{n} \sum_{i=1}^{n} p_i = O\left(\frac{1 + \ln n}{n} \sum_{1}^{n} h_i i^3 + \sum_{n+1}^{\infty} h_i i^2\right), \tag{42}$$

and by (25) as $n \to \infty$

$$\bar{p}_n \to 0 \quad \text{and} \quad \frac{\ln n}{n} = O(\bar{p}_n) . \tag{43}$$

From (40), (41) and (43) we obtain

$$L_k = 1/(1-z) + \alpha k + kO(\bar{p}_k) ,$$

which plus (34) and (43) imply (39).

LEMMA 4. For Q_n we have the representation

$$Q_n = \frac{1 + O(\bar{p}_n)}{\alpha n + 1} . \tag{44}$$

Proof. From (1) it follows that there exists k_0 such that $\forall n \geq k_0 \, |Q_n^{-1} \frac{1}{n}| < 2\alpha$. By Lemma 3 for $z=0$ this relation proves (44).

LEMMA 5. There exists $k_6 > 0$ such that for any integer $s > 0$ and any fixed $k_1 > 0$ we have

$$\sum_{k=1}^{(s+2)/2} \left(\frac{1}{(s+2-k)^2} - \frac{2\mu}{(s+2-k)^3} - \frac{k_1 \bar{p}_{s+2-k}}{(s+2-k)^3}\right) f_k \geqslant$$
$$\geqslant \frac{1}{(s+2)^2}\left(1 - k_6 \frac{1}{s+2} \bar{p}_{s+2}\right). \tag{45}$$

Proof. We express the left-hand side of (45) as

$$\sum_{k=1}^{(s+2)/2}\left(-\frac{1}{(s+2)^2}+\frac{1}{(s+2-k)^2}-\frac{2\mu}{(s+2-k)^3}-\frac{k_1}{(s+2-k)^3}\right.$$
$$\left.-\bar{p}_{s+2-k}\right)f_k+\frac{1}{(s+2)^2}\left(1-\sum_{(s+2)/2+1}^{\infty}f_k\right) \tag{46}$$

and estimate the resulting expression by parts. Next, by $s>0$ we mean an integer. Note that

$$\bar{p}_{s-k}=\frac{1}{s-k}\sum_{1}^{s-k}p_i\leqslant\frac{s}{s-k}\bar{p}_s.$$

This inequality implies the existence of a constant $k_2>0$ such that $\forall s>0$

$$\sum_{k=1}^{(s+2)/2}\frac{k_1}{(s+2-k)^3}\bar{p}_{s+2-k}f_k\leqslant k_2\bar{p}_{s+2}\frac{1}{(s+2)^3}.$$

By the condition $M\eta^3<\infty$, $\forall s>0\ \exists k_3>0$ such that

$$\sum_{k=1}^{(s+2)/2}\frac{2\mu}{(s+2-k)^3}f_k\leqslant\sum_{k=1}^{(s+2)/2}\frac{2\mu}{(s+2)^3}f_k-\sum_{k=1}^{(s+2)/2}\frac{2f_k\mu}{(s+2)^3(s+2-k)^3}\times$$
$$\times\left(3(s+2)k^2-3(s+2)^3k-k^3\right)\leqslant\frac{2\mu}{(s+2)^3}+k_3\frac{1}{(s+2)^4}.$$

Similarly $\exists k_4>0$ and $\exists k_5>0$ such that

$$\sum_{k=1}^{(s+2)/2}\left[\frac{1}{(s+2-k)^2}-\frac{1}{(s+2)^3}\right]f_k\geqslant\sum_{k=1}^{(s+2)/2}\left[\frac{2f_kk}{(s+2-k)^2(s+2)}-\right.$$
$$\left.-\frac{k^2f_k}{(s+2-k)^2(s+2)^2}\right]\geqslant\sum_{k=1}^{(s+2)/2}\frac{2f_kk}{(s+2)^3}-k_4\frac{1}{(s+2)^4}\geqslant$$
$$\geqslant\frac{2\mu}{(s+2)^3}-k_5\frac{1}{(s+2)^4}.$$

Since $\sum_{(s+2)/2}^{\infty}f_k=o(s^{-2})$ and (43) hold, then (46) and the estimates which follow prove Lemma 5. All the needed auxiliary results are obtained. Next we prove Lemma 2.

Proof of Lemma 2. We show that

$$h'(P_n) = 1 - BQ_n + O\left(\frac{1}{n} p_n\right). \tag{47}$$

Indeed,

$$\frac{1 - h'(P_n) - BQ_n}{Q_n} = \frac{\sum_1^\infty h_k k \left(1 - P_n^{k-1}\right) - BQ_n}{Q_n} =$$

$$= \sum_2^\infty h_k \left[k \sum_0^{k-2} P_n^i - k(k-1) \right] = \sum h_k k \sum_0^{k-2} \left(1 - P_n^i\right) =$$

$$= O\left(\sum_1^n h_k k^3 (1 - P_n) + \sum_{n+1}^\infty h_k k^2 \right) = O(p_n).$$

In the expansion of $h(x)$ and $P_k(x)$ in Taylor series all the coefficients are positive, therefore, $h'(x)$ and $P_k(x)$ are monotone increasing functions on $0 \le x \le 1$, which enables us to write:

$$h(P_k(x')) - h(P_k(x)) = \int_{P_k(x)}^{P_k(x')} h'(y)\,dy \geqslant h'(P_k) D_k. \tag{48}$$

Using (5) and (48), we obtain

$$D_n = \sum_{k=1}^n [h(P_{n-k}(x')) - h(P_{n-k})(x))] f_k + (x' - x) q_n \geqslant$$

$$\geqslant \sum_{k=1}^n D_{n-k} h'(P_{n-k}) f_k + (x' - x) q_n. \tag{49}$$

Then, by the fact that $D_0 = x' - x \neq 0$ for $x' \neq x$ and $h'(P_n) > 0$, it follows that $D_n/(x'-x) > 0$ for any n. Therefore, for any n there exist $\beta_n > 0$ such that

$$D_n \geqslant (x' - x)\beta_n 1/(n+1)^2. \tag{50}$$

By (47), (44), (25) and the definition of $\bar{p}_n$ there exists k_1 such that

$$h'(P_n)=1-BQ_n+O\left(\frac{1}{n}p_n\right)\geqslant 1-\frac{2\mu}{n+1}-\frac{k_1}{n}\bar{p}_n. \qquad (51)$$

We assume that in (45) k_1 from (51) is used. Let n_0 be arbitrarily large so that the right-hand sides in (45) and (51) are positive for s and k larger than n_0. For $n\leq n_0$ let $\beta_n>0$ be such that they satisfy the condition (50) and $\beta_{n_0}\leq\beta_n$.

We now choose β_n, $n\geq n_0$, such that they satisfy (50), and (50) implies (38). We define β_{n+1} for $n\geq n_0$ by the recurrence relation $\beta_{n+1}=\beta_n(1-k_6\frac{1}{n+2}\bar{p}_{n+2})$. We note that β_n does not increase in n. Assume that for β_n thus chosen the condition (50) is satisfied for k such $n_0\leq k\leq s$. Then from (49) and (51) there follows

$$D_{s+1}\geqslant\sum_{k=1}^{(s+2)/2}D_{s+1-k}\left(1-\frac{2\mu}{s+2-k}-\frac{k_1}{s+2-k}\bar{p}_{s+2-k}\right)f_k\geqslant$$

$$\geqslant(x-x')\beta_s\left(\sum_{k=1}^{(s+2)/2}\frac{1}{(s+2-k)^2}\left(1-\frac{2\mu}{s+2-k}-\frac{k_1\bar{p}_{s+2-k}}{s+2-k}\right)f_k\right)$$

Using Lemma 5, we then have

$$D_{s+1}\geqslant(x'-x)\beta_s\frac{1}{(s+2)^2}\left(1-k_6\frac{1}{s+2}\bar{p}_{s+2}\right). \qquad (52)$$

The inequality (52) shows that the condition (50) is satisfied for the given sequence β_n.

From the estimate

$$\sum_{n_0+1}^{\infty}\frac{1}{s}\bar{p}_s=O\left(\sum_{1}^{\infty}\frac{1+\ln s}{s^2}\sum_{k=1}^{s}h_kk^3+\sum_{1}^{\infty}\frac{1}{s}\sum_{s+1}^{\infty}h_kk^2\right)=$$

$$=O\left(\sum_{k=1}^{\infty}h_kk^3\sum_{k}^{\infty}\frac{\ln s}{s^2}\right)+O\left(\sum h_kk^2\ln k\right)=O\left(\sum_{1}^{\infty}h_k(\ln k)k^2\right)<\infty \qquad (53)$$

obtained by virtue of (42), it follows that the product $\prod_{n_0}^{\infty}\left(1-k_6\frac{1}{s+2}\bar{p}_{s+2}\right)\neq 0$ converges, i.e., by the obvious inequality $\beta_n\geqslant\beta_{n_0}\prod_{n_0}^{\infty}\left(1-k_6\frac{\bar{p}_{s+2}}{s+2}\right)$ we have proved Lemma 2.

THEOREM 3. Let $|z|=1$. Then there exists a constant $k_8>0$ and an index $\bar{n}_0$ such that for any $n\geq\bar{n}_0$ we have the inequality

$$|Q_n(z)|\geqslant k_8/(n+1/|1-z|). \tag{54}$$

Proof. First we prove the inequality (54) for z close to 1. Note that we can assume $\bar{n}_0=0$ up to and including Lemma 8.

From [4] it follows that

$$Q_n(x)(n\alpha+1/(1-x))\to 1 \tag{55}$$

as $n\to\infty$ uniformly in $0\leq x<1$.

As was mentioned earlier, $P_n'(1)=1$; therefore

$$|Q_n(z)|=\left|\int_z^1 P_n'(z)\,dz\right|\leqslant|1-z|. \tag{56}$$

From (23), (42), (56) and the equality $Q_0(z)=1-z$ we obtain

$$Q_n(z)=Q_0(z)-\alpha\sum_0^{n-1}Q_i^2(z)+(1-z)^2\sum_0^{n-1}O(p_k)=$$

$$=(1-z)-\alpha\sum_0^{n-1}Q_i^2(z)+(1-z)^2O(n\bar{p}_n). \tag{57}$$

Square (23) yields $Q_{n+1}^2(z)=Q_n^2(z)+O_n(|1-z|^3)$ and, next,

$$Q_n^2(z)=Q_{n-k}^2(z)+kO_n(|1-z|^3). \tag{58}$$

From (58) and (57) it follows that

$$Q_n(z)=1-z-\alpha n(1-z)^2+O_n(n^2|1-z|^3)+O_n(|1-z|^2n\bar{p}_n).$$

It is seen that $1-e^{it} = -it + O(t^2)$.

Since $\bar{p}_n \to 0$ as $n\to\infty$,

$$Q_n(e^{it}) = -it + \alpha nt^2 + O_n(n^2t^3) + o_n(nt^2), \tag{59}$$

or

$$P_n(e^{it}) = 1 + it - \alpha nt^2 + O_n(n^2t^3) + o_n(nt^2). \tag{60}$$

Since

$$|1+it-\alpha nt^2| = \sqrt{(1-\alpha nt^2)^2+t^2} = \\ = \sqrt{1-2\alpha nt^2+O_n(n^2t^4)+t^2} = 1-\alpha nt^2 + O_n(n^2t^4) + O_n(t^2),$$

from (60) for $|nt|>1$ there follows

$$|P_n(e^{it})| = 1-\alpha nt^2 + O_n(n^2t^3) + o_n(nt^2) \leqslant 1- \\ - t[\alpha nt(1+O_n(nt)+o_n(1))],$$

i.e., for sufficiently small ε_1 and ε_2 $(0<\varepsilon_1<\varepsilon_2<1)$ for n and t such that $\varepsilon_1 \leq |nt| \leq \varepsilon_2$, where $n \geq N_0$, and N_0 is sufficiently large, it is possible to find $\eta_0 = \eta_{\varepsilon_1,\varepsilon_2,N_0} > 0$ such that

$$|P_n(e^{it})| \leqslant 1 - \eta_0|1-e^{it}| = \rho_0(t) = \rho_0 < 1. \tag{61}$$

Eq. (59) implies $|Q_n(e^{it})| = t(1+o_n(1))$ as $|nt|\to 0$, which together with the obvious estimate

$$\frac{1}{n+1/|1-e^{it}|} = \frac{t+O_n(t^2)}{n\left(t+O_n(t^2)\right)+1} \quad \text{as } |nt|\to 0$$

establishes the following lemma.

LEMMA 6. There exists ε_2' such that Theorem 3 holds for $|tn|<\varepsilon_2'$.

In proving Theorem 3, we assume that ε_1, ε_2, ε_2', N_0 and η_0 defined before are fixed, letting without loss of generality $\varepsilon_2=\varepsilon_2'<1$.

Let $n_1=\min\{n: \varepsilon_1\le|tn|, |t|\le\delta\}$, and $n_2=\max\{n: |tn|\le\varepsilon_2, |t|\le\delta\}$. Then for any n_0 there exists δ such that $0<\delta\le\varepsilon_1/N_0$ and $n_2-n_1\ge n_0$.

Suppose n_0 is such that for $n\ge n_0$ we have the inequality

$$1-(n+1)^2q_n/d > 0 , \tag{62}$$

where d is the same as in the formulation of Lemma 2. This is possible because the inequality $M\eta^3<\infty$ implies $q_n n^2\to 0$ as $n\to\infty$. Using n_0 we choose and fix $\delta>0$ such that $n_2-n_1\ge n_0$ and $\delta\le\varepsilon_1/N_0$.

We prove now by induction that there exists a non-increasing sequence $\eta_k>0$, $k\ge 0$, such that for fixed t ($t\ne 0$ and $|t|\le\delta$)

$$\left|P_{n_1+k}(e^{it})\right|\leqslant P_k(\rho_k), \tag{63}$$

where $\rho_k=\rho_k(t)=1-\eta_k|1-e^{it}|$ and $\inf\eta_k>0$. We note that η_k does not depend on t.

We need the following lemma.

<u>LEMMA 7</u>. For any δ^2 and ρ such that $0\le\rho\le\rho+\delta^2\le 1$ we have the inequalities

$$P_n(\rho)\le P_{n+1}(\rho)\le P_{n+1}(\rho+\delta^2) . \tag{64}$$

<u>Proof</u>. The proof of (64) is immediate because $P_n(x)=\sum_0^\infty p_{nk}x^k$, where $p_{nk}\ge 0$. To prove the first inequality of (43), we apply the induction. It is known that $h(x)$ increases in $0\le x\le 1$ and $h(x)\ge x$ (see [1]). Hence $P_1(x)=h(x)f_1+xq_1\ge x=P_0(x)$.

Let $P_k(x)\le P_n(x)$ for $k\le n\le n_3$. We prove that $P_{n_3+1}(x)\ge P_{n_3}(x)$.

Eq. (5) and the inequalities written above imply

$$P_{n_3+1}(x) = \sum_{k=1}^{n_3+1} h\left(P_{u_3+1-k}(x)\right) f_k + xq_{n_3+1} \geqslant$$

$$\geqslant \sum_{k=1}^{n_3} h\left(P_{n_3-k}(x)\right) f_k + x\left(f_{n_3+1} + q_{n_3+1}\right) = P_{n_3}(x),$$

which completes proving Lemma 7.

We proceed to derive (63). By definition, n_0 from (64) and (61) for $k \leq n_0$ we obtain

$$\left|P_{n_1+k}(e^{it})\right| \leqslant \rho_0 = P_0(\rho_0) \leqslant P_k(\rho_0),$$

i.e., $\eta_k = \eta_0$ for $0 \leq k \leq n_0$.

We define η_{n_0+1}. It is seen that

$$\left|P_{n_0+n+1}(e^{it})\right| \leqslant \sum_{k=1}^{n_0+n+1} h\left(\left|P_{n_1+n_0+1-k}(e^{it})\right|\right) f_k + q_{n_0+n_1+1} \leqslant$$

$$\leqslant \sum_{k=1}^{n_0+1} h\left(P_{n_0+1-k}(\rho_0)\right) f_k + q_{n_0+1} = P_{n_0+1}(\rho_0) + (1 - \rho_0) q_{n_0+1},$$

where the last equality rests on (5).

By Lemma 2

$$P_{n_0+1}\left(\rho_0 + \frac{(1-\rho_0)(n_0+2)^2}{d} q_{n_0+1}\right) - P_{n_0+1}(\rho_0) \geqslant (1-\rho_0) q_{n_0+1},$$

whence

$$\left|P_{n_0+n+1}(e^{it})\right| \leqslant P_{n_0+1}\left(1 - \eta_0\left|1 - e^{it}\right|\left(1 - \frac{(n_0+2)^2}{d} q_{n_0+1}\right)\right). \quad (65)$$

By (62)

$$\eta_{n_0+1} = \eta_0\left(1 - \frac{(n_0+2)^2}{d} q_{n_0+1}\right)$$

satisfies the assumption concerning induction. For $k=s+n_0$ let this assumption be satisfied. We prove that it holds also for $k=s+n_0+1$. The following inequality is easily obtainable from the same considerations as those for (65):

$$\left|P_{n_0+s+1+n_1}(e^{it})\right| \leqslant P_{n_0+s+1}(\rho_{s+n_0}) + (1-\rho_{s+n_0})\, q_{n_0+s} \leqslant P_{n_0+s+1}(\rho_{s+n_0+1}),$$

where

$$\rho_{s+n_0+1} = 1 - \eta_{n_0+s}\left(1 - \frac{(s+n_0+2)^2}{d}\, q_{s+n_0+1}\right)\left|1-e^{it}\right|,$$

thus proving (63).

Since $\sum_{n_0}^{\infty} n^2 q_n \leq \sum n^3 f_n < \infty$, then the product

$$\prod_0^{\infty}\left(1 - \frac{(n_0+k+2)^2}{d} \times q_{n_0+k+1}\right)$$

is convergent and positive, leading, by (63) and (64), to the estimate

$$\left|P_{n_1+k}(e^{it})\right| \leqslant P_k(\rho), \tag{66}$$

where

$$\rho = 1 - \eta_0 \prod_0^{\infty}\left(1 - \frac{(n_0+k+2)^2}{d}\, q_{n_0+k+1}\right)\left|e^{it}-1\right| = 1 - \bar{\eta}\left|e^{it}-1\right|.$$

The estimates (55) and (66) imply the existence of $k_9>0$ such that

$$\left|Q_{n_1+k}(e^{it})\right| \geqslant 1 - \left|P_{n_1+k}(e^{it})\right| \geqslant 1 - P_k(\rho) \geqslant \frac{k_9}{\alpha k + 1/\bar{\eta}\left|1-e^{it}\right|} \geqslant$$

$$\geqslant \frac{k_9}{\alpha(n_1+k) + 1/\bar{\eta}\left|1-e^{it}\right|} \geqslant \frac{k_9(1-1/\bar{\eta})^{-1}}{\alpha(n_1+k) + 1/\left|1-e^{it}\right|}.$$

From this estimate and Lemma 6 we have the following lemma.

<u>LEMMA 8</u>. There exists $\delta>0$ such that for $|t|<\delta$ Theorem 3 holds (with $n_0=0$).

It remains only to investigate $P_n(z)$ for $|1-z|>\delta_1$

for an arbitrary fixed $\delta_1>0$. If there exists $\bar{n}_0$ such that for $z\neq 1$ for any $n\geq\bar{n}_0$

$$|P_n(z)| \leq 1 , \qquad (67)$$

then $\forall\, n_0'>0$ it is possible to find $\gamma_{n_0'}=\gamma_{n_0'\bar{n}_0'\delta_1}<1$ such that $|P_n(z)|<\gamma_{n_0'}$ for $\bar{n}_0\leq n\leq\bar{n}_0+n_0'$ and $|1-z|>\delta_1$ and therefore, also $\rho_{n_0'}=1-\eta_{n_0'}$ such that $\eta_{n_0'}>0$ and $|P_n(z)|\leq P_{n-\bar{n}_0}(\rho_{n_0'})$ $(\bar{n}_0\leq n\leq\bar{n}_0+n_0')$. Hence, repeating almost word-for-word the proof of (63) and (66) with n_1 equal to $\bar{n}_0$, $|1-e^{it}|$ equal to 1, and the same n_0, we obtain that there exists $\rho<1$ such that $|P_n(z)|\leq P_{n-\bar{n}_0}(\rho)$. By (55) and Lemma 8 the next lemma completes proving the Theorem.

LEMMA 9. If conditions "a" and "b" of Section 1 are satisfied, the inequality (67) is satisfied beginning from some $\bar{n}_0$.

Proof. First we note that the properties of characteristic functions used in what follows can be found in [10, p. 115]. From condition "a" it follows that if only one f_k is not zero, i.e., $f_k=1$, then $k=1$. Then, from condition "a" and the properties of characteristic functions for lattice random variables it follows that $|h(z)|<1$ for $|z|=1$ and $z\neq 1$. But $P_1(z)=h(z)$, and $P_k(z)=h(P_{k-1}(z))$. Therefore, in this case we can assume $\bar{n}_0=1$.

It remains only to investigate the case where there exist at least two $f_k\neq 0$ and condition "b" of Section 1 is satisfied.

Since $h(1)=1$, $h'(1)=1$ and $h''(1)>0$, we obtain $h(0)>0$. This means that $h(e^{it})$ is the characteristic function of a lattice random variable and 0 belongs to the lattice. This immediately implies that if $|h(z)|=1$, then $h(z)=1$, i.e., the equality $h(z)=z$ for $|z|=1$ is possible only for $z=1$. We call it property (*).

Let $S=\{k|f_k \neq 0\}$, $j=\sup\{k|k \in S\}$. In the sequel $S \setminus j=S$, if $j=\infty$. From the property (*), the inequality $|h(z)|<1$ for $|z|<1$, the system (5), and the inequality $|az_1+bz_2|<1$ holding for $a>0$, $b>0$, $a+b=1$, $|z_1|=|z_2|=1$ and $z_1 \neq z_2$ and <u>a fortiriori</u> for $|z_1| \leq 1$, $|z_2|<1$, there follows $|P_n(z)|<1$ for $|z|=1$, $z \neq 1$ and $n \in S \setminus j$. From the last inequality and the considerations used to have obtained it, it follows that this inequality holds as well for n of the form

$$\alpha_j j + \sum_{\substack{k \in S \setminus j \\ \alpha_k \geqslant 0}} k\alpha_k,$$

where $\alpha_j \equiv 0$, if $j=\infty$, and α_k are integers and not all of them are equal to zero for $k \neq j$. The assertion obtained together with the next Lemma 10 prove Lemma 9.

We draw the reader's attention to the following: from the set S defined above it is possible to choose a definite subset S' such that the greatest common divisor of $k \in S'$ is 1.

<u>LEMMA 10</u>. There exists $\bar{n}_0$ such that any integer $n \geq \bar{n}_0$ can be represented as

$$n = \sum_{\substack{k \in S' \\ \alpha_k > 0}} \alpha_k k.$$

This lemma has been proved in [10, p. 207]. Although the formulation of the lemma therein does not contain $\alpha_k \geq 0$, the author notes in the proof that the assertion holds true for

$\alpha_k > 0 \ \forall k$.

REMARK. For small n (54) does not, in general, hold. For example, $h(z)=1/2+z^2/2$, $f_4=f_3=1/2$. It is not hard to show that $P_0(z)=P_1(z)=P_2(z)=z$, $P_3(z)=\frac{1}{2}(z+h(z))$, $P_4(h)=h(z)$ and $P_4(-1)=1$, although conditions "a" and "b" of Section 1 are satisfied.

COROLLARY. There exist $\bar{n}_0>0$ and constant k_{10} such that for any $|z|\leq 1$ and $n\geq\bar{n}_0$

$$|Q_n(z)| \geqslant \frac{k_{10}}{|n\alpha+(1-z)^{-1}|}. \tag{68}$$

Proof. We begin by noting that for $|z|<1$, first, by definition of $Q_n(z)$ and the fact that $Q_n(1)=0$ $(P_n(1)=1)$, we have $Q_n(z)\neq 0$; second, $n\alpha+1/(1-z)\neq 0$ $(\mathrm{Re}(n\alpha+(1-z)^{-1})>0)$. We define $Q_n(z)(n\alpha+(1-z)^{-1})$ more precisely for $z=1$ in continuity by unity, which is possible by (27). This function is analytic in $|z|<1$; therefore, if we prove (68) for $|z|=1$, (68) will hold for $|z|\leq 1$ by the maximum modulus principle. The needed estimate follows from (54) and the inequality

$$\frac{1}{n\alpha+|1-z|^{-1}} \geqslant \frac{1}{2}\left|\frac{1}{n\alpha+(1-z)^{-1}}\right|$$

for $|z|=1$. The inequality is equivalent to $n\alpha|1-z|+1\leq$ $\leq 2|n\alpha(1-z)+1|$, which follows from the obvious inequalities $|n\alpha(1-z)|\leq|n\alpha(1-z)+1|$ and $1\leq|n\alpha(1-z)+1|$.

The proved corollary justifies the further references to Lemma 3 or the formula (39).

THEOREM 4. There exists $\bar{m}_0$ such that for $n\geq\bar{m}_0$

$$Q_n(z) = \frac{1+\tilde{O}(\bar{p}_n)}{n\alpha + (1-z)^{-1}}. \tag{69}$$

Proof. If we had used the estimate

$$\|[(1/(1-z)+\alpha n)Q_n(z)]^{-1}\| \leqslant k_{11} < \infty, \; n \geqslant \bar{m}_0 \geqslant \bar{n}_0, \tag{70}$$

in deriving (39), we could replace everywhere O by Õ in proving Lemma 3; we could thus arrive at (69).

Therefore, to prove Theorem 4 it suffices to establish (70).

If in proving (47) instead of P_n we put s satisfying the conditions $|s|\leq 1$ and $|1-s|\leq \alpha'/n$, where $\alpha'>0$ is fixed, we obtain the estimate

$$|1-h'(s)-B(1-s)| \leqslant \frac{1}{n} O(p_n). \tag{71}$$

Making use of the properties of the integral and of the estimates (27), (37) and (71), we derive

$$\left| \int\limits_{P_{n-k-1}(z)}^{P_{n-k}(z)} (1-h'(z)-B(1-z))\,dz \right| \leqslant |Q_{n-k}(z)-Q_{n-k-1}(z)|\, O\left(\frac{p_{n-k}}{n-k}\right). \tag{72}$$

From (23), (25a), (39), (72), (25), (27) and the obvious identity

$$h(P_{n-k}(z)) + Q_{n-k}(z) - h(P_{n-k-1}(z)) - Q_{n-k-1}(z) = \frac{B}{2} Q^2_{n-k}(z) - $$
$$- \frac{B}{2} Q^2_{n-k-1} - \int\limits_{P_{n-k-1}(z)}^{P_{n-k}(z)} (1-h'(z)-B(1-z))\,dz$$

it follows that

$$h(P_{n-k}(z)) + Q_{n-k}(z) - h(P_{n-k-1}(z)) - Q_{n-k-1}(z) =$$
$$= \frac{B}{2}\left(Q^2_{n-k}(z) - Q^2_{n-k-1}(z)\right) + \frac{O(p_{n-k})}{(n-k)\left(\alpha(n-k) + (1-z)^{-1}\right)^2} =$$
$$= -\frac{B}{2} 2\alpha Q^3_{n-k-1}(z) + \frac{O(p_{n-k})}{(n-k)\left(\alpha(n-k)+(1-z)^{-1}\right)^2}. \qquad (73)$$

Using (29), (26) and (73), we obtain

$$Q_{n-1}(z) - Q_n(z) = -(1 - h(P_{n-1}(z)) - Q_{n-1}(z))/\mu +$$
$$+ \frac{B}{\mu}\alpha Q^3_{n-1}(z) + \frac{O(p_n)}{n(\alpha n + (1-z)^{-1})^2} - B\alpha \sum_{k=0}^{N} Q^3_{n-k-1}(z)\, b_k +$$
$$+ \sum_{k=0}^{N} \frac{O(p_{n-k})}{(n-k)(n-k+(1-z)^{-1})^2} b_k + \frac{B}{2} \sum_{k=N+2}^{n} Q^2_{n-k}(z) \times$$
$$\times \left(1 + \tilde{O}(p_{n-k})\right)(b_k - b_{k-1}) + b_{N+1} Q^2_{n-N-1}(z)\left(1 - \tilde{O}(p_{n-N+1})\right).$$

From (24), (25) and (27) there follows $Q^3_{n-1}(z) = Q^2_{n-1}(z)\tilde{O}(p_n)$; and from (24), (15) and (28) there follows

$$\sum_{k=0}^{N} \frac{O(p_{n-k})\, b_k}{(n-k)\left((n-k)\alpha + (1-z)^{-1}\right)^2} = \sum_{k=0}^{N} \frac{O(p_n)}{n(n\alpha + (1-z)^{-1})^2} b_k =$$
$$= \frac{O(p_n)}{n(n\alpha + (1-z)^{-1})^2}.$$

The last equalities plus the estimates (33), (25) and (26) imply

$$Q_{n-1}(z) - Q_n(z) = \alpha Q^2_{n-1}(z)\left(1 + \tilde{O}(p_n)\right) -$$
$$- B\alpha \sum_{k=0}^{N} Q^3_{n-k-1}(z)\, b_k + \frac{B}{2} \sum_{k=N+2}^{n} Q^2_{n-k}(z)(b_k - b_{k-1}) \times$$
$$\times \left(1 + \tilde{O}(p_{n-k})\right) + \frac{O(p_n)}{n(\alpha n + (1-z)^{-1})^2}. \qquad (74)$$

From (23) it is easy to obtain for $k \leq N$

$$Q_{n-k}^3(z) = Q_n^3(z) + k\frac{\widetilde{O}(1)}{(\alpha n + (1-z)^{-1})^4}. \tag{75}$$

Indeed, cubing (23) and noting (27) and (28), we have

$$Q_n^3(z) = Q_{n-1}^3(z) + \frac{\widetilde{O}(1)}{(\alpha n + (1-z)^{-1})^4}.$$

The sum of k equalities of this kind is (75).

From (16), (25), (37) and (75) there follows the estimate

$$\sum_{k=0}^{N} Q_{n-k-1}^3(z)\, b_k = Q_{n-1}^3(z)\sum_{k=0}^{N} b_k + \frac{\widetilde{O}(1)}{(\alpha n + (1-z)^{-1})^4}\sum_{k=0}^{N} |kb_k| =$$
$$= Q_{n-1}^2(z)\,\widetilde{O}(p_n) + \frac{\widetilde{O}(p_n)}{n(\alpha n + 1/(1-z))^2}. \tag{76}$$

Using (27), (28) and (14), we conclude that

$$\sum_{n=4}^{\infty}\left\| \frac{B}{2}\sum_{k=N+2}^{n} Q_{n-k}^2(z)(n\alpha + 1/(1-z))^2(b_k - b_{k-1})\right\| =$$
$$= O\left(\sum_{n=4}^{\infty}\sum_{k=N+2}^{n}\left(\frac{n}{n-k+1}\right)^2 |b_k - b_{k-1}|\right) =$$
$$= O\left(\sum_{k=3}^{\infty} |b_k - b_{k-1}|\cdot\sum_{n=k}^{2k}\frac{n^2}{(n-k+1)^2}\right) =$$
$$= \left(\sum_{k=3}^{\infty}|b_k - b_{k-1}|\,k^2\sum_{1}^{\infty}\frac{1}{i^2}\right) = O(1). \tag{77}$$

The inequality $\sum p_n/n<\infty$ (follows from (46), (53) and the definition of p_n) together with the estimates (76) and (77) justify us in asserting that there exist $r_n>0$ such that

$$-B\alpha\sum_{k=0}^{N} Q_{n-k-1}^3(z)\, b_k + \frac{B}{2}\sum_{k=N+2}^{n} Q_{n-k}^2(z)(b_k - b_{k-1})(1+\widetilde{O}(p_{n-k})) =$$
$$= Q_{n-1}^2(z)\,\widetilde{O}(p_n) + \frac{\widetilde{O}(r_n)}{(n\alpha + 1/(1-z))^2}, \tag{78}$$

and $\sum r_n<\infty$.

Therefore, (78) enables us to transform (74) into

$$Q_n(z) = Q_{n-1}(z) - \alpha Q_{n-1}^2(z)(1 + \tilde{O}(p_n)) + \frac{\tilde{O}(r_n)}{(n\alpha + 1/(1-z))^2}. \qquad (79)$$

We note that this equality makes Lemma 1 more precise.

From Lemma 3 by (68) for sufficiently large n there follows

$$Q_n^{-1}(z)\frac{1}{n\alpha + 1/(1-z)} = O_n(1).$$

However, until now we cannot say anything definite about the norm of the last expression, i.e., we have no reason to substitute $\tilde{O}(1)$ for $O_n(1)$. Hence we will write only O instead of $\tilde{O}$. As before, $L_n = Q_n^{-1}(z)$.

It follows from (79) and (39) that for $n > \bar{m}_0$

$$L_n = L_{n-1}\left(1 - \alpha Q_{n-1}(z)(1 + \tilde{O}(p_n)) + O(r_n)\left(n\alpha + (1-z)^{-1}\right)^{-1}\right)^{-1} =$$
$$= L_{n-1}(1 - \alpha Q_{n-1}(z)(1 + \tilde{O}(p_n)) + \frac{O(r_n)}{n\alpha + 1/(1-z)} =$$
$$= L_{n-1} + \alpha(1 + \tilde{O}(p_n)) + O(r_n).$$

Summing up the resulting equalities from $\bar{m}_0+1$ to n, we obtain by (40)

$$L_n = (1-z)^{-1} + \alpha n + \sum_{k=\bar{m}_0}^{n} \tilde{O}(p_k) + \sum_{k=\bar{m}_0}^{n} O(r_k) + \tilde{O}(1) =$$
$$= (1-z)^{-1} + \alpha n(1 + \tilde{O}(\bar{p}_n)) + K(z) + O\left(\sum_n^{\infty} r_k\right),$$

where $K(z) = \sum_{k=\bar{m}_0}^{\infty} O(r_k)$ does not depend on n, is bounded because $\sum r_k < \infty$ and analytic, since it is the limit of a uniformly convergent sequence of analytic functions. Hence $K(z)$ can be viewed as the Fourier series for $|z|=1$. The same can be said of the last term in the equality considered (however, for fixed n).

Taking the inverse quantities from the both sides of the last equality and taking into account (43) and (37), we have

$$Q_n(z) = \frac{1+\tilde{o}(\bar{p}_n)}{\alpha n + (1-z)^{-1}} + \frac{K(z)+o_n(1)}{(n\alpha + (1-z)^{-1})^2}, \tag{80}$$

where by $o_n(1)$ we mean the function expressed as the Fourier series for $|z|=1$ according to the foregoing considerations. We write this series as $\sum_{k=0}^{\infty} c_{nk} z^k$.

Now we prove that the quantity

$$\left\| Q_n(z)\left(\alpha n + (1-z)^{-1}\right) - 1 + \sum_{k=1}^{tn} \alpha n p_{nk} z^k - \sum_{k=1}^{tn} \sum_{i=k+1}^{\infty} p_{ni} z^k \right\| =$$

$$= |1 - Q_n(\alpha_n + 1| + \left\| \sum_{k=tn+1}^{\infty} \alpha n p_{nk} z^k - \sum_{k-tn+1}^{\infty} \sum_{i=k+1}^{\infty} p_{ni} z^k \right\| \tag{81}$$

can be made arbitrarily small by fixing an integer t and taking a sufficiently large n. Indeed, by (44) the first term on the right side of (81) tends to zero as $n\to\infty$.

On the other hand (see [2, p. 295]), $\sum_1^{\infty} k^2 p_{nk} = O(n)$, which in turn yields, by the Chebyshev inequality,

$$\alpha n \sum_{k=tn+1}^{\infty} p_{nk} = \frac{n}{(tn)^2} O_t(n) \to 0 \quad \text{as} \quad t\to\infty$$

and

$$\sum_{k=tn+1}^{\infty} \sum_{i=k+1}^{\infty} p_{ni} = O_t\left(\sum_{k=tn+1}^{\infty} k p_{nk}\right) = \frac{O_t(n)}{tn} \quad \text{as} \quad t\to\infty,$$

proving thereby that the right side of (81) can be made arbitrarily small.

The resulting assertion implies, by (43) and (80), the

existence of t such that

$$\sum_{k=tn+1}^{\infty}\left|\mathbf{C}_k\left(\frac{K(z)+o_n(1)}{n\alpha+(1-z)^{-1}}\right)\right|<\varepsilon. \tag{82}$$

Let $\tau_k=C_k(K(z))$. From the properties of Fourier coefficients and the equality $\sum_0^{\infty} c_{nk}z^k=o_n(1)$ it follows that $\tau_k\to 0$ as $k\to\infty$ and $c_{nk}=o_n(1)$ for any k. Then by the equality (37) for any fixed t we obtain

$$\left\|\sum_0^{tn}\mathbf{C}_k\left(\frac{K(z)+o_n(1)}{n\alpha+(1-z)^{-1}}\right)z^k\right\|\leqslant\left\|\frac{1}{n\alpha+(1-z)^{-1}}\right\|\left(\sum_0^{tn}|\tau_k|+\sum_0^{tn}|c_{nk}|\right)=$$

$$=\frac{2}{n\alpha+1}\left(\sum_0^{tn}|\tau_k|+\sum_0^{tn}|c_{nk}|\right)=o_n(1). \tag{83}$$

Therefore, by (37), (82) and (83) plus the arbitrariness of $\varepsilon>0$ we have the estimate

$$\left\|\frac{K(z)+o_n(1)}{n\alpha+(1-z)^{-1}}\right\|=o_n(1) \quad (n>m_0),$$

which by virtue of (80) immediately implies (70) thus completing proving Theorem 4.

<u>LEMMA 11</u>. Under the conditions stated in this section we have the representation

$$Q_n(z)-Q_n=(Q_{n-1}(z)-Q_{n-1})(1-\alpha(Q_{n-1}(z)+Q_{n-1})+$$
$$+\tilde{O}(r_n))+\tilde{O}(r_n/n^2), \tag{84}$$

where $r_n>0$ denotes a sequence such that $\sum r_n<\infty$.

<u>Proof</u>. We shall need in the sequel several sequences of positive numbers. These sequences (we denote them by r_n) must majorize up to a constant factor the corresponding expressions and also be summable. Hence we can assume without loss of generality that these sequences are equal.

As we mentioned earlier, it is easy to show by analogy with (53) that $\sum p_n/n<\infty$; this justifies the further substitution of r_n for p_n/n.

Let

$$R_n^{(k)}(z)=Q_n^k-Q_n^k(z);\quad R_n(z)=R_n^{(1)}(z);$$
$$\bar{h}(R_n(z))=h(P_n(z))-h(P_n).$$

From (29) we obtain

$$R_{n-1}(z)-R_n(z)=-(\bar{h}(R_n(z))-R_n(z))/\mu+\sum_{k=0}^{N}(-h(R_{n-k}(z))+$$
$$+\bar{h}(R_{n-k-1}(z))+R_{n-k}(z)-R_{n-k-1}(z))b_k-\sum_{k=N+2}^{n}[\bar{h}(R_{n-k}(z))-$$
$$-R_{n-k}(z))]\times(b_k-b_{k-1})-b_{N+1})[\bar{h}(R_{n-N-1}(z))-R_{n-N-1}(z)]. \quad (85)$$

As before, we put $N=[n/2]$.

It is seen that

$$\|R_n(z)\|=\left\|\sum_1^\infty p_{nk}z^k\right\|=R_n(1)=Q_n=O\left(\frac{1}{n}\right) \quad (86)$$

and

$$\|R_n^{(2)}(z)\|=\|(Q_n(z)-Q_n)(Q_n(z)+Q_n)\|\leqslant 3Q_n^2=O\left(\frac{1}{n^2}\right). \quad (87)$$

By analogy with (47) it is easy to obtain

$$h''(P_n)-B=O(p_n)\ . \quad (88)$$

Indeed,

$$\sum_1^\infty h_k(k-1)kP_n^k-B=\sum_1^\infty h_kk(k-1)(P_n^k-1)=$$
$$=\sum_1^n h_kk(k-1)(P_n-1)\sum_{i=0}^{k-1}P_n^i+\sum_{k=n+1}^{\infty}h_kk(k-1)=O(p_n).$$

By the definition of $\bar{h}(R_n(z))$ we have

$$\overline{h}(R_n(z)) - R_n(z) + (B/2)R_n^{(2)}(z) = \sum_0^\infty h_k\left((R_n(z)+P_n)^k - P_n^k\right) -$$

$$- R_n(z) + (B/2)R_n(z)(2Q_n - R_n(z)) = \sum_3^\infty h_k \sum_{i=0}^{k-3} C_k^i P_n^i R_n^{k-i}(z) +$$

$$+ \sum_1^\infty h_k R_n(z) P_n^{k-1} k + \frac{1}{2}\sum_2^\infty h_k R_n^2(z) P_n^{k-2} k(k-1) - R_n(z) +$$

$$+ \frac{B}{2} R_n(z)(2Q_n - R_n(z)) = \sum_3^\infty h_k \sum_{i=0}^{k-3} C_k^i P_n^i R_n^{k-i}(z) +$$

$$+ R_n(z)(h'(P_n) - 1 + BQ_n) + \frac{1}{2} R_n^2(z)(h''(P_n) - B).$$

Using the estimates (89) for $z=1$ plus (86), we obtain

$$\left\| \frac{\sum_3^\infty h_k \sum_{i=0}^{k-3} C_k^i P_n^i R_n^{k-i}(z)}{R_n(z)} \right\| = \sum_3^\infty h_k \sum_{i=0}^{k-3} C_k^i P_n^i Q_n^{k-i-1} =$$

$$= \frac{\sum_0^\infty h_k\left((Q_n + P_n)^k - P_n^k\right) - Q_n + (B/2)Q_n^2 - Q_n(h'(P_n) - 1 + BQ_n)}{Q_n} -$$

$$- \frac{Q_n^2(h''(P_n)/2 - B/2)}{Q_n} = \frac{1 - h(P_n) - Q_n + \frac{B}{2}Q_n^2}{Q_n} -$$

$$-(h'(P_n) - 1 + BQ_n) - \frac{1}{2} Q_n(h''(P_n) - B).$$

The last representation for the norm together with (26), (47) and (88) enables us to write:

$$\sum_3^\infty h_k \sum_{i=0}^{k-3} C_k^i P_n^i R_n^{k-i}(z) = R_n(z)\widetilde{O}\left(\frac{p_n}{n}\right). \tag{90}$$

Applying the estimates (47), (86), (88) and (90) to (89), we obtain

$$\overline{h}(R_n(z)) - R_n(z) + \frac{B}{2} R_n^{(2)}(z) = R_n(z)\widetilde{O}\left(\frac{p_n}{n+1}\right). \tag{91}$$

The estimate (91) enables us to transform (85) to

$$R_{n-1}(z) - R_n(z) = \alpha R_n^{(2)}(z) + R_n(z)\widetilde{O}\left(\frac{p_n}{n}\right) + \sum_{k=0}^{N} \frac{B}{2}\left(R_{n-k}^{(2)}(z) - \right.$$

$$\left. - R_{n-k-1}^{(2)}(z)\right) b_k + \sum_{k=N+1}^{n} \left[\frac{B}{2} R_{n-k}^{(2)}(z) - R_{n-k}(z)\widetilde{O}\left(\frac{p_{n-k}}{n-k+1}\right)\right] \times$$

$$\times (b_k - b_{k-1}) + b_{N+1}\left[\frac{B}{2} R_{n-N-1}^{(2)}(z) - R_{n-N-1}(z)\widetilde{O}\left(\frac{p_{n-N-1}}{n-N-1}\right)\right] +$$

$$+ \sum_{k=0}^{N}\left[R_{n-k}(z)\widetilde{O}\left(\frac{p_{n-k}}{n-k}\right) - R_{n-k-1}(z)\widetilde{O}\left(\frac{p_{n-k-1}}{n-k-1}\right)\right] b_k. \qquad (92)$$

By (86), (87), (77), (25) and (1) we have

$$\sum_n \left\| \sum_{k=N+2}^{n} \frac{B}{2}\left(R_{n-k}^{(2)}(z) + R_{n-k}(z)\widetilde{O}\left(\frac{p_{n-k}}{n-k+1}\right)\right)(b_k - b_{k-1})\right\| n^2 \leqslant$$

$$\leqslant \sum_n \sum_{k=N+2}^{n} n^2 \frac{B}{2}\left[Q_{n-k}^2 + \widetilde{O}\left(\frac{p_{n-k}}{n-k+1}\right) Q_{n-k}\right] |b_k - b_{k-1}| =$$

$$= O\left(\sum_n \sum_{k=N+2}^{n} \frac{n^2}{(n-k+1)^2} |b_k - b_{k-1}|\right) < \infty. \qquad (93)$$

From (13), (86) and (87) we have

$$\left\| b_{N+1}\left(\frac{B}{2} R_{n-N-1}^{(2)}(z) + R_{n-N-1}(z)\widetilde{O}\left(\frac{p_{n-N-1}}{n-N-1}\right)\right)\right\| = O\left(\frac{1}{n^4}\right). \qquad (94)$$

By the estimates (93) and (94) there exist $r_n > 0$ such that $\sum r_n < \infty$ and

$$\sum_{k=N+2}^{n} \left[\frac{B}{2} R_{n-k}^{(2)}(z) + R_{n-k}(z)\widetilde{O}\left(\frac{p_{n-k}}{n-k+1}\right)\right](b_k - b_{k-1}) +$$

$$+ b_{N+1}\left(\frac{B}{2} R_{n-N-1}^{(2)}(z) + R_{n-N-1}(z)\widetilde{O}\left(\frac{p_{n-N-1}}{n-N-1}\right)\right) = \widetilde{O}\left(\frac{r_n}{n^2}\right), \qquad (95)$$

which enables us to transform (92) to

$$R_{n-1}(z) - R_n(z) = \alpha R_n^{(2)}(z) - \sum_{k=0}^{N} \frac{B}{2}\left(R_{n-k}^{(2)}(z) - R_{n-k-1}^{(2)}(z)\right) b_k +$$

$$+ \widetilde{O}\left(\frac{r_n}{n^2}\right) + \sum_{k=0}^{N} \widetilde{O}\left(\frac{p_{n-k}}{n-k}\right) R_{n-k}(z) b_k + R_n(z)\widetilde{O}\left(\frac{p_n}{n}\right), \qquad (96)$$

whence by (86), (87) and (25) we obtain

$$\|R_{n-1}(z)-R_n(z)\| = O(n^{-2}) + O(n^{-2}) \sum |b_k| ,$$

or by (15)

$$R_n(z) = R_{n-1}(z) + \tilde{O}(n^{-2}) \tag{97}$$

yielding

$$R_{n-k}(z) = R_n(z) + k\tilde{O}(n^{-2}) . \tag{98}$$

By virtue of (16), (98) and (24) we can write:

$$\sum_{k=0}^{N} \tilde{O}\left(\frac{p_{n-k}}{n-k}\right) R_{n-k}(z)\, b_k = R_n(z) \sum_k \tilde{O}\left(\frac{p_{n-k}}{n-k}\right) b_k + \\ + \tilde{O}(p_n/n^3) \sum |kb_k| = R_n(z)\, \tilde{O}\left(\frac{p_n}{n}\right) + \tilde{O}(r_n/n^2). \tag{99}$$

By definition,

$$R_n^{(2)}(z) = R_n(z)(R_n(z)+2Q_n) . \tag{100}$$

Squaring both sides in (97) and taking into account (86), we have

$$R_n^2(z) = R_{n-1}^2(z) + \tilde{O}(n^{-3}) . \tag{101}$$

Multiplying both sides of (97) by Q_n and taking into account (23) for $z=0$ and the estimates (86) and (1), we obtain

$$R_n(z)Q_n = R_{n-1}(z)Q_{n-1} + R_{n-1}(z)(Q_n - Q_{n-1}) + \\ + \tilde{O}(n^{-3}) = R_{n-1}(z)Q_{n-1} + \tilde{O}(n^{-3}). \tag{102}$$

Using (100), (101) and (102), we obtain

$$R_n^{(2)}(z) - R_{n-1}^{(2)}(z) = \tilde{O}(n^{-3}) . \tag{103}$$

Using (99) and (103), we transform (96) to

$$R_{n-1}(z) - R_n(z) = \alpha R_n^{(2)}(z) + \sum_{k=0}^{N} \tilde{O}\left(\frac{b_k}{(n-k)^3}\right) + \tilde{O}\left(\frac{r_n}{n^2}\right) + R_n(z)\, \tilde{O}\left(\frac{p_n}{n}\right).$$

From (15) we have $\sum_{k=0}^{N} \tilde{O}\left((n-k)^{-3} b_k\right) = \tilde{O}\left(n^{-3}\right)$, which plus the estimates (27), (37) and (103) makes it possible to obtain from the last but one equality:

$$R_n(z) = R_{n-1}(z)(1 + \tilde{O}(p_n/n)) - \alpha R_{n-1}(z)(Q_{n-1} - Q_{n-1}(z)) + \tilde{O}(n^{-3}) + \\ + \tilde{O}(r_n/n^3) = R_{n-1}(z)(1 + \tilde{O}(n^{-1})) + \tilde{O}(n^{-3}) + \tilde{O}(r_n/n^2). \tag{104}$$

Making the same operations with (54) as we did with (97) we arrive at

$$R_n^2(z) = R_{n-1}^2(z)\left(1 + \tilde{O}(1/n)\right) + \tilde{O}\left(n^{-4}\right) + \tilde{O}(r_n/n^3)$$

and

$$R_n(z)\, Q_n = R_{n-1}(z)\, Q_{n-1}\left(1 + \tilde{O}(1/n)\right) + \tilde{O}\left(n^{-4}\right) + \tilde{O}(r_n/n^3),$$

which are analogs of (101) and (102). Summing these equalities and noting (1), we obtain an analog of (103):

$$R_n^{(2)}(z) = R_{n-1}^{(2)}(z) + R_{n-1}(z)\, \tilde{O}\left(n^{-2}\right) + \tilde{O}\left(n^{-4}\right) + \tilde{O}(r_n/n^3). \tag{105}$$

Using (105), we estimate the second term in (96):

$$\sum_{k=0}^{N} \frac{B}{2}\left(R_{n-k}^{(2)}(z) - R_{n-k-1}^{(2)}(z)\right) b_k = \sum_{k=0}^{N} R_{n-k-1}(z)\, \tilde{O}\left((n-k)^{-2}\right) b_k + \\ + \sum_{k=0}^{N} \tilde{O}\left((n-k)^{-4}\right) b_k + \sum_{k=0}^{N} \tilde{O}\left(\frac{r_{n-k}}{(n-k)^3}\right) b_k. \tag{106}$$

Using (98) and (16) we obtain

$$\sum_{k=0}^{N} R_{n-k-1}(z)\, \tilde{O}\left((n-k)^{-2}\right) b_k = R_{n-1}(z)\, \tilde{O}\left(n^{-2}\right) \sum_{k=0}^{N} |b_k| + \\ + n^{-2} \sum \tilde{O}\left(k(n-k)^{-2}\right) b_k = R_{n-1}(z)\, \tilde{O}\left(n^{-2}\right) + \tilde{O}\left(n^{-4}\right). \tag{107}$$

It is seen that

$$\sum_{k=0}^{N} \tilde{O}\left((n-k)^{-4}\right) b_k = \tilde{O}\left(n^{-4}\right) \sum |b_k| = \tilde{O}\left(n^{-4}\right). \tag{108}$$

Now we prove that

$$\sum_{n=4}^{\infty} n^2 \left\| \sum_{k=0}^{N} O\left(\frac{r_{n-k}}{(n-k)^3}\right) b_k \right\| < \infty. \qquad (109)$$

We estimate the right-hand side of (109) from above by

$$\sum_{n=4}^{\infty} n^2 \sum_{k=0}^{N} O\left(\frac{r_{n-k}}{(n-k)^3}\right) b_k = \sum_{n=4}^{\infty} \frac{1}{n} \sum_{k=0}^{N} O(r_{n-k})\, b_k =$$
$$= \sum_{n=4}^{\infty} \frac{1}{n} \sum_{k=N+1}^{n} O(r_k)\, b_{n-k} = \sum_{n=4}^{\infty} O(r_n)\, \Sigma |b_k| < \infty.$$

The last inequality is due to (15) and the definition of r_n. This proves (109).

From (106)-(109) there follows the existence of $r_n > 0$ such that

$$\Sigma \frac{B}{2}\left(R^{(2)}_{n-k}(z) - R^{(2)}_{n-k-1}(z)\right) b_k = R_{n-1}(z)\, \tilde{O}\left(n^{-2}\right) + \tilde{O}\left(r_n/n^2\right) \qquad (110)$$

and $\sum r_n < \infty$.

By (96), (105), (110), (99) and the remarks concerning r_n and p_n we obtain

$$R_{n-1}(z) - R_n(z) = \alpha R^{(2)}_{n-1}(z) + \tilde{O}\left(r_n/n^2\right) + R_{n-1}(z)\, \tilde{O}(r_n),$$

which is equivalent to the assertion of Lemma 11.

4. THE PROOF OF THEOREM 1

At this final stage of discussion, we take the approach suggested in [6]. The remarks concerning r_k made while proving Lemma 11, remain valid in this section as well. However, we substitute r_k for p_k/k and $\bar{p}_k/k$ without stipulation.

It follows from (23), (25) and (1) that

$$Q_n = Q_{n-1} - \alpha Q_{n-1}^2 + O(p_{n-1})/n^2 = Q_{n-1}(1 - \alpha Q_{n-1} + O(r_{n-1})) =$$
$$= Q_{n-1}(1 + \alpha Q_{n-1} + O(r_{n-1}))^{-1}. \qquad (111)$$

Next, we divide the left and right sides of (84), respectively, by the left and right sides of (111). Then by (1) we obtain

$$\frac{R_k(z)}{Q_k} = \frac{R_{k-1}(z)}{Q_{k-1}}\left(1 - \alpha Q_{k-1}(z) + \tilde{O}(r_{k-1})\right) + O\left(\frac{r_k}{k}\right),$$

or

$$\frac{R_k(z)}{Q_k} = \frac{R_{k-1}(z)}{Q_{k-1}}(1 - \alpha Q_{k-1}(z) + \Psi_{k-1}(z)) + \varphi_k(z), \qquad (112)$$

where $\Psi_{k-1}(z)=\tilde{O}(r_{k-1})$ and $\phi_k(z)=\tilde{O}(r_k/k)$.

Let $\phi_k^{(1)}(z)=(1-R_{k-1}(z)/Q_{k-1})\phi_k(z)$, and let $\Psi_k^{(1)}(z)=\Psi_k(z)+\phi_{k+1}(z)$. Then by (86) we have $\Psi_k^{(1)}(z)=\tilde{O}(r_k)$ and $\phi_k^{(1)}(z)=\tilde{O}(r_k/k)$; since $R_{k-1}(z)=Q_{k-1}$ and $Q_k(1)=0$, then $\phi_k^{(1)}(1)=0$ and $\Psi_k^{(1)}(1)=0$: therefore, we assume without loss of generality that $\phi_k(1)=\Psi_k(1)=0$.

We define $\theta_k(z)$ and $\pi_k(z)$ from the equalities

$$\theta_k(z) = 1 - \alpha Q_k(z) + \Psi_k(z) = \left(1 - \frac{\alpha}{\alpha k + (1-z)^{-1}}\right)(1 + \pi_k(z)).$$

Using the representation (69), the equalities (37) and

$$\left(1 - \frac{\alpha}{\alpha n + (1-z)^{-1}}\right)^{-1} = \frac{\alpha n + (1-z)^{-1}}{\alpha(n-1) + (1-z)^{-1}} = \tilde{O}(1)$$

(the last equality holds by (28)), we obtain

$$\pi_k(z) = \left(1 - \frac{\alpha + O(\bar{p}_k)}{\alpha k + (1-z)^{-1}} + \Psi_k(z)\right)\left(1 - \frac{\alpha}{\alpha k + (1-z)^{-1}}\right)^{-1} - 1 =$$
$$= \tilde{O}(r_k) + \tilde{O}(\|\Psi_k(z)\|) = \tilde{O}(r_k), \quad k > \bar{m}_0. \qquad (113)$$

Therefore, by the definition of $\theta_k(z)$

$$\frac{R_k(z)}{Q_k} = \frac{R_{k-1}(z)}{Q_{k-1}}\theta_{k-1}(z) + \varphi_k(z). \tag{114}$$

By the equalities $\phi_k(1)=\Psi_k(1)=0$ it follows from (114) and (113) that $\theta_k(1)=1$ and $\pi_k(1)=0$.

From (113) we have the convergence of $\prod_{k=M}^{\infty}(1+\pi_k(z))$ in the norm, where $M\geq\bar{m}_0+1$ is fixed arbitrarily. Therefore, we have the relations ($\phi_n(z)=\tilde{o}_n(1)$, if $\|\phi_n(z)\|\to 0$ as $n\to\infty$)

$$\prod_M^{\infty}(1+\pi_k(z)) = \tilde{O}(1) \tag{115}$$

and

$$\prod_M^{n-1}(1+\pi_k(z)) = \prod_M^{\infty}(1+\pi_k(z)) + \tilde{o}_n(1). \tag{116}$$

The definition of $\theta_i(z)$ yields

$$\prod_{i=k}^{n-1}\theta_i(z) = \frac{\alpha(k-1)+(1-z)^{-1}}{\alpha(n-1)+(1-z)^{-1}}\prod_{i=k}^{n-1}(1+\pi_i(z)). \tag{117}$$

Multiplying the respective parts of the equalities (117) and (114) and summing the resulting equalities in k from M through n-1 (for k=n-1, (117) is trivial 1=1), we obtain

$$\frac{R_n(z)}{Q_n} = \frac{R_{M-1}(z)}{Q_{M-1}}\prod_{k=M}^{n-1}\theta_k(z) + \sum_{k=M}^{n-1}\frac{\alpha k+(1-z)^{-1}}{\alpha(n-1)+(1-z)^{-1}}\tilde{O}\left(\frac{r_k}{k}\right). \tag{118}$$

It is seen that

$$\mathbf{C}_n(\varphi(z)\Psi(z)) \leqslant \|\varphi(z)\|\|\Psi(z)\|, \tag{119}$$

which by (28) implies

$$\mathbf{C}_s\left(\sum_M^{n-1}\frac{\alpha k+(1-z)^{-1}}{\alpha(n-1)+(1-z)^{-1}}\tilde{O}\left(\frac{r_k}{k}\right)\right) = \frac{1}{n}O\left(\sum_M^{n-1}r_k\right). \tag{120}$$

By (116)

$$\frac{R_{M-1}(z)}{Q_{M-1}} \prod_{M}^{n-1} \theta_k(z) = \frac{R_{M-1}(z)}{Q_{M-1}} \prod_{k=M}^{\infty} (1 + \pi_k(z)) \frac{\alpha(M-1) + (1-z)^{-1}}{\alpha(n-1) + (1-z)^{-1}} + \\ + \tilde{o}_n(1) \frac{R_{M-1}(z)}{Q_{M-1}(z)} \cdot \frac{\alpha(M-1) + (1-z)^{-1}}{\alpha(n-1) + (1-z)^{-1}}. \quad (121)$$

By analogy with (120), by virtue of (119), (28) and (86) we have the estimate

$$\mathrm{C}_k \left(\tilde{o}_n(1) \frac{R_{M-1}(z)}{Q_{M-1}} \frac{\alpha(M-1) + (1-z)^{-1}}{\alpha(n-1) + (1-z)^{-1}} \right) = o_n\left(\frac{1}{n}\right) \quad (122)$$

If (115) is taken into account, then by analogy with (122) the first term on the right side of (121) has a coefficient equal to $O(1/n)$ for z^r; this plus the equalities (122), (121), (120) and (118) implies

$$C_k(R_n(z)/Q_n) = O(1/n) .$$

This relation, by the definition of $R_n(z)$ and p_{nk}, is equivalent for $k>0$ to $\sup\limits_{\substack{n \ge m_0 \\ k \ge 1}} n^2 p_{nk} < \infty$.

Since the last inequality is obvious for $n \le m_0$, Eq. (3) is proved.

From (28) it is easy to obtain

$$\frac{\alpha(M-1) + (1-z)^{-1}}{\alpha(n-1) + (1-z)^{-1}} = \frac{1 + (M-1)\alpha}{1 + (n-1)\alpha} + \\ + \frac{(n-M)}{(1 + (n-1)\alpha)(n-1)} \sum_{k=1}^{\infty} \left(\frac{(n-1)\alpha}{1 + (n-1)\alpha} \right)^k z^k. \quad (123)$$

Let

$$\prod_{M}^{\infty} (1 + \pi_k(z)) \frac{R_{M-1}(z)}{Q_{M-1}} = \sum_{i=0}^{\infty} t_i z^i = \Psi(z). \quad (124)$$

From (123) and (124) via elementary computations we obtain

$$\mathbf{C}_k\left(\frac{R_{M-1}(z)}{Q_{M-1}}\prod_{M}^{\infty}(1+\pi_i(z))\frac{\alpha(M-1)+(1-z)^{-1}}{\alpha(n-1)+(1-z)^{-1}}\right)=t_k\frac{1+(M-1)\alpha}{1+(n-1)\alpha}+$$
$$+\frac{n-M}{(1+(n-1)\alpha)(n-1)}\sum_{i=0}^{k-1}t_i\left(\frac{(n-1)\alpha}{1+\alpha(n-1)}\right)^{k-i}. \qquad (125)$$

Since $\|\Psi(z)\|<\infty$, then

$$\sum_{s}^{\infty}|t_i| = o_s(1) . \qquad (126)$$

For fixed s

$$\sum_{i=0}^{s-1}t_i\left(\frac{(n-1)\alpha}{1+(n-1)\alpha}\right)^{k-i}=\left(\frac{(n-1)\alpha}{1+(n-1)\alpha}\right)^{k}\left\{\sum_{i=0}^{s-1}t_i+\sum_{i=0}^{s-1}\left(t_i\left(\frac{(n-1)\alpha}{1+(n-1)\alpha}\right)^{-i}-\right.\right.$$
$$\left.\left.-t_i\right)\right\}=\left(\frac{(n-1)\alpha}{1+(n-1)\alpha}\right)^{k}\left(\sum_{i=0}^{s-1}t_i+o_n(1)\right). \qquad (127)$$

We write $\phi(n,k)=\hat{o}((n\alpha)^{-1}e^{-k/n\alpha})$, if $k/n<c$ and $n\phi(n,k)\to 0$ as $n\to\infty$ and $k\to\infty$.

It follows from (126) that for any fixed M

$$t_k\frac{1+(M-1)\alpha}{1+(n-1)\alpha}+\frac{n-M}{(1+(n-1)\alpha)(n-1)}\sum_{i=s}^{k-1}t_i\left(\frac{(n-1)\alpha}{1+(n-1)\alpha}\right)^{k-i}=$$
$$=\frac{1}{n}O\left(\sum_{s}^{k}|t_i|\right). \qquad (128)$$

From (124) it is easy to obtain

$$\Psi(1)=\sum_{0}^{\infty}t_i=R_N(1)/Q_N=1. \qquad (129)$$

It follows from (126) and (129) that one can choose s_0 such that for $s>s_0$ the expressions $\sum_{s}^{\infty}|t_i|$ and $\left|\sum_{i=0}^{s}t_i-1\right|$ can be arbitrarily small. Then it follows that by the appropriate choice of s_0 for sufficiently large n and k one can have (by (124), (125), (127) and (128)) that the inequality

$$\left| \mathbf{C}_k\left(\Psi(z)\frac{\alpha(M-1)+(1-z)^{-1}}{\alpha(n-1)+(1-z)^{-1}}\right)-\left(\frac{(n-1)\alpha}{1+(n-1)\alpha}\right)^k\frac{1}{(n-1)\alpha}\right| \leqslant \frac{\varepsilon_1}{n},$$

where $\varepsilon_1>0$ is fixed arbitrarily, is safisfied.

Since ε_1 is arbitrary, the last inequality can be written as

$$\mathbf{C}_k\left(\Psi(z)\frac{\alpha(M-1)+(1-z)^{-1}}{\alpha(n-1)+(1-z)^{-1}}\right)=\frac{1}{n\alpha}e^{-k/n\alpha}+\hat{o}\left(\frac{1}{n\alpha}e^{-k/n\alpha}\right), \tag{130}$$

where $M\geq\bar{m}_0$ is fixed.

Applying the estimates (122) and (130) to (121), we obtain

$$\mathbf{C}_k\left(\frac{R_{M-1}(z)}{Q_{M-1}}\prod_{M}^{n-1}\theta_n(z)\right)=\frac{1}{n\alpha}e^{-k/n\alpha}+\hat{o}\left(\frac{1}{n\alpha}e^{-k/n\alpha}\right). \tag{131}$$

It follows from (120) that for any $\varepsilon>0$ there exists $M_0>\bar{m}_0$ such that

$$\left|\mathbf{C}_r\left(\sum_{k=M_0}^{n-1}\frac{\alpha k+(1-z)^{-1}}{\alpha n+(1-z)^{-1}}\tilde{O}\left(\frac{r_k}{k}\right)\right)\right|<\varepsilon\frac{1}{n},$$

which together with the equality (118) enables us to obtain

$$\left|\mathbf{C}_k\left(\frac{R_n(z)}{Q_n}\right)-\mathbf{C}_k\left(\frac{R_{M_0-1}(z)}{Q_{M_0-1}}\prod_{M_0}^{n-1}\theta_i(z)\right)\right|\leqslant\varepsilon\frac{1}{n}.$$

Using the arbitrariness of ε and the estimate (131), we then have

$$\mathbf{C}_k\left(\frac{R_n(z)}{Q_n}\right)=\frac{1}{n\alpha}e^{-k/n\alpha}+\hat{o}\left(\frac{1}{n\alpha}e^{-k/n\alpha}\right).$$

This is the first part of the assertion of Theorem 1. The second part has been proved earlier.

REFERENCES

[1] T.E. Harris. *The Theory of Branching Processes.* Berlin: Springer-Verlag, 1963.

[2] B.A. Sevastyanov. *Branching Processes.* Moscow: Nauka, 1971. (In Russian.)

[3] B.A. Sevastyanov. "Age-dependent Branching Processes." *Theory Prob. Applications,* 4 (1964): 521-537.

[4] M.I. Goldstein. "Critical Age-dependent Branching Processes: Single and Multi-type." *Z. Wahrscheinlichkeitstheorie und verw. Gebiete,* 17, 4 (1971): 74-88.

[5] H. Kesten, P. Ney, and F. Spitzer. "The Galton-Watson Process With Mean and Finite Invariance." *Theory Prob. Applications,* 4 (1966): 513-540.

[6] S.V. Nagaev and R. Mukhamedkhanova. "Certain Limit Theorems in the Theory of Branching Random Processes." In *Limit Theorems and Statistical Inferences,* 90-113. Tashkent: FAN, 1966. (In Russian.)

[7] S.V. Nagaev. Transition Phenomena for Age-dependent Branching Processes With Discrete Time," I. *Sibirskij matemat. zh.,* 15, 2 (1974): 261-281; "Transfer Effects for Age-dependent Discrete-time Branching Processes," II. *Sibirskij matemat. zh.,* 15, 3 (1974): 408-415.

[8] B.A. Rogozin. "An Estimate of the Remainder Term in Limit Theorems of Renewal Theory." *Theory Prob. Applications,* 4 (1973): 662-677.

[9] M.A. Naimark. *Normed Rings*. P. Noordhoff: Groningen, 1970.

[10] A.A. Borovkov. *Probability Theory*. Moscow: Nauka, 1972. (In Russian.)

THE ESTIMATES OF THE RATE OF CONVERGENCE IN A LOCAL LIMIT THEOREM FOR THE SQUARE OF THE NORM IN ℓ_2

V.I. Chebotarev

1. BASIC ASSUMPTIONS. THE FORMULATION OF THE RESULTS

Let $X_1, X_2, \ldots, X_n$ be a sequence of independent identically distributed random vectors with values in ℓ_2, $\mathbf{E}X_1 = 0$, $\mathbf{E}\|X_1\|^2 = 1$, $\beta = \mathbf{E}\|X_1\|^3 < \infty$. Here $\|\cdot\|$ denotes the norm in ℓ_2. Let $X_i = \{X_{i1}, X_{i2}, \ldots\}$, $i=1,2,\ldots,n$, $\sigma_j^2 = \mathbf{E}X_{1j}^2$, $\beta_j = \mathbf{E}|X_{1j}|^3$, $j=1,2,\ldots$. Furthermore, let $Z=\{Z_1, Z_2, \ldots\}$ be a Gaussian random vector in ℓ_2, satisfying the conditions $EZ=0$, and let the covariance operator of the vector Z coincide with the covariance operator of the random vector X_1. This implies that $EZ_j^2 = \sigma_j^2$. We can assume without loss of generality that $\sigma_j \geq \sigma_{j+1}$ for all $j=1,2,\ldots$.

Let

$$S_n = n^{-1/2} \sum_{i=1}^{n} X_i, \; S_{nj} = n^{-1/2} \sum_{i=1}^{n} X_{ij}.$$

Under the assumptions:

a) the coordinates X_{11}, X_{12} have absolutely continuous distributions with bounded densities and, in addition,

b) the coordinates of each vector among $X_1, X_2, \ldots, X_n$, Z are mutually independent, we estimate

$$\delta_n = \sup_x \left| \frac{d}{dx} \mathbf{P}\{\|S_n\|^2 < x\} - \frac{d}{dx} \mathbf{P}\{\|Z\|^2 < x\} \right|.$$

We shall use the following notation in formulating and proving the basic results of our research: $\phi_j(\cdot)$ is the probability density of the random variable $\sigma_j^{-1} X_{1j}$; $\phi_{nj}(\cdot)$ is the density of $\sigma_j^{-1} S_{nj}$; $\phi(\cdot)$ is the density of $\sigma_j^{-1} Z_j$; $A_j = \sup_x \phi_j(x)$, $j=1,2$; $A_{1,2} = \max_{j=1,2} \max \{1, A_j^3\}$; C is an absolute constant not necessarily the same in different formulas.

To express the results in a simpler way, we use the following assumptions without loss of generality:

c) $n^{-1/2} A_{1,2} \sigma_1^{-3} \beta_1 < 1$;

c') $n^{-1/2} \left(\prod_{j=3}^{5} \sigma_j \right)^{-1} \sum_{j=6}^{\infty} \beta_j < 1$.

THEOREM 1. If the conditions "a", "b", and "c" are satisfied, then

$$\delta_n \leqslant n^{-1/2} C A_{1,2} (\sigma_1 \sigma_2)^{-1} \left[\left(\prod_{j=3}^{6} \sigma_j \right)^{-3/4} \sum_{j=7}^{\infty} \beta_j + \sum_{j=1}^{6} \sigma_j^{-3} \beta_j \right].$$

THEOREM 2. If the conditions "a", "b", "c" and "c'" are satisfied, then

$$\delta_n \leqslant n^{-1/2} C A_{1,2} (\sigma_1 \sigma_2)^{-1} \left\{ \left(\prod_{j=3}^{5} \sigma_j \right)^{-1} \left(\sum_{j=6}^{\infty} \beta_j \right) \times \right.$$

$$\left. \times \left[1 + \ln\left(n^{1/2} \prod_{j=3}^{5} \sigma_j \left(\sum_{j=6}^{\infty} \beta_j \right)^{-1} \right) \right] + \sum_{j=1}^{5} \sigma_j^{-3} \beta_j \right\}.$$

THEOREM 3. If the conditions "a", "b" and "c" are satisfied, then

$$\delta_n \leqslant n^{-1/3} C A_{1,2} (\sigma_1\sigma_2)^{-1} \left[(\sigma_3\sigma_4)^{-1} \left(\sum_{j=5}^{\infty} \beta_j \right)^{2/3} + n^{-1/6} \sum_{j=1}^{4} \sigma_j^{-3} \beta_j \right].$$

THEOREM 4. If the conditions "a", "b" and "c" are satisfied, then

$$\delta_n \leqslant n^{-1/6} C A_{1,2} (\sigma_1\sigma_2)^{-1} \left[\sigma_3^{-1} \left(\sum_{j=4}^{\infty} \beta_j \right)^{1/3} + n^{-1/3} \sum_{j=1}^{3} \sigma_j^{-3} \beta_j \right].$$

We note that the convergence of the series $\sum_{j=1}^{\infty} \beta_j$ follows from the existence of the third moment β.

The magnitude of the characteristics in the estimates given in Theorems 1-4 determines the accuracy of each (see [1]. If $A_{1,2} \leq C$, $\sigma_6 \geq C$, $\beta \leq C$, then Theorem 1 yields the estimate $\delta_n = O(n^{-\frac{1}{2}})$.

2. THE PROOF OF THEOREMS 1-4

Let

$$\Delta_n^{(k)}(x) = \mathbf{P}\left\{ \sum_{j=k+1}^{\infty} S_{nj}^2 < x \right\} - \mathbf{P}\left\{ \sum_{j=k+1}^{\infty} Z_j^2 < x \right\};$$

$$\Delta_n^{(k)} = \sup_x |\Delta_n^{(k)}(x)|, \quad k = 0, 1, 2, \ldots, \quad \Delta_n^{(0)}(x) \equiv \Delta_n(x);$$

p_{nj} denotes the probability density of S_{nj}^2; p_j denotes the density of Z_j^2; $\cdot * \cdot$ denotes convolution. We shall assume from now on that conditions "a" and "b" are satisfied.

To prove Theorems 1-4 we need five lemmas. Lemma 1 provides the estimate δ_n as the sum of the "infinite-dimensional" and "finite-dimensional" parts.

LEMMA 1. For all n

$$\delta_n \leqslant \sup_x \left| \int_{0-}^{x+} p_1 * p_2 (x-y)\, d\Delta_n^{(2)}(y) \right| + \sup_x | p_{n1} * p_{n2}(x) - p_1 * p_2 (x) |. \quad (1)$$

Proof. The estimate (1) follows from the representation

$$\frac{d}{dx}\Delta_n(x) = \int_{0-}^{x+} p_1 * p_2 (x-y)\, d\Delta_n^{(2)}(y) +$$

$$+ \int_{0-}^{x+} (p_{n1} * p_{n2} - p_1 * p_2)(x-y)\, d\mathbf{P}\left\{\sum_{j=3}^{\infty} S_{nj}^2 < x\right\}.$$

We estimate the first supremum in (1).

LEMMA 2. We have the estimate

$$\sup_x \left| \int_{0-}^{x+} p_1 * p_2 (x-y)\, d\Delta_n^{(2)}(y) \right| \leqslant (\sigma_1\sigma_2)^{-1}\Delta_n^{(2)}.$$

Proof. Integrating by parts, we obtain

$$\sup_x \left| \int_{0-}^{x+} p_1 * p_2 (x-y)\, d\Delta_n^{(2)}(y) \right| = \sup_x | p_1 * p_2 (x-y)\, \Delta_n^{(2)}(y) \Big|_{0-}^{x+} -$$

$$- \int_{0-}^{x+} \Delta_n^{(2)}(y)\, d(p_1 * p_2)(x-y) | = \sup_x \left| \int_{0-}^{x+} \Delta_n^{(2)}(y)\, d(p_1 * p_2)(x-y) \right| \leqslant$$

$$\leqslant \Delta_n^{(2)} \int_0^{\infty} | d(p_1 * p_2)(y) |. \quad (2)$$

Since for $x>0$

$$p_1 * p_2 (x) = \frac{1}{2\pi\sigma_1\sigma_2} \int_0^x \frac{1}{\sqrt{(x-t)t}} \exp\left\{-\frac{x-t}{2\sigma_1^2} - \frac{t}{2\sigma_2^2}\right\} dt =$$

$$= \frac{1}{2\pi\sigma_1\sigma_2} \int_0^1 \frac{1}{\sqrt{(1-y)y}} \exp\left\{-\frac{x}{2}\left(\frac{1-y}{\sigma_1^2} + \frac{y}{\sigma_2^2}\right)\right\} dy, \quad (3)$$

the derivative $\frac{d}{dx}(p_1 * p_2)(x)$ is negative for all $x>0$. Therefore

$$\int_{0-}^{\infty} |d(p_1 * p_2)(y)| = (p_1 * p_2)(0+) + \int_{0+}^{\infty} |d(p_1 * p_2)(y)| = (p_1 * p_2)(0+) - $$
$$- \int_{0+}^{\infty} d(p_1 * p_2)(y) = 2(p_1 * p_2)(0+). \qquad (4)$$

From (2)-(4) there follows the estimate

$$\sup_x \left| \int_{0-}^{x+} p_1 * p_2(x-y)\, d\Delta_n^{(2)}(y) \right| \leqslant 2\Delta_n^{(2)}(p_1 * p_2)(0+) = (\sigma_1\sigma_2)^{-1}\Delta_n^{(2)}.$$

//

To estimate the second term in (1) we make use of the result obtained by N. Shakhaidarova [2], which we formulate in the terms used in Section 1.

LEMMA 3. For all n

$$\sup_x |\varphi_{nj}(x) - \varphi(x)| \leqslant n^{-1/2} C \max\{1, A_j^3\} \sigma_j^{-3}\beta_j, \qquad j = 1, 2.$$

LEMMA 4. For all n

$$\sup_x |(p_{n1} * p_{n2} - p_1 * p_2)(x)| \leqslant n^{-1/2} C A_{1,2} (\sigma_1\sigma_2)^{-1} \Big(\sum_{j=1}^{2} \sigma_j^{-3}\beta_j +$$
$$+ n^{-1/2} A_{1,2} \sigma_1^{-3}\beta_1 \sigma_2^{-3}\beta_2 \Big).$$

Proof. We have

$$\sup_x |p_{n1} * p_{n2}(x) - p_1 * p_2(x)| \leqslant \sup_x |(p_{n1} - p_1) * (p_{n2} - p_2)(x)| +$$
$$+ \sup_x |(p_{n1} - p_1) * p_2(x)| + \sup_x |(p_{n2} - p_2) * p_1(x)|. \qquad (5)$$

We have the following relations:

$$p_{nj}(x) = (2\sqrt{x}\,\sigma_j)^{-1} \left(\varphi_{nj}(\sqrt{x}\,\sigma_j^{-1}) + \varphi_{nj}(-\sqrt{x}\,\sigma_j^{-1})\right);$$
$$p_j(x) = (\sqrt{x}\,\sigma_j)^{-1} \varphi(\sqrt{x}\,\sigma_j^{-1}); \qquad x > 0,\ j = 1, 2. \qquad (6)$$

From (6) and Lemma 3 it follows that

$$\sup_x |(p_{n1}-p_1)*(p_{n2}-p_2)(x)| = (4\sigma_1\sigma_2)^{-1}\sup_x \left|\int_0^x [(x-t)\,t]^{-1/2} \times \right.$$
$$\times \left[\varphi_{n1}(\sqrt{x-t}\,\sigma_1^{-1}) + \varphi_{n1}(-\sqrt{x-t}\,\sigma_1^{-1}) - \varphi(\sqrt{x-t}\,\sigma_1^{-1}) - \right.$$
$$\left. - \varphi(-\sqrt{x-t}\,\sigma_1^{-1})\right]\left[\varphi_{n2}(\sqrt{t}\,\sigma_2^{-1}) + \varphi_{n2}(-\sqrt{t}\,\sigma_2^{-1}) - \varphi(\sqrt{t}\sigma_2^{-1}) - \right.$$
$$\left.\left. - \varphi(-\sqrt{t}\,\sigma_2^{-1})\right]dt\right| \leqslant n^{-1}C\max\{1; A_1^3\}\max\{1, A_2^3\}\,\sigma_1^{-4}\beta_1\sigma_2^{-4}\beta_2. \tag{7}$$

In the similar way we obtain the inequality

$$\sup_x |(p_j - p_{nj})*p_m(x)| \leqslant n^{-1/2}C\max\{1, A_j^3\}\,\sigma_m^{-1}\sigma_j^{-4}\beta_j, \tag{8}$$

where $j,m=1,2$. From (5), (7) and (8) we have the estimate obtained in Lemma 4.

LEMMA 5. For all n

$$\delta_n \leqslant (\sigma_1\sigma_2)^{-1}\left[\Delta_n^{(2)} + n^{-1/2}CA_{1,2}\left(\sum_{j=1}^{2}\sigma_j^{-3}\beta_j + n^{-1/2}A_{1,2}\sigma_1^{-3}\beta_1\sigma_2^{-3}\beta_2\right)\right]. \tag{9}$$

Proof. The estimate (9) follows immediately from Lemmas 1, 2 and 4.

The assertions of Theorems 1-4 follow from Lemma 5 and Theorems 1-4 [3], respectively.

REMARK. An example illustrating the presence of the factor σ_2^{-1} in the estimates δ_n is available.

REFERENCES

[1] S.V. Nagaev and V.I. Chebotarev. "Estimates of the Rate of Convergence in the Central Limit Theorem in ℓ_2 for Independent Coordinates." In *Proceedings of the Second Vilnius Conference on Probability Theory and Mathematical Statistics*, 68-69. Vol. 2. Vilnius, 1977. (In Russian.)

[2] N. Shakhaidarova. "Uniform, Local and Global Theorems for Densities." *Izv. AN Uzbek. S.S.R., Ser. Phys.-Math. Sciences*, 5 (1966): 90-91. (In Russian.)

[3] S.V. Nagaev and V.I. Chebotarev. "On Estimates of the Rate of Convergence in the Central Limit Theorem for Random Vectors With Values in a Space ℓ_2." In *Mathematical Analysis and Related Problems in Mathematics*, 153-182. Novosibirsk: Nauka, 1978. (In Russian.)

HOMOGENIZATION OF NON-DIVERGENT RANDOM ELLIPTIC OPERATORS

V.V. Yurinskij

In this article we consider the limit behavior of $\varepsilon \to 0$ of solutions of boundary value problems for the equations

$$L_\varepsilon u_\varepsilon = f, \ \frac{\partial u_\varepsilon}{\partial t} = L_\varepsilon u_\varepsilon + f,$$

where

$$L_\varepsilon v = \sum_{i,j} a_{ij}(\bar{x}, \bar{x}/\varepsilon) \frac{\partial^2 v}{\partial x^{(i)} \partial x^{(j)}} + \sum_i b_i(\bar{x}, \bar{x}/\varepsilon) \frac{\partial v}{\partial x^{(i)}} + c(\bar{x}, \bar{x}/\varepsilon)\, v;$$

$$v = v(\bar{x}); \ \bar{x} = \{x^{(0)}, \ldots, x^{(n)}\} \in R^{n+1},$$

and the coefficients $a_{ij}(\bar{x}, \bar{y}), \ldots$ are smooth random fields, periodic in the variables $y^{(1)}, \ldots, y^{(n)}$ and satisfying in the variable $y^{(0)}$ the strong mixing condition.

The basic results of this article consist in the theorems on the approximation of solutions u_ε of boundary value problems for the operator L_ε by solutions U of the homogenized problems for the operator L_0 with smooth nonrandom coefficients in the sense that $\sup |u_\varepsilon(\bar{x}-U(\bar{x})| \to 0$ in probability. The formulations and proofs of these theorems are given in Section 3.

The homogenization of non-divergent equations with nonrandom periodic rapidly oscillating coefficients is dealt with in [1] (this result has been obtained in [2] via other techniques). The method of homogenization for equations in the divergent form has been discussed, for example, in [2, 3-10].

In what follows we use the method of [2, 3, 10]. The main difficulty related to the application of this method consists in the construction of rapidly oscillating correctors to the solution of the homogenized problem, which are solutions of auxiliary problems for the main part of the operator L_ε. These correctors are constructed in Sections 1 and 2. In Section 4 the results will be used to consider the weak convergence of distributions of diffusion processes generated by the operator L_ε to a measure in the space of continuous functions, generated by the homogenized operator.

The author is grateful to T.I. Zelenyak and S.V. Uspenskij for the useful discussions.

1. ESTIMATES OF PERIODIC SOLUTIONS OF AUXILIARY EQUATIONS

Let $a=\|a_{ij}(\bar{y})\|$, $i,j=0,\dots,n$, $\bar{y}=\{y^{(0)},\dots,y^{(n)}\}$, be a symmetric matrix satisfying the condition

$$\forall \bar{y}, \bar{\xi} \quad 0 < a_- |\bar{\xi}|^2 \leqslant a_{ij}(\bar{y})\,\xi^{(i)}\xi^{(j)} \leqslant a_+ |\bar{\xi}|^2 < \infty, \tag{1}$$

where $|\bar{\xi}|^2 = \xi^{(i)}\xi^{(i)}$, and the recurrence of the superscripts implies here and below the summation from 0 to n.

The coefficients a_{ij} are assumed to be three times continuously differentiable and satisfy for all $\bar{y}$

$$|\nabla_k a_{ij}(\bar{y})| < A, \quad |\nabla_k \nabla_l a_{ij}(\bar{y})| < A, \quad |\nabla_k \nabla_l \nabla_m a_{ij}(\bar{y})| < A, \tag{2}$$

where $\nabla_k \phi = \partial\phi/\partial y^{(k)}$.

The functions periodic with the period 1 only in the variables $y^{(1)},\dots,y^{(n)}$ are called below 1-periodic; the 1-periodic functions which are also periodic functions with a period T in $y^{(0)}$ are called below (T,1)-periodic:

$$\varphi(\bar{y}) = \varphi(y^{(0)} + \varepsilon_0 T,\ y^{(1)} + \varepsilon_1, \dots),\ \varepsilon_i = 0, \pm 1.$$

If a_{ij} are (T,1)-periodic, the equation

$$\nabla_i \nabla_j (a_{ij}\mu) = 0 \tag{3}$$

has (see, for example, [11]) a unique (T,1)-periodic solution satisfying the condition

$$\int_0^T dy \int' \mu(y) = c,$$

with the reduction

$$\int' \varphi(y) = \int_{[0,1]^n} \varphi\left(y, y^{(1)} \dots y^{(n)}\right) dy^{(1)} \dots dy^{(n)}.$$

We use next the identity, which holds for (T,1)-periodic frunctions,

$$\int_0^T dy \int' \nabla_i \nabla_j (a_{ij}\mu)\,\varphi(y) = \int_0^T dy \int' \mu a_{ij} \nabla_i \nabla_j \varphi(y)$$

in the case where ϕ does not depend on $y^{(1)},\dots,y^{(n)}$; it is easy to see that $\int' a_{00}\mu(y)$ does not depend on y. This enables us to use in what follows a more convenient condition for normalization

$$\int' \mu a_{00}(y) = 1 , \qquad (4)$$

which also separates uniquely the $(T,1)$-periodic solution of (3).

The Harnack inequality (see, for example, [12]) allows us to infer that the $(T,1)$-periodic solution of the probelm (3), (4) admits the estimates

$$1/c \le \mu(\bar{y}) \le c . \qquad (5)$$

The smoothness of coefficients allows us to complete (5) with the inequalities

$$|\nabla_j \mu(\bar{y})| < c ; \quad |\nabla_i \nabla_j \mu(\bar{y})| < c , \qquad (6)$$

using the Schauder estimation [13]. Here and below c denotes any positive constant whose value is given only by the dimension and constants from conditions (1), (2), but does not depend on T so that $c+c=c$, $c \cdot c=c$, etc.

One can also see that (5) holds if one uses the well-known theorems (see, for example, [14]) on fundamental solutions of the equation $u_t = \nabla_i \nabla_j (a_{ij}\mu)$, which describe the transition densities of a random walk on the torus $[0,T] \times [0,1]^n$ with the infinitesimal operator $a_{ij}\nabla_i \nabla_j$, since μ is proportional to the invariant density of this walk.

In what follows the variable $y^{(0)}$ will play a special role. In the cases where we need to emphasize this fact we shall use the short form

$\bar{y} = \{y, 'y\}$, $y\ (= y^{(0)}) \in R^1$, $'y\ (= \{y^{(1)}, \dots, y^{(n)}\}) \in R^n$.

LEMMA 1. Let $\mu^{(k)}$, $k=1,2$, be 1-periodic functions such that for $0<y<S$

$$\nabla_i \nabla_j a_{ij}^{(k)} \mu^{(k)} (\bar{y}) = 0, \quad \int' a_{00}^{(k)} \mu^{(k)} (y) = 1,$$

with $\mu^{(k)}$ admitting the estimates (5), (6), and the 1-periodic coefficients $a_{ij}^{(k)}$ satisfy (1), (2) and for $1<y<S-1$, $\alpha \leq 1$

$$| a_{ij}^{(1)} (\bar{y}) - a_{ij}^{(2)} (\bar{y}) | \leqslant \alpha, \quad | \nabla_k a_{ij}^{(1)} (\bar{y}) - \nabla_k a_{ij}^{(2)} (\bar{y}) | \leqslant \alpha.$$

Then

$$\int_0^S dy \int' | \mu^{(1)} - \mu^{(2)} | (y) \leqslant c \left[\alpha S + S^{1/2} \right].$$

Proof. Let $\rho(y)=0$ outside $[1,S-1]$, $0 \leq \rho(y) \leq 1$, $\rho(y)=1$ for $y \in [2,S-2]$, $|\rho'(y)| + |\rho''(y)| \leq c$. The function $u(\bar{y}) = \rho(y)(\mu^{(1)}(\bar{y}) - \mu^{(2)}(\bar{y}))/\mu^{(1)}(\bar{y})$ satisfies the equality

$$\int_0^S dy \int' u \nabla_i \nabla_j \left(a_{ij}^{(1)} \left(\mu^{(1)} - \mu^{(2)} \right) \rho \right) = \int_0^S dy \int' u \nabla_i \nabla_j \left(a_{ij}^{(1)} \mu^{(1)} u \right) =$$

$$= - \int_0^S dy \int' \mu^{(1)} a_{ij}^{(1)} \nabla_i u \nabla_j u - \int_0^S dy \int' \nabla_j \left(a_{ij}^{(1)} \mu^{(1)} \right) \nabla_i (u^2/2) =$$

$$= - \int_0^S dy \int' \mu^{(1)} a_{ij}^{(1)} \nabla_i u \nabla_j u$$

obtained by integrating by parts with the 1-periodicity and finiteness of u taken into account.

The left-hand side of the previous equality can be transformed into

$$\int_0^S dy \int' u \nabla_i \nabla_j \left(a_{ij}^{(1)} \left(\mu^{(1)} - \mu^{(2)} \right) \rho \right) = I_1 + I_2,$$

where

$$I_1 = \int_0^S dy \int' u \nabla_i \nabla_j \left[\rho \left(a_{ij}^{(2)} - a_{ij}^{(1)}\right) \mu^{(2)}\right];$$

$$I_2 = \int_0^S dy \int' u \nabla_i \nabla_j \left[\rho \left(a_{ij}^{(1)} \mu^{(1)} - a_{ij}^{(2)}\right) \mu^{(2)}\right].$$

The first term admits the estimate

$$|I_1| = \left| -\int_0^S dy \int' \nabla_i u \cdot \nabla_j \left(\rho \left(a_{ij}^{(2)} - a_{ij}^{(1)}\right) \mu^{(2)}\right) \right| \leqslant \left(\int_0^S dy \int' u_{\bar{y}}^2 \right)^{1/2} \times$$

$$\times \left(\int_0^S dy \int' \left[\rho \left(a_{ij}^{(2)} - a_{ij}^{(1)}\right) \mu^{(2)}\right]_{\bar{y}}^2 \right)^{1/2} \leqslant c \left(\int_0^S dy \int' u_{\bar{y}}^2 \right)^{1/2} (S\alpha^2)^{1/2},$$

where $\phi_{\bar{y}}^2 = \nabla_i \phi \nabla_i \phi$. This estimate follows from the proximity of $a_{ij}^{(k)}$, $\nabla_\ell a_{ij}^{(k)}$ for $\rho \neq 0$ and by (5), (6).

The second term satisfies the inequality

$$|I_2| = \left| \int_{\rho \neq 0} dy \int' u \cdot \{\rho'' \left(a_{00}^{(1)} \mu^{(1)} - a_{00}^{(2)} \mu^{(2)}\right) + 2\rho' \nabla_j \left[a_{0j}^{(1)} \mu^{(1)} - a_{0j}^{(2)} \mu^{(2)}\right]\} \right| \leqslant$$

$$\leqslant c \left(\int_{0<\rho<1} dy \int' u^2 \right)^{1/2} \left(\int_{0<\rho<1} dy \int' 1 \right)^{1/2},$$

since $\mu^{(1)}$, $\mu^{(2)}$ are solutions of Eq. (3). Since $u(\bar{y})$ turns into zero for $y=0,S$,

$$\int_{0<\rho<1} dy \int' u^2 \leqslant c \int_{0<\rho<1} dy \int' u_{\bar{y}}^2$$

by the Poincaré-Friedrichs estimate.

These estimates show that

$$\int_0^S dy \int' u_{\bar{y}}^2 \leqslant c\,(S\alpha^2 + 1).$$

The normalization condition and proximity of the coefficients $a_{ij}^{(k)}$ ensure the inequality $|\bar{u}(y)| \leq c\alpha$, $1 \leq y \leq S-1$, to hold, where

$$\bar{u}(y) = \int' a_{00}^{(1)} \mu^{(1)} u(y) \Big/ \int' a_{00}^{(1)} \mu^{(1)}(y).$$

Let $\hat{u}(\bar{y}) = u(\bar{y}) - \bar{u}(y)$. Then

$$\int' \hat{u}^2(y) \leqslant c \int' \hat{u}^2_{\cdot y}(y); \; v^2_{\cdot y} = \sum_{i=1}^{n} (\nabla_i v)^2$$

(this version of the Poincaré-Freidrichs inequality for periodic functions with a bounded period and zero mean in the bounded density is easily obtained by, for example, computations similar to the derivation of the Erling inequalities in [15]).

Since the "cross-sectional" derivatives u and $\hat{u}$ coincide, $\hat{u}^2_{\cdot y} = u^2_{\cdot y} \le u^2_{\bar{y}}$. Hence

$$\int_1^{S-1} dy \int' |\hat{u}|(y) \leqslant \left(\int_1^{S-1} dy \int' \hat{u}^2 \right)^{1/2} \left(\int_1^{S-1} dy \int' 1 \right)^{1/2} \leqslant$$

$$\leqslant c S^{1/2} \left(\int_1^{S-1} dy \int' u^2_{\bar{y}} \right)^{1/2} \leqslant c\left[S^{1/2} + S\alpha\right].$$

With the estimate for $|\bar{u}|$ this leads to the inequality

$$\int_1^{S-1} dy \int' |u|(y) \leqslant c\left(S^{1/2} + S\alpha\right),$$

which, in turn, implies the estimate of the Lemma, if we take into account (5), (6) and the inequality

$$\int_{0<\rho<1} dy \int' |\mu^{(1)} - \mu^{(2)}|(y) \leqslant c.$$

On can verify (repeating, for instance, the arguments of Section 5 of Chapter III in [13]) for the space of $(T,1)$-periodic functions with norm $(\int\int' u^2 + u^2_{\bar{y}})^{\frac{1}{2}}$, that the solvability condition of an equation with the $(T,1)$-periodic coefficients

$$a_{ij} \nabla_i \nabla_j \phi = g \;, \tag{7}$$

adjoint to (3), is

$$\int_0^T dy \int' g\mu(y) = 0, \tag{8}$$

where μ is the (T,1)-periodic solution to Eqs. (3) and (4). In this case the unique solution can be separated, for example, by

$$\int\int' \phi = 0 . \tag{9}$$

If g satisfies the Hölder condition, ϕ is the classical solution of (7), (9). We shall use in the sequel a corollary of the lemma concerning the growth of solutions (Section 4 of Chapter 1 in [16]).

<u>LEMMA 2</u>. Let the 1-periodic function ϕ satisfy Eq. (7) for $|\bar{y}-\bar{y}_0| \leq 4\ell$:

$$\Phi(l) = \varphi_+(l) - \varphi_-(l); \quad \varphi_+(l) = \sup_{|\bar{y}-\bar{y}_0|<l} \varphi(y),$$

$$\varphi_-(l) = \inf_{|\bar{y}-\bar{y}_0|<l} \varphi(\bar{y}).$$

There exist positive constants x, κ (their values are given only by the dimension and the constants in (1), (2)) such that

$$\Phi(l) \leqslant cl^2 G; \quad G = \sup_{|\bar{y}-\bar{y}_0|<4l} |g(\bar{y})|,$$

if only $\Phi(4\ell) \leq \Phi(\ell)(1+\kappa)$.

<u>Proof</u>. Let G_1 be a constant such that for $|\bar{y}-\bar{y}_0| \leq 4\ell$

$$a_{ij}\nabla_i\nabla_j(G_1|\bar{y}|^2 \pm \phi) > 0 ;$$

let $\phi_0 = (\phi_+ + \phi_-)/2$. The quantity G_1 can, obviously, be chosen such that $0<G_1<cG$. It suffices to give the proof only for the case $\bar{y}_0=0$, $\Phi(\ell)>4\ell^2G_1/\kappa$ (restrictions on the choice of κ will be listed later).

Let $s>0$ be so large that the s-capacity of C_s is an upper capacity for the operator (7). In the case where

$$C_s(H_0) \geqslant l^s/2,\ H_0 = \{\bar{y} : \varphi(\bar{y}) \leqslant \varphi_0,\ |\bar{y}| \leqslant l\},$$

we apply the lemma on the growth of solutions [16] to the function

$$u(\bar{y}) = \varphi(\bar{y}) - \varphi_0 - \Phi(l)\varkappa/4 + G_1|\bar{y}|^2.$$

Indeed,

$$H = \{\bar{y} : u(\bar{y}) \leqslant 0,\ |\bar{y}| \leqslant l\} \supseteq H_0.$$

Hence $C_s(H)\geq \ell^s/2$ and by the lemma on the growth of solutions, we have

$$\sup_{|\bar{y}|\leqslant 4l} u(\bar{y}) \geqslant (1+c_1) \sup_{|\bar{y}|\leqslant l} u(\bar{y}).$$

By hypothesis of the Lemma

$$\sup_{|\bar{y}|\leqslant 4l} u(\bar{y}) \leqslant \sup_{|\bar{y}|\leqslant 4l} \varphi(\bar{y}) - \varphi_0 - \Phi(l)\varkappa/4 + 4^2l^2G_1 \leqslant$$
$$\leqslant (1/2+2\varkappa)\Phi(l) + 16G_1l^2;$$

at the same time

$$\sup_{|\bar{y}|\leqslant l} u(\bar{y}) \geqslant \Phi(l)/2 - \Phi(l)\varkappa/4.$$

Therefore, for $\kappa<\kappa_0$ (the value of κ_0 depends on the constant c_1 from the lemma on the growth of solutions)

$$16l^2G_1 \geqslant \Phi(l)[(1+c_1)(1/2-\varkappa/4) - (1/2+\varkappa 7/4)] \geqslant \Phi(l)/c,$$

which is the estimate sought.

If $C_s(H_0)<\ell^s/2$, it suffices to repeat the computations for the function $\bar{u}=G_1|\bar{y}|^2-\Phi(\ell)\kappa/4+\phi_0-\phi(\bar{y})$, since in this case

$$\bar{H}=\{\bar{y}: u(\bar{y})\leqslant 0,\ |\bar{y}|\leqslant l\}\supseteq\{\bar{y}:\varphi_0\leqslant\varphi(\bar{y}),\ |\bar{y}|\leqslant l\}\supseteq\{\bar{y}:|\bar{y}|\leqslant l\}\setminus H_0$$

and $C_s(\bar{H})\geq\ell^s-C_s(H_0)\geq\ell^s/2$.

COROLLARY 1. If the (T,1)-periodic function ϕ satisfies Eq. (7) with the (T,1)-periodic right-hand side g, with $g\neq 0$ in an "interval" $Q=\{\bar{y}:\ 0\leq y\leq N\}$ for $0\leq y\leq T$, then

$$\Phi=\sup_{\bar{y}}\varphi(\bar{y})-\inf_{\bar{y}}\varphi(\bar{y})\leqslant cN^2\sup_{\bar{y}}|g(\bar{y})|.$$

Indeed, according to the maximum principle, ϕ attains the maximum and minimum on the set, where $g\neq 0$. Hence for any point $\bar{y}\in\{\bar{y}:\ 0\leq y\leq N\}$

$$\Phi(\sqrt{N^2+n})=\Phi(4\sqrt{N^2+n})=\Phi,$$

which is the estimate sought.

COROLLARY 2. Let ϕ be the 1-periodic solution of Eq. (7), with $g(\bar{y})=0$ for $|\bar{y}-y_0|\leq M$ (M>1).

Then

$$|\nabla_j\varphi(\bar{y}_0)|\leqslant cM^{-\varkappa_1}\Phi,\quad |\nabla_i\nabla_j\varphi(\bar{y}_0)|\leqslant cM^{-\varkappa_1}\Phi,$$

where $\Phi=\sup\limits_{|\bar{y}-\bar{y}_0|<M}\varphi-\inf\limits_{|\bar{y}-\bar{y}_0|<M}\varphi$, and κ_1 depends only on the dimension and the ellipticity constants of the problem.

Proof. Let $\Phi(\ell)$ be defined in the same way as in Lemma 2. For $4^k<M$

$$\Phi(4^k)\geqslant(1+\varkappa)\Phi(4^{k-1}),$$

since the converse assumption leads via Lemma 2 to the equality $\Phi(4^{k-1})=0$ and a contradiction. Hence

$$\Phi(1)\leqslant c(1+\varkappa)^{-\log_4 M}\Phi=cM^{-\varkappa_1}\Phi,\quad \varkappa_1>0.$$

The function ϕ+const is also a solution to Eq. (7). This enables us, if we use the smoothness of coefficients, to obtain the needed result from the Schauder estimate [13].

2. ERGODIC PROPERTIES OF PERIODIC SOLUTIONS OF AUXILIARY EQUATIONS

We shall use $a_{ij}(\bar{y})=a_{ij}(\bar{y},\omega), g(\bar{y})=g(\bar{y},\omega)$, $\bar{y}=\{y,'y\}\in R^{n+1}$, $\omega\in\mathcal{R}$, to denote 1-periodic smooth random fields on a probability space $<\mathcal{R},\mathcal{U},P>$. We assume in addition that for $y\in(\alpha,\beta)$ the random variables $a_{ij}(\bar{y})$, $g(\bar{y})$, $\bar{y}=\{y',y\}$ are measurable with respect to the σ-algebra M_α^β, the family $\{M_\alpha^\beta\}$ permitting the inclusion $M_\alpha^\beta\subseteq M_\gamma^\delta$ for $(\alpha,\beta)\subset(\gamma,\delta)$ and satisfying the strong mixing condition

$$|\mathbf{E}\xi\eta-\mathbf{E}\xi\mathbf{E}\eta)\leqslant\alpha(\tau), \tag{10}$$

where ξ, $|\xi|\le 1$, is M_α^β-measurable, η, $|\eta|\le 1$, is $M_{\beta+\tau}^\delta$-measurable.

With respect to the variable y, the fields a_{ij}, g are assumed to be homogeneous (it suffices to assume that the mutual distributions of the values of a_{ij}, g for a set of points $\bar{y}_1,\dots,\bar{y}_N$ do not change for an integer shift of $\bar{y}_i$ along the first coordinate).

The field a_{ij} satisfies the constraints (1), (2) with probability 1; the field g with probability 1 admits the estimates

$$|g(\bar{y})|+\sum_j|\nabla_j g(\bar{y})|<c \quad \forall\bar{y}. \tag{11}$$

Let $\rho(y)$ be a nonrandom function such that

$$0 \leqslant \rho(y) \leqslant 1, \ |\rho'(y)| + |\rho''(y)| \leqslant c, \ \rho(\bar{y}) = 0$$

in the neighborhood $y=0$, $\rho(y)=1$ for $|y|>1$. The transformation of the field a_{ij} in the neighborhood of hyperplanes $y=kT$ by the formulas

$$\hat{a}_{ij}(\bar{y}) = \rho(y-kT)a_{ij}(\bar{y}) + (1-\rho(y-kT))a_{-}\delta_{ij} \tag{12}$$

retains, obviously, all the properties described above, including the homogeneity (for shifts by multiples of T). In this case the field $\hat{a}$ permits a $(T,1)$-periodic extension from each of the intervals $[kT,(k+1)T]\times[0,1]^n$. Later on, when we speak of periodic extensions of a_{ij} or g, we assume that a, g are a priori transformed by (12) if necessary.

LEMMA 3. Let μ_T be the $(T,1)$-periodic solution of (3), (4), corresponding to the $(T,1)$-periodic extension of a_{ij} from $[0,T]\times[0,1]^n$. The limits exist

$$G_i = \lim_{T\to\infty} \mathbf{E}\gamma^{(i)}(T), \quad i=1,2,$$

where

$$\gamma^{(i)}(T) = T^{-1}\int_0^T dy \int' g^{(i)}\mu_T(y), \ g^{(1)}(\bar{y}) = 1, \ g^{(2)}(\bar{y}) = g(\bar{y}).$$

If the mixing coefficient in the condition (10) admits the estimate

$$\alpha(\tau) \leqslant c_1 \exp(-c_2\tau), \tag{13}$$

then for any integer $S<T$

$$\mathbf{P}(|\gamma^{(2)}(T)/\gamma^{(1)}(T) - \mathbf{E}\gamma^{(2)}(S)/\mathbf{E}\gamma^{(1)}(S)| \geqslant c'(S^{-1/2} + S/T + x)) \leqslant$$
$$\leqslant c'' \exp\{-x(T/S)^{1/2}\ln^{-1/2}(T/S)\},$$

where the constants c', c'' do not depend on T or S.

Proof. Let $T=kS+M$, $0\le M<S$, where k, S are integers. Let $\hat{\mu}(\bar{y})$, $\bar{y}=\{y,'y\}$ for $\ell S<y\le(\ell+1)S$ coincide with the $(S,1)$-periodic solution of (3), (4), corresponding to the $(S,1)$-periodic extension of a_{ij} from the "interval" $Q_\ell=[S\ell,S(\ell+1)]\times[0,1]^n$. Then by Lemma 1 (with $\alpha=0$)

$$\gamma^{(i)}(T)=(kS)^{-1}\int_0^{Sk}dy\int' g^{(i)}\hat{\mu}(y)+r_1^{(i)},$$

where with probability 1, $|r_1^{(i)}|\le c(S/T+S^{-\frac{1}{2}})$. But

$$(kS)^{-1}\int_0^{Sk}dy\int' g^{(i)}\hat{\mu}(y)=\sum_{l=1}^{k}\xi_l^{(i)}/k,$$

where the random variables $\xi_\ell^{(i)}$ are $M_{\ell S}^{(\ell+1)S}$-measurable and (by the homogeneity property) distributed as $\gamma^{(i)}(S)$. The existing estimates for g and μ ensure that the inequalities

$$1/c\leqslant\xi_l^{(1)}\leqslant c;\quad |\xi_l^{(2)}|\leqslant c$$

be satisfied with probability 1. Hence, in particular,

$$\mathbf{E}\gamma^{(i)}(T)=\mathbf{E}\gamma^{(i)}(S)+O(S^{-1/2}+S/T),$$

which readily implies the existence of the limits G_i. Since the sequence $\xi_\ell^{(i)}$ satisfies the strong mixing condition with the coefficient $\hat{\alpha}(\ell)\le c\exp(-S\ell)$, the estimates for probabilities of the large deviations

$$\mathbf{P}\left(\left|\sum_{l=1}^{k}\xi_l^{(i)}/k-\mathbf{E}\xi_1^{(i)}\right|\geqslant c_1^{(i)}x\right)\leqslant c_2^{(i)}\exp\{-x\sqrt{k/\ln k}\}$$

follows from the results obtained in [17].

REMARK. If the random fields $a_{ij}(\bar{x},\bar{y})$, $g(\bar{x},\bar{y})$ depend smoothly on the parameter $\bar{x}\in D\subset R^{n+1}$ in the sense that a_{ij}, g satisfy (1), (2), (11) with probability 1 for all $\bar{x}$ with common constants and for all $\bar{y}$

$$|a_{ij}(\bar{x},\bar{y}) - a_{ij}(\bar{x}',\bar{y})| \leqslant c|\bar{x}-\bar{x}'|^{\delta};$$
$$|\nabla_k a_{ij}(\bar{x},\bar{y}) - \nabla_k a_{ij}(\bar{x}',\bar{y})| \leqslant c|\bar{x}-\bar{x}'|^{\delta};$$
$$|g(\bar{x},\bar{y}) - g(\bar{x}',\bar{y})| \leqslant c|\bar{x}-\bar{x}'|^{\delta}; \quad \delta > 0,$$

by Lemma 1 the mean values of Lemma 3 admit with probability 1 the estimate

$$|\gamma_{\bar{x}}^{(i)}(T) - \gamma_{\bar{x}'}^{(i)}(T)| \leqslant c|\bar{x}-\bar{x}'|^{\delta} + O(T^{-1/2}).$$

This, in particular, implies that the limit values $G_i(\bar{x})$ satisfy the Hölder condition. Furthermore, the convergence of the mean values $E\gamma_{\bar{x}}^{(i)}(T)$ to their limits is uniform on any compact.

LEMMA 4. Let $T=LN$ (L, N are integers), and let a_{ij}, g be $(T,1)$-periodic. There exists a $(T,1)$-periodic function ϕ such that with probability 1

$$\sup|\varphi(\bar{y})| \leqslant cTN, \quad \sup|\nabla_j\varphi(\bar{y})| \leqslant cN^{1+\varkappa_1}T^{1-\varkappa_1};$$
$$\sup|\nabla_i\nabla_j\varphi(\bar{y})| \leqslant cN^{1+\varkappa_1}T^{1-\varkappa_1},$$

where κ_1 is the constant from Corollary 2; also, if the mixing coefficient (10) admits the estimate (13), then for any integer $S<N$

$$\mathbf{P}\{\sup|a_{ij}\nabla_i\nabla_j\varphi - g + G_2/G_1| \geqslant |G_2/G_1 - \mathbf{E}\gamma^{(2)}(S)/\mathbf{E}\gamma^{(1)}(S)| +$$
$$+ c(S^{-1/2} + (N/S)^{-1/4})\} \leqslant c(T/N)\exp\{-(N/S)^{1/4}\ln^{-1/2} N/S\}.$$

Proof. Let the $(T,1)$-periodic functions $\zeta_\ell(y)=0$ outside the 1-neighborhood of the interval $\Delta_\ell=[\ell N,(\ell+1)/N]$, $|\zeta_\ell'(y)|+|\zeta_\ell''(y)|\leq c$, $0\leq\zeta_\ell\leq 1$, $\sum_{\ell=1}^{L}\zeta_\ell\equiv 1$. Let $\hat{g}_\ell$ be constants such that

$$\int_0^T dy \int' \zeta_l (g - \hat{g}_l)\, \mu_T = 0,$$

where μ_T is the (T,1)-positive solution of the adjoint equation (3). Then the problem

$$a_{ij} \nabla_i \nabla_j \varphi_l = \zeta_l (g - \hat{g}_l), \quad \int_0^T dy \int' \varphi_l = 0$$

is uniquely solvable in the class of (T,1)-periodic functions and by Corollary 1 admits the estimate $\sup |\phi_\ell(\bar{y})| \leq cN^2$.

Corollary 2 yields for the derivatives the estimate

$$|\nabla_i \varphi_l (\bar{y})| \leqslant cN^2 [1 + \min \{(lN - x)^+, (x - (l+1) N)^+\}]^{-\varkappa_1},$$

where $a^+ = \max(a,0)$. Hence $\phi = \sum_{\ell=1}^{L} \phi_\ell$ satisfies the inequalities

$$|\varphi(\bar{y})| \leqslant cN^2 T/N = cTN; \quad |\nabla_j \varphi (\bar{y})|, \quad |\nabla_i \nabla_j \varphi (\bar{y})|$$

$$\leqslant cN^2 \sum_{l=1}^{T/N} (1 + lN)^{-\varkappa_1} \leqslant cN^2 \left(1 + N^{-\varkappa_1} (T/N)^{1-\varkappa_1}\right)$$

at each point $\bar{y}$.

Let the (T,1)-periodic solution μ_T of the adjoint problem be separated by the condition (4). Then the estimates of Lemma 1 yield the equality

$$\hat{g}_l = \int_0^N dy \int' g\mu_N \Big/ \int_0^N dy \int' \mu_N + r_l,$$

where μ_N is the (N,1)-periodic solution of (4), (3), corresponding to the (N,1)-periodic extension (possibly, modified by (12)) of the coefficients a_{ij} from the "interval" $[\ell N, (\ell+1)N] \times [0,1]^n$, and the remainder is estimated with probability 1 by the inequality $|r_\ell| \leq cN^{-\frac{1}{2}}$. However, by Lemma 3 for integer $S<N$ and $x = (N/S)^{-\frac{1}{4}}$

$$\mathbf{P}\left\{\left|\int_0^N dy \int' g\mu_N \Big/ \int_0^N dy \int' \mu_N - \mathbf{E}\gamma^{(2)}(S)/\mathbf{E}\gamma^{(1)}(S)\right| \geqslant \right.$$
$$\left. \geqslant c'\left(S^{-1/2} + (S/N)^{1/4} + S/N\right\} \leqslant c'' \exp\{-(N/S)^{1/4}\ln^{-1/2} N/S\},$$

thus completing the proof.

3. THEOREMS ON HOMOGENIZATION

Let the random fields $a_{ij}(\bar{x},\bar{y})$, $b_i(\bar{x},\bar{y})$, $c(\bar{x},\bar{y})$, $f(\bar{x},\bar{y})$, $\bar{x}=\{x^{(0)},\ldots,x^{(n)}\}\in\bar{G}\subset R^{n+1}$, $\bar{y}=\{y^{(0)},\ldots,y^{(n)}\}\in R^{n+1}$ for fixed $\bar{x}$ be 1-periodic in the variables $y^{(1)},\ldots,y^{(n)}$, and let them satisfy the ellipticity conditions (1) and of smoothness in $\bar{y}$ (2), (11) (where $g=b_1,c,f$) with probability 1 for all $\bar{x}\in\bar{G}$, and also the conditions of homogeneity and of strong mixing (10), (13) in the variable $y^{(0)}$. Furthermore, let with probability 1 for all $\bar{y}$

$$\sum_{i,j}|a_{ij}(\bar{x},\bar{y}) - a_{ij}(\bar{x}',\bar{y})| + \sum_{i,j,k}|\nabla_k a_{ij}(\bar{x},\bar{y}) - \nabla_k a_{ij}(\bar{x}',\bar{y})| \leqslant c|\bar{x}-\bar{x}'|^{\alpha};$$

$$|g(\bar{x},\bar{y}) - g(\bar{x}',\bar{y})| \leqslant c|\bar{x}-\bar{x}'|^{\alpha}; \quad 0<\alpha\leqslant 1, \tag{14}$$

where $\nabla_j=\partial/\partial y^{(j)}$ is the operator of differentiation in the "rapid" variable $\bar{y}$.

The constraints imposed above and the condition

$$c(\bar{x},\bar{y}) \leq 0 \qquad \forall\ \bar{x},\ \bar{y} \tag{15}$$

ensure with probability 1 the unique solvability [13] in the class $C^{2+\alpha}(G)$ of the Dirichlet problem

$$L_\varepsilon u_\varepsilon = a_{ij}(\bar{x},\bar{x}/\varepsilon)D_iD_ju_\varepsilon + b_i(\bar{x},\bar{x}/\varepsilon)D_iu_\varepsilon + c(\bar{x},\bar{x}/\varepsilon)u_\varepsilon =$$
$$= f(\bar{x},\bar{x}/\varepsilon), \quad u_\varepsilon|_{\partial G} = \Phi(\bar{x}), \tag{16}$$

where Φ is the nonrandom function of the class $C^{2+\alpha}(\partial G)$;

G denotes a bounded domain with the boundary of the class $C^{2+\alpha}$; $D_i = \partial/\partial x^{(i)}$ denotes the operator of "slow" differentiation.

In what follows we denote by U the solution of the homogenized Dirichlet problem

$$L_0 U = \bar{a}_{ij} D_i D_j U + \bar{b}_i D_i U + \bar{c} U = \bar{f}; \qquad (17)$$
$$U/\partial G = \Phi,$$

where the coefficients $\bar{a}_{ij}(\bar{x})$, $\bar{b}_i(\bar{x})$, ... are obtained by averaging a_{ij},... via the following procedure.

Let $\mu_T^{\bar{x}}(\bar{y}) \geq 0$ be the $(T,1)$-periodic (see Section 1) solution of the equation

$$\nabla_i \nabla_j \left(a_{ij}(\bar{x}, \bar{y}) \mu_T^{\bar{x}} \right) = 0,$$

corresponding to the $(T,1)$-periodic extension of the coefficients $a_{ij}(\bar{x},\bar{y})$ with a "frozen" argument $\bar{x}$ (if necessary, a_{ij} are modified by the formula (12) as to make a periodic extension possible). The mean of the function g is defined by the passage to the limit by the formulas

$$\bar{g}(\bar{x}) = G_2(\bar{x})/G_1(\bar{x}); \quad G_1(\bar{x}) = \lim_{T\to\infty} \mathrm{E} T^{-1} \iint\limits_{[0,T]\times[0,1]^n} \mu_T^{\bar{x}}(\bar{y})\, d\bar{y};$$
$$G_2(\bar{x}) = \lim_{T\to\infty} \mathrm{E} T^{-1} \iint\limits_{[0,T]\times[0,1]^n} g(\bar{x}, \bar{y})\, \mu_T^{\bar{x}}(\bar{y})\, d\bar{y}.$$

The limits involved in the defining of the mean $a_{ij}(\bar{x})$, $b_i(\bar{x})$,..., exist by Lemma 3 and Remark 1 in Section 2. The averaged coefficients for $\bar{x} \in \bar{G}$ satisfy the Hölder condition with an exponential α. Hence the averaged Dirichlet problem (17) is also uniquely solvable in the class $C^{2+\alpha}$.

Under the constraints imposed above we have the following

theorem.

THEOREM 1. As $\varepsilon\to 0$ the random variable

$$\zeta_\varepsilon = \sup_{\bar{x}\in G} |u_\varepsilon(\bar{x}) - U(\bar{x})| \to 0$$

in probability.

Proof. The proof is based on the representation of the solution (compare with [2,3,6,10]) in the form $u_\varepsilon = U+\varepsilon^2\phi(\bar{x},\bar{x}/\varepsilon)+w_\varepsilon$, where the rapidly oscillating corrector ϕ is chosen to eliminate the principal term on the right side of the equation for w_ε, and U denotes the solution of the averaged problem.

In the case where only the coefficients a_{ij} are rapidly oscillating, the corrector ϕ can be constructed, for example, in the following way. Let $\zeta_k(\bar{x})$, $0\le\zeta_k(\bar{x})\le 1$ be finite smooth functions composing the decomposition of unity in G $(\sum\zeta_k\equiv 1)$: the supports of ζ_k have the diameter not greater than εT ($T=T(\varepsilon)$ will be chosen later), the multiplicity of the crossing of the supports of ζ_k is bounded by a constant not depending on ε, and the derivatives of the ℓth order admit the estimate $|D^{(\ell)}\zeta_k|\le c(\varepsilon T)^{-\ell}$.

Let $\bar{x}_k$ be a point of the support of ζ_k. Also, let $\hat{\phi}^{(k)}_{\ell m}(\bar{y})$ be the $(2T,1)$-periodic function (described in Lemma 4) constructed on the basis of the $(2T,1)$-periodic extension of the coefficients $a_{ij}(\bar{x}_k,\bar{y})$, $g(\bar{y})=-a_{\ell m}(\bar{x}_k,\bar{y})$ (possibly, modified by (12)) from the interval $[a,a+2T]\times[0,1]^n$, the basis of which covers the projection of the support of $\zeta_k(\bar{x}/\varepsilon)$ onto the axis $y^{(0)}$.

We can use as ϕ the function

$$\varphi = \sum_k \zeta_k(\bar{x})\,\hat{\varphi}^{(k)}_{lm}(\bar{x}/\varepsilon)\,D_l D_m U(\bar{x}_k).$$

Indeed, if

$$\varepsilon^2 TN + \varepsilon N + \varepsilon N^{1+\varkappa_1} T^{1-\varkappa_1} \to 0 \tag{18}$$

are satisfied, the expression $b_i D_i(\varepsilon^2\phi)+c_i\varepsilon^2\phi$ containing only the function ϕ and its first-order derivatives, tends to zero with probability 1 uniformly in $\bar{x}\in\bar{G}$.

The principal term $L_\varepsilon(\varepsilon^2\phi)$ is representable as

$$\sum_k \zeta_k a_{ij}(\bar{x}, \bar{x}/\varepsilon)\,\nabla_i\nabla_j\hat{\varphi}^{(k)}_{lm}(\bar{x}/\varepsilon)\,D_l D_m U(\bar{x}_k) + r_1 =$$
$$= \sum_k \zeta_k a_{ij}(\bar{x}_k, \bar{x}/\varepsilon)\,\nabla_i\nabla_j\hat{\varphi}^{(k)}_{lm}(\bar{x}/\varepsilon)\,D_l D_m U(\bar{x}_k) + r,$$

where with probability 1

$$|r| \leqslant c\left\{N/T + N^{1+\varkappa_1}/T^{\varkappa_1} + (\varepsilon T)^{\alpha} N^{1+\varkappa_1} T^{1-\varkappa_1}\right\}. \tag{19}$$

In this case if $N(\varepsilon)$ is the number of ζ_k in the decomposition of unity of G, then with probability not smaller than

$$1 - c\mathfrak{N}(\varepsilon)(T/N)\exp(-(N/S)^{1/4}\ln^{-1/2} N/S), \tag{20}$$

for all k $(S<N<T)$ the remainder r_k in the representation

$$a_{ij}(\bar{x}_k, \bar{x}/\varepsilon)\,\nabla_i\nabla_j\hat{\varphi}^{(k)}_{lm}(\bar{x}/\varepsilon) = -a_{lm}(\bar{x}_k, \bar{x}/\varepsilon) + \bar{a}_{lm}(\bar{x}_k) + r_k(\bar{x})$$

admits the estimate

$$\sup_{k,\bar{x}} |r_k(\bar{x})| \leqslant c\left[S^{-1/2} + (S/N)^{1/4}\right] + \left|\bar{a}_{lm}(\bar{x}_k) - \mathbf{E}\gamma^{(2)}(S)/\mathbf{E}\gamma^{(1)}(S)\right|, \tag{21}$$

where, as was noted in Remark 1 (Section 2), the last term tends to zero uniformly in $\bar{x}\in G$. Furthermore, obviously, $N(\varepsilon)\le c(\varepsilon T)^{-(n+1)}$.

Therefore, taking into account the smoothness of coefficients of the primary problem in $\bar{x}$, (14) and the averaged solution U for

$$S^{-1}+S/N+N/T+N^{1+\varkappa_1}T^{-\varkappa_1}+\varepsilon^{\alpha}T^{1+\alpha-\varkappa_1}N^{1+\varkappa_1}+$$
$$+(\varepsilon T)^{-n-1}(T/N)\exp\left[-(N/S)^{1/4}\ln^{-1/2}(N/S)\right]\to 0 \qquad (22)$$

we see that the remainder in the relation

$$a_{ij}(\bar{x},\bar{x}/\varepsilon)D_iD_j(\varepsilon^2\varphi)=-\left(a_{lm}(\bar{x},\bar{x}/\varepsilon)-\bar{a}_{lm}(\bar{x})\right)D_lD_mU(\bar{x})+\rho_\varepsilon$$

satisfies the relation $\sup_{\bar{x}}|\rho_\varepsilon(\bar{x})|\to 0$ in probability.

When Eqs. (18), (22) are satisfied at the same time, the equation for the residual w_ε (we assume that the minor terms have no rapidly oscillating coefficients) becomes

$$L_\varepsilon w_\varepsilon=\bar{f}-a_{ij}D_iD_jU-\bar{b}_iD_iU-\bar{c}U+\left(a_{lm}-\bar{a}_{lm}\right)D_lD_mU+\rho_\varepsilon=\rho_\varepsilon.$$

On the boundary, $\sup_{\bar{x}}|w_\varepsilon|\le c\varepsilon^2NT$ tends to zero with probability 1. Hence $\sup_{\bar{G}}|w_\varepsilon|\to 0$ in probability.

It remains to note that (18), (22) can be satisfied by $T\sim\varepsilon^{-\delta}$, $N\sim T^{\delta}$, $S\sim N^{\delta}$, where δ is small positive.

In the case where the coefficients in minor derivatives and the free term are also rapidly oscillating, the corrector involves terms of the form

$$\Sigma\zeta_k\hat{\varphi}_{bl}^{(h)}D_lU(\bar{x}_k),\ \Sigma\zeta_k\hat{\varphi}_c^{(h)}U(\bar{x}_k),\ \Sigma\zeta_k\hat{\varphi}^{(h)},$$

where $\hat{\phi}$ are constructed on the basis of $g=b_i,\ldots$.

To complete the proof of Theorem 1 is suffices to note that $\sup_{\bar{x}\in G}|\varepsilon^2\phi|\to 0$ with probability 1.

The next theorem can be proved in the similar way.

<u>THEOREM 2</u>. As $\varepsilon\to 0$ the solution of the boundary problem

$$\frac{\partial u_\varepsilon}{\partial t}=L_\varepsilon u_\varepsilon+f_\varepsilon;\ (\bar{x},t)\in G\times[0,\tau];$$
$$\lim_{t\downarrow 0}u_\varepsilon(\bar{x},t)=\Phi_1;\ u_\varepsilon|_{\partial G}=0$$

approaches the solution of the averaged problem

$$\frac{\partial U}{\partial t} = L_0 U + \bar{f};\ U|_{t=0} = \Phi_1;\ U|_{\partial G} = 0$$

in the sense that $\sup_{(t,\bar{x})} |u_\varepsilon - U| \to 0$ in probability.

4. THE AVERAGING OF STOCHASTIC DIFFERENTIAL EQUATIONS

The results of the preceding section can be applied to investigating the limit behavior of $\varepsilon \to 0$ of distributions in the space of continuous functions generated by solutions of stochastic differential equations of the form

$$dx_\varepsilon^{(i)}(t) = A_j^{(i)}(\bar{x}_\varepsilon(t), \bar{x}_\varepsilon(t)/\varepsilon)\, dw^{(j)}(t) + b^{(i)}(\bar{x}_\varepsilon(t), \bar{x}_\varepsilon(t)/\varepsilon)\, dt;\ x_\varepsilon^{(i)}(0) = x^{(i)},$$

where the iteration of the indices implies the summation and $\bar{x}_\varepsilon = \{\bar{x}_\varepsilon^{(0)}, \dots, x_\varepsilon^{(n)}\}$; the independent Wiener processes $w^{(i)}$, $i=0,\dots,n$, do not depend on the random coefficients $A_j^{(i)}(\bar{x},\bar{y},\omega)$, $b^{(i)}(\bar{x},\bar{y},\omega)$; the coefficients $A_j^{(i)}$, $b^{(i)}$ are periodic in $y^{(i)},\dots,y^{(n)}$ and satisfy the mixing conditions in $y^{(0)}$ mentioned in Section 2. Furthermore, the coefficients $A_j^{(i)}$, $b^{(i)}$ are bounded smooth functions of $\bar{y}$, $a_{ij} = A_k^{(i)} A_k^{(j)}$ satisfying the strong ellipticity condition (1); as functions of $\bar{x}$ these coefficients and their derivatives in $y^{(j)}$ satisfy a Lipschitz condition uniform in $\bar{y}$, ω.

The boundedness of the coefficients that we assumed implies the estimate

$$\mathbf{M}|\bar{x}_\varepsilon(t+h) - \bar{x}_\varepsilon(t)|^4 \leqslant ch^2$$

with probability 1 (M will designate the conditional averaging provided the random coefficients $A_j^{(i)}$, $b^{(i)}$ are fixed). This estimate shows (see, for example, Theorem 2 in Section 2 in Chapter IX in [18]) that with probability 1 the family of distributions generated by the processes $\bar{x}_\varepsilon$ in a space of continuous functions is weakly compact.

Let G be any bounded domain with a smooth boundary, let ϕ be a smooth function finite in G, and let τ_ε be the instant of the first exit time from G. Then from the results of Section 3 and also from [19] it follows that for all x from G

$$\mathbf{M}_{\bar{x}}\varphi\left(\bar{x}_\varepsilon(t)\right) I_{\{\tau_\varepsilon<t\}} \to U\left(\bar{x}\right)$$

in probability, where $U(\bar{x})$ is a solution of the averaged boundary problem

$$\frac{\partial U}{\partial t} = \frac{1}{2}\bar{a}_{ij}\frac{\partial^2 U}{\partial x^{(i)}\partial x^{(j)}} + \bar{b}^{(i)}\frac{\partial U}{\partial x^{(i)}};$$

$$U|_{t=0} = \varphi\left(\bar{x}\right);\ U|_{\partial G} = 0.$$

This readily implies that for any $t>0$, $\bar{x}\in R^{n+1}$ the distribution $\bar{x}_\varepsilon(t)$ in R^{n+1} in probability converges to the distribution of $\bar{x}_0(t)$ that is a solution of the stochastic equation

$$d\bar{X}_0^{(i)}(t) = \left(\sqrt{\bar{a}}\right)_{ij} dw^{(j)}(t) + \bar{b}^{(i)}dt,$$

where $\sqrt{\bar{a}}$ is a positive definite symmetric matrix such that $\sqrt{\bar{a}}\sqrt{\bar{a}}=\bar{a}$. The convergence of distributions in probability means:

$$\mathfrak{L}\left(F_\varepsilon^{\bar{x},t}, F_0^{\bar{x},t}\right) \to 0 \tag{23}$$

in probability, where $\mathfrak{L}$ is the Lévy-Prokhorov distance bebetween the distributions F_ε of $\bar{x}_\varepsilon(t)$ and F_0 of $\bar{x}_0(t)$

in R^{n+1}.

Choosing from the arbitrary sequence $\varepsilon_n \downarrow 0$ the subsequence $\varepsilon_{n'}$ having the property that with probability 1

$$\mathfrak{L}\left(F_{\varepsilon_{n'}}^{\bar{x}_k, t_l}, F_0^{\bar{x}_k, t_l}\right) \to 0$$

for any $\bar{x}_k \in \bar{X}$, $t_\ell \in \bar{T}$, where $\bar{X}$, $\bar{T}$ are everywhere countable dense sets in R^{n+1}, $R_+ = [0,\infty)$, using the weak compactness with probability 1 of the family of distributions $x_\varepsilon(t)$ (in $C_{[0,t]}$, $t<\infty$), we can verify that $\mathfrak{L}_\pi\left(\mathfrak{F}_{\varepsilon_{n'}}^t, \mathfrak{F}_0^t\right) \to 0$ with probability 1, where L_π designates the Lévy-Prokhorov distance of the distributions in $C_{[0,t]}$. This implies that $\mathfrak{L}_\pi\left(\mathfrak{F}_\varepsilon^t, \mathfrak{F}_0^t\right) \to 0$ in probability.

REFERENCES

[1] M.I. Freidlin. "Dirichlet's Problem for an Equation With Periodic Coefficients Depending on a Small Parameter." *Theory Prob. Applications*, 1 (1964): 121-125.

[2] S.M. Kozlov. "Averaging Differential Operators With Almost Periodic, Rapidly Oscillating Coefficients." *Math. USSR, Sbornik*, 35, 4 (1979): 481-498.

[3] N.S. Bakhvalov. "Averaged Characteristics of Bodies with Periodic Structure." *Soviet Phys. Dokl.*, Vol. 19 (1974-75): 650-651.

[4] N.S. Bakhvalov. "Averaging of Partial Differential Equations With Rapidly Oscillating Coefficients." *Soviet Math. Dokl.*, 16, 2 (1975): 351-355.

[5] N.S. Bakhvalov and A.A. Zlotnik. "Coefficient Stability of Differential Equations and Averaging of Equations With Random Coefficients." *Soviet Math.,Dokl.,* 19, 5 (1978): 1171-1175.

[6] V.L. Berdichevskij. "Spatial Averaging of Periodic Structures," *Soviet Phys. Dokl.,* 20, 5 (1975): 334-335.

[7] S.M. Kozlov. "Averaging Random Structures." *Soviet Math.Dokl.,* 19, 4 (1978): 950-954.

[8] V.V. Yurinskij. "On the Application of the Homogenization Method to Elliptic Equations With Random Coefficients." In *Mathematical Analysis and Related Problems in Mathematics,* 330-339. Novosibirsk: Nauka, 1978. (In Russian.)

[9] V.V. Yurinskij. *The International Symposium on Stochastic Differential Equations, Summary of Reports,* 54-56. Vilnius, 1978. (In Russian.)

[10] A. Bensoussan, J.-L. Lions, and G. Papanicolaou. "Sur quelques phénomènes asymptotiques stationnaires." *CR Ac. Sci., Ser. A,* 281, 2-3 (1975): A89-A94.

[11] R.Z. Khas'minskij. "Ergodic Properties of Recurrent Diffusion Processes and Stabilization of the Cauchy Problem for Parabolic Equations." *Theory Prob. Applications,* 2 (1960): 196-214. (In Russian.)

[12] D.G. Aronson and J. Serrin. "Local Behavior of Solutions of Quasilinear Parabolic Equations." *Arch. Ration. Mech. and Anal.,* 25, 2 (1967): 81-122.

[13] O.A. Ladyzhenskaya and N.N. Ural'tseva. *Linear and Quasilinear Elliptic Equations.* New York: Academic Press, 1968.

[14] O.A. Ladyzhenskaya, V.A. Solonnikov, and N.N. Ural'tseva. *Linear and Quasilinear Equations of Parabolic Type*. Providence: American Mathematical Society, 1968.

[15] K. Maurin. *Methods of Hilbert Spaces*. Warszawa, Poland: Polish Scientific Publishers, 1967.

[16] E.M. Landis. *Second-order Equations of Elliptic and Parabolic Type*. Moscow: Nauka, 1971. (In Russian.)

[17] J. Sunklodas. "Some Bounds for the Distribution of Sums of Weakly Independent Random Variables." *Lithuanian Math. J.*, 18, 1 (2978): 144-154.

[18] I.I. Gikhman and A.V. Skorokhod. *Introduction to the Theory of Random Processes*. Philadelphia: W.B. Saunders and Co., 1969; 2nd rev. ed., Mowcow: Nauka, 1977. (In Russian.)

[19] E.B. Dynkin. *Markov Processes*. New York: Academic Press, 1965.

Part 3

PROPERTIES OF DISTRIBUTIONS AND APPLIED PROBLEMS

ON THE DISTRIBUTION OF THE SUPREMUM OF A RANDOM WALK ON A MARKOV CHAIN

K. Arndt

1. INTRODUCTION

Let $\{\kappa_n\}_{n\geq 0}$ be a homogeneous irreducible noncyclic Markov chain with a finite number of states $N=\{1,2,\ldots,N\}$, with the matrix of transition probabilities $P=\|p_{kj}\|_{k,j\in N}$ and with a uniquely defined stationary distribution $\pi=\{\pi_1,\pi_2,\ldots,\pi_N\}$, $\pi_k>0$, $k\in N$. Next, let $\{\xi_{kj}^n\}_{k,j\in N}^{n\geq 1}$ be the family of independent random variables independent of $\{\kappa_n\}$ and identically distributed in each sequence $\{\xi_{kj}^n\}^{n\geq 1}$, $F_{kj}(x)=P(\xi_{kj}^n\leq x)$, $k,j\in N$. We consider a two-dimensional Markov process $\{Y_n,\kappa_n\}_{n\geq 0}$ the evolution of which is given by the initial value $\{0,\kappa_0\}$ and the relation $Y_{n+1}=Y_n+\xi_{\kappa_n\kappa_{n+1}}^{n+1}$. The distribution $\{Y_n,\kappa_n\}_{n\geq 0}$ is completely defined if the distribution κ_0 and the matrix $A(\mu)=\left\|p_{kj}\int_{-\infty}^{\infty}e^{i\mu x}dF_{kj}(x)\right\|$ are specified. We call $\{Y_n\}_{n\geq 0}$ a random walk on the Markov chain $\{\kappa_n\}$.

Everywhere in the sequel we will assume that

$$\mathbf{M}_\pi Y_1\equiv\sum_{k,j\in N}\pi_k p_{kj}\times\int_{-\infty}^{\infty}x\,dF_{kj}(x)<0.$$

Then the random variable $\bar{Y} \equiv \sup_{n \geq 0} Y_n$ is finite with probability 1 for any initial state κ_0 of the chain $\{\kappa_n\}$ [1].

Let V_n be the ring of quadratic matrices of the order N, the elements of which are Fourier-Stieltjes transforms of the right continuous functions of bounded variation on a real axis. We denote by V_{N+} (V_{N-}) the subring of matrices

$$\left\| \int_{-\infty}^{\infty} e^{i\mu x} db_{kj}(x) \right\|_{k,j \in \mathcal{N}} \in V_N,$$

such that $\operatorname{Var}_{(-\infty,0]} b_{kj}(x)=0$ $(\operatorname{Var}_{(0,\infty)} b_{kj}(x)=0)$.

Let I be the identity matrix of the order N.

DEFINITION. The matrix $I-B \in V_N$ permits a right factorization if there exist elements $B_- \in V_{N-}$ and $B_+ - I \in V_{N+}$ such that $I-B=B_-B_+$. If there exist also inverse $(B_-)^{-1} \in V_{N-}$, $(B_+)^{-1} - I \in V_{N+}$, we say that I-B permits a right canonical factorization. The element B_+ (B_-) is said to be a positive (negative) component of the factorization. For brevity, we will write in the sequel $P_k(\cdot)$ instead of $P(\cdot|\kappa_0=k)$ and $M_k(\cdot)$ instead of $M(\cdot|\kappa_0=k)$, $k \in \mathcal{N}$.

ASSERTION 1. (See [4]). The matrix $I-A(\mu) \in V_N$ permits the right factorization

$$I - A(\mu) = A_-(\mu)A_+(\mu) , \tag{1}$$

where the components are representable as

$$A_+(\mu) = I - \left\| \int_0^{\infty} e^{i\mu x} da_{kj+}(x) \right\|, \; a_{kj+}(x) = \sum_{n=1}^{\infty} P_k(\bar{y}_{n-1} \leq 0, \; Y_n \leq x, \; \varkappa_n = j);$$

$$A_-(\mu) = I - \left\| \int_{-\infty}^{0} e^{i\mu x} da_{kj-}(x) \right\|, \; a_{kj-}(x) = \sum_{n=1}^{\infty} P_k(\bar{y}_{n-1} < Y_n \leq x, \; \varkappa_n = j),$$

where $\bar{y}_n = \max_{1 \le m \le n} Y_m$, $n \ge 1$, $\bar{y}_0 = -\infty$.

We note that $\det A_-(\mu) \ne 0$ for $\operatorname{Im} \mu < 0$, and by $M_\pi Y_1 < 0$ $\det A_+(\mu) \ne 0$ for $\operatorname{Im} \mu \ge 0$.

ASSERTION 2. (See [1]). For $\operatorname{Im} \mu \ge 0$

$$\left\| \mathbf{M}_k e^{i\mu \bar{Y}} \right\|_{k \in \mathcal{N}} = (A_+(\mu))^{-1} A_+(0)\, \mathbf{1}, \tag{2}$$

where $\mathbf{1}$ designates the column vector of the order N, the elements of which are units.

The question arises whether it is possible to find the distribution $\bar{Y}$ in the explicit form from the formula (2), if the matrix $A(\mu)$ is given. The relations determining the factorization components (1) (see Assertion 1) cannot, in general, be an effective tool for finding the matrices $A_+(\mu)$, $(A_+(\mu))^{-1}$. We denote by

$$F(x) = \| p_{kj} F_{kj}(x) \|_{k,j \in N} \tag{3}$$

the matrix of conditional distributions, completely defining the matrix $A(\mu)$. Furthermore, we specify the metric ρ in the following way: If $F^{(1)}(x) = \| p_{kj}^{(1)} F_{ij}^{(1)}(x) \|$ and $F^{(2)}(x) = \| p_{kj}^{(2)} F_{kj}^{(2)}(x) \|$, then

$$\rho\left(F^{(1)}, F^{(2)}\right) = \max_{k,j \in \mathcal{N}} L\left(F_{kj}^{(1)}, F_{kj}^{(2)}\right) + \max_{k,j \in \mathcal{N}} \left| p_{kj}^{(1)} - p_{kj}^{(2)} \right|,$$

where $L(\cdot,\cdot)$ is the Lévy distance.

In this paper we discuss two problems:

1. the class R of matrices of the form (3), everywhere dense in V_N in the sense of ρ-convergence on the set of all matrices of the form (3), such that the matrix $A_+(\mu)$ can be found in the explicit form for $F \in R$. The theorem on the con-

tinuous dependence between $A_+(\mu)$ and $A(\mu)$ complementing this approach can be found in [3];

2. the application of the results obtained in the first part of the paper to the case of a queueing system. For the case $N=1$ corresponding to a random walk with independent identically distributed terms, the strongest results are due to A.A. Borovkov [2].

2. ON CONDITIONS FOR EXPLICIT SOLVABILITY OF THE FACTORIZATION

We will need the following conditions concerning the matrices $F(x)$ (or $A(\mu)$):

A_1) the distributions $F_{kj}(x)$ have no singular components, and at least one of these distributions corresponding to $p_{kj}>0$ has an absolutely continuous component;

A_2) $|M\xi_{kj}^n|<\infty$ for all $k,j\in N$;

A_3) all distributions $F_{ij}(x)$ are absolutely continuous.

The class R of matrices of the form (3) is described as the following: $F\in R$, if conditions A_2, A_3 are satisfied and at least one of the matrices $A^{\pm}(\mu)$ in the expansion $A(\mu)=A^+(\mu)+A^-(\mu)$,

$$A^+(\mu)=\left\| p_{kj}\int_0^{\infty} e^{i\mu x} dF_{kj}(x)\right\|;\ A^-(\mu)=\left\| p_{kj}\int_{-\infty}^{0} e^{i\mu x} dF_{kj}(x)\right\|,$$

is a rational matrix (i.e., a matrix whose elements are rational functions).

For $A^+(\mu)$ to be rational, it is necessary and sufficient that $1-F_{kj}(x)$ for $x>0$ be representable as

$$1-F_{kj}(x)=\sum_{m} P_{kj,m}(x)\exp\{-\alpha_{kj,m}x\}, \tag{4}$$

where $P_{kj,m}(x)$ are the polynomials of x; $\alpha_{kj,m}>0$; $k,j\in N$ (see [2, p. 139]). Following [2], we call expressions of the form (4) exponential polynomials.

ASSERTION 3. It is possible to approach closely to any matrix of the form (3) in the sense of ρ-convergence, using matrices from the class R. This assertion follows immediately from the respective assertion for the case $N=1$ in [2, pp. 139, 140].

Suppose that along with (1) we have the right factorization of the matrix $I-A(\mu)$:

$$\left.\begin{array}{r} I-A(\mu)=B_{-}(\mu)B_{+}(\mu), \\ B_{+}(i\infty)=I,\ \det B_{-}(\mu)\neq 0 \text{ for } \operatorname{Im}\mu<0, \\ \det B_{+}(\mu)\neq 0 \text{ for } \operatorname{Im}\mu\geqslant 0. \end{array}\right\} \tag{5}$$

LEMMA 1. Let conditions A_1, A_2 be satisfied. Then the factorization (5) is unique and $B_{\pm}(\mu)=A_{\pm}(\mu)$, which is a mere corollary from Theorem 2 in [3].

THEOREM 1. For $A_{+}(\mu)$ to be a rational matrix, it is necessary and, if conditions A_2 and A_3 are satisfied, sufficient that $A^{+}(\mu)$ be a rational matrix. If also $A^{+}(\mu)=\left\|\frac{q^{+}_{kj}(\mu)}{r^{+}_{kj}(\mu)}\right\|$, where $\frac{q^{+}_{kj}(\mu)}{r^{+}_{kj}(\mu)}$ are irredundant relations of polynomials, $k,j\in N$, $r(\mu)$ denotes the least common factor of all $r^{+}_{kj}(\mu)$, R is the degree of the polynomial $r(\mu)$, then the function $\det\{(I-A(\mu))r(\mu)\}$ has in the domain $\operatorname{Im}\mu<0$ exactly RN zeros and $A_+(\mu)=(r(\mu))^{-1}B(\mu)$, where $B(\mu)$ is the matrix of polynomials (i.e., the matrix the elements of which are polynomials). The methods for finding the matrix $B(\mu)$ are contained in the proof.

Proof. Necessity. It follows from the identity (1) that for $x>0$

$$p_{kj}(1-F_{kj}(x)) = (a_{kj+}(\infty)-a_{kj+}(x)) - \\ -\sum_{m=1}^{N}\int_{-\infty}^{0} da_{km-}(t)\{a_{mj+}(\infty)-a_{mj+}(x)\}. \quad (6)$$

It remains only to note that by assumption the right-hand side of (6) for all $k,j\in N$ is an exponential polynomial, thereby $p_{kj}(1-F_{kj}(x))$ for $x>0$ is an exponential polynomial. Therefore,

$$A^+(\mu) = \left\| -p_{kj}\int_0^\infty e^{i\mu x} d(1-F_{kj}(x))\right\|$$

is a rational matrix.

Sufficiency. We will prove the sufficiency in several stages.

1. If $A^+(\mu)\equiv 0$, then $A_+(\mu)\equiv I$ and the Theorem is then proved. Now let $A^+(\mu)\not\equiv 0$. Then the matrix $A^+(\mu)$ permits the representation $A^+(\mu)=(r(\mu))^{-1}\|q_{kj}(\mu)\|$, where $q_{kj}(\mu)$ are polynomials. Under the assumptions made the matrix

$$\mathscr{V}(\mu) = J(\mu)\cdot(I-A(\mu)), \quad (7)$$

where

$$J(\mu) = \left\|\frac{i\mu+1}{i\mu}\delta_{1k}+\delta_{kj}(1-\delta_{1j})\right\|\cdot \operatorname{diag}(\pi_1,\ldots,\pi_N),$$

δ_{kj} denotes the Krönecker delta, permits the right canonical factorization with the components $J(\mu)A_-(\mu)$ and $A_+(\mu)$ (see Theorem 2 in [3]). We define $\operatorname{ind} G \equiv \frac{1}{2\pi}\int_{-\infty}^{\infty} d \arg \det G(\mu)$ for $G \in V_N$. Since $\operatorname{ind}(I-A(\mu))=0$ (see [4]) and $\operatorname{ind} J(\mu)=0$, then $\operatorname{ind} \mathcal{V}=0$. We denote by $\mathcal{V}_1$ and $\mathcal{V}_2$ the factors in the product

$$\mathcal{V}(\mu) = \{I(\mu)\left[r(\mu)I - \|q_{kj}(\mu)\| - r(\mu)A^-(\mu)\right](i\mu+1)^{-R}\}\left\{\frac{(i\mu+1)^R}{r(\mu)}I\right\}.$$

Then

$$\operatorname{ind} \mathcal{V}_1 = -RN \quad . \tag{8}$$

Indeed, since all the zeros of $r(\mu)$ lie in the domain $\operatorname{Im}\mu<0$, then $\mathcal{V}_2 - I \in V_{N+}$, $\det \mathcal{V}_2 - 1 \in V_{1+}$. Furthermore, $\det \mathcal{V}_2(\mu)$ has RN zeros in the domain $\operatorname{Im}\mu>0$ and $\det \mathcal{V}_2(\mu) \to i^{RN}$ as $|\mu|\to\infty$. This means that $\operatorname{ind} \mathcal{V}_2 = RN$. The assertion (8) follows from the equality $\operatorname{ind}\mathcal{V} = \operatorname{ind}\mathcal{V}_1 + \operatorname{ind}\mathcal{V}_2$.

2. Now we show that $\mathcal{V}_1(\mu) \in V_{N-}$, $\det \mathcal{V}_1(\mu) \in V_{1-}$ and $\det \mathcal{V}_1(\mu)$ has exactly RN zeros in the domain $\operatorname{Im}\mu<0$. We decompose $A^{-1}(\mu)$ as $A^-(\mu) = A^-(0) + (J(\mu))^{-1}D(\mu)$. Here $D(\mu) \equiv -J(\mu)\{A^-(0) - A^-(\mu)\} \in V_{N-}$. Indeed,

$$J(\mu) = (I + Z(\mu))T \operatorname{diag}(\pi_1, \ldots, \pi_N),$$

where

$$T = \|\delta_{kj} + \delta_{1k}(1-\delta_{1j})\|;\ Z(\mu) = \left\|\frac{-i}{\mu}\delta_{1k}\delta_{1j}\right\|.$$

The multiplication on the left by $(I+Z(\mu))$ changes in the matrix $T \operatorname{diag}(\pi_1, \ldots, \pi_N)\{A^-(0) - A^-(\mu)\}$ only the first row the elements of which are

$$-\sum_{m=1}^{N} \pi_m p_{mj} \int_{-\infty}^{0} (e^{i\mu x} - 1)\, dF_{mj}(x) = i\mu \int_{-\infty}^{0} e^{i\mu x}\left(\sum_{m=0}^{N} \pi_m p_{mj} F_{mj}(x)\right) dx, \quad j \in \mathcal{N},$$

so that in the first row of the matrix $D(\mu)$

$$(i\mu+1)\int_{-\infty}^{0} e^{i\mu x}\left(\sum_{m=1}^{N}\pi_m p_{mj}F_{mj}(x)\right)dx, \qquad j\in\mathcal{N},$$

which, as $\mu\to 0$, tend to the limits

$$\sum_{m=1}^{N}\pi_m p_{mj}\int_{-\infty}^{0}F_{mj}(x)\,dx = -\sum_{m=1}^{N}\pi_m p_{mj}M\left(\xi_{mj}^{n}\right)^{-} < \infty.$$

Therefore, $D(\mu)\in V_{N-}$.

We will show that the elements of the matrix $J(\mu)\{r(\mu)I-\|q_{kj}(\mu)\|-r(\mu)A^{-}(0)\}$ are polynomials. The multiplication on the right by $I+Z(\mu)$ changes in the matrix of the polynomials

$$T\,\mathrm{diag}\,(\pi_1,\ldots,\pi_N)\{r(\mu)I-\|q_{kj}(r)\|-r(\mu)A^{-}(0)\} \tag{9}$$

only the first row, with zeros for $\mu=0$, since $\pi(I-A(0))=\pi(I-P)=0$. Hence the polynomials in the first row of the matrix (9) are divisible by $i\mu$.

The matrix $\mathcal{V}_1(\mu)$ permits the representation

$$\mathcal{V}_1(\mu)=J(\mu)\{r(\mu)I-\|q_{kj}(\mu)\|-r(\mu)A^{-}(0)\}$$
$$(i\mu+1)^{-R}-D(\mu)r(\mu)(i\mu+1]^{-R},$$

so that $\mathcal{V}_1(\mu)\in V_{N-}$ and $\det \mathcal{V}_1(\mu)\in V_{1-}$.

3. Let $A^{-}(\mu)=\|f_{kj}^{-}(\mu)\|$. Since, as $|\mu|\to\infty$ and for $\mathrm{Im}\,\mu\le 0$, $|f_{kj}^{-}(\mu)|\le p_{kj}$,

$$\det \mathcal{V}_1(\mu)\sim\prod_{m=1}^{N}\pi_m \det\left(I-A^{-}(\mu)\right)i^{-R\cdot N}$$

and, in addition, there exists a pair of subscripts (k_0,j_0) such that $|f_{k_0j_0}^{-}(\mu)|<p_{k_0j_0}$, then it follows from $\mathrm{ind}\,\mathcal{V}_1=-RN$ that $\det \mathcal{V}_1(\mu)$ has in the domain $\mathrm{Im}\,\mu<0$ exactly RN zeros which we denote by $\mu_1,\mu_2,\ldots,\mu_{RN}$. It is seen that the function $\det \{(I-A(\mu))r(\mu)\}$ has in the domain $\text{Im}\,\mu<0$ the same zeros as $\det V_1(\mu)$ has.

4. Since $\det V_1(\mu_1)=0$, there exists a column vector $c_1=(c_{11},\ldots,c_{1N})^T\neq 0$ such that $V_1(\mu_1)c_1=0$. Let j_1 be such that $c_{1j_1}\neq 0$, and let

$$a_1(\mu) = \{c_{1j_1}(\mu-\mu_1)\}^{-1}V_1(\mu)c_1 .$$

Furthermore, we replace in $V_1(\mu)$ the j_1^{th} column by $a_1(\mu)$. We find a new matrix $B_1(\mu)=V_1(\mu)C_1(\mu)$, where $C_1(\mu)$ results from the substitution of the vector $\{c_{1j_1}(\mu-\mu_1)\}^{-1}c_1$ for the jth column in I. The function $\det B_1(\mu)$ has in the domain $\text{Im}\,\mu<0$ exactly $RN-1$ zeros. Next, for a column vector $c_2=(c_{21},\ldots,c_{2N})^T\neq 0$ we have $B_1(\mu_2)c_2=0$; and we can find the matrix $C_2(\mu)$ from I by substituting the vector $\{c_{2j_2}(\mu-\mu_2)\}^{-1}c_2$ for the j_2^{th} column, where j_2 is such that $c_{2j_2}\neq 0$. Let $B_2(\mu)=B_1(\mu)C_2(\mu)$. Continuing this construction, we obtain the matrix $B_{RN}(\mu)=V_1(\mu)C_1(\mu),\ldots,C_{RN}(\mu)$ for which $\det B_{RN}(\mu)$ in the domain $\text{Im}\,\mu<0$ has no zeros, $B_{RN}(\mu)\in V_{N-}$.

5. Let $B(\mu)=\{C_1(\mu)\ldots C_{RN}(\mu)\}^{-1}$. We have $(r(\mu))^{-1}B(\mu)-I\in V_{N+}$, $\det\{(r(\mu))^{-1}B(\mu)\}-1\in V_{1+}$, $\det\{(r(\mu))^{-1}B(\mu)\to 1\}$ as $|\mu|\to\infty$. The form of the matrix $(r(\mu))^{-1}B(\mu)$ depends on the numeration of zeros of the function $\det V_1(\mu)$ for $\text{Im}\,\mu<0$ and on the choice of the subscripts j_n in the construction of the matrices $C_n(\mu)$. By Lemma 1 it is possible to find a set of matrices $C_1(\mu),\ldots,C_{RN}(\mu)$, such that $(r(\mu))^{-1}B(\mu)\to I$ as $\mu\to i\infty$, i.e.,

$(r(\mu))^{-1}B(\mu)=A_+(\mu)$. //

THEOREM 2. For $A_+(\mu)$ to be representable in the form

$$A_+(\mu) = R(\mu)(I-A(\mu))c ,$$

where $R(\mu)$ is a rational matrix, c is a constant, it is necessary and, if conditions A_2 and A_3 are satisfied, sufficient that $A^-(\mu)$ be a rational matrix. If in this case $A^-(\mu)=\frac{q^-_{kj}(\mu)}{r^-_{kj}(\mu)}$, where $\frac{q^-_{kj}(\mu)}{r^-_{kj}(\mu)}$ are irredundant relations of polynomials, $k,j\in N$, $r(\mu)$ is the least common factor of all $r^-_{kj}(\mu)$, R is the degree of the polynomial $r(\mu)$, then the function $\det\{(I-A(\mu))r(\mu)\}$ has in the domain $\operatorname{Im}\mu>0$ exactly $RN-1$ zeros and

$$R(\mu) = B(\mu)\|\delta_{kj}(1-\delta_{1j})\pi_j + \delta_{1k}\pi_j(i\mu)^{-1}\|r(\mu),$$

where $B(\mu)$ is a rational matrix. The methods for finding the matrix $B(\mu)$ are contained in the proof.

Proof. The proof of Theorem 2 follows closely the proof of Theorem 1. Hence we make it short.

Necessity. From the identity (1) we find for $x<0$ the equalities

$$p_{kj}F_{kj}(x) = a_{kj-}(x) - \sum_{m=1}^{N}\int_0^\infty a_{km-}(x-t)\,da_{mj+}(t), \qquad k,j\in\mathcal{N},$$

which, as in the proof of Theorem 1, implies the following.

Sufficiency. The matrix $A^-(\mu)$ permits the representation $A^-(\mu)=(r(\mu))^{-1}\|q_{kj}(\mu)\|$, where $q_{kj}(\mu)$ are polynomials. Denoting by V_1 and V_2 the factors in the product

$$\mathcal{V}(\mu) = \{(r(\mu))^{-1}(i\mu - 1)^R I\}\{J(\mu)[r(\mu)I - r(\mu)A^+(\mu) - \\ - \|q_{kj}(\mu)\|](i\mu - 1)^{-R}\},$$

we obtain ind $V_1 = -RN$, ind $V_2 = RN$. It is not hard to show that $V_2(\mu)-I\in V_{N+}$, $\det V_2(\mu)-1\in V_{1+}$. Therefore, by the argument principle the function $\det V_2(\mu)$ has in the domain $\mathrm{Im}\,\mu>0$ exactly RN zeros $\mu_0,\mu_1,\ldots,\mu_{RN-1}$. But Y det J($\mu$) is the unique zero $\mu_0=i$, so that $\det\{(I-A(\mu))\times r(\mu)\}$ has for $\mathrm{Im}\,\mu>0$ RN-1 zeros $\mu_0,\mu_1,\ldots,\mu_{RN-1}$. Following the proof of Theorem 1, we construct similarly the sequence of rational matrices $C_0(\mu)$, $C_1(\mu),\ldots$, $C_{RN-1}(\mu)$ such that the function $\det\{B(\mu)C_0(\mu)V_2(\mu)\}$, where $B(\mu)=C_{RN-1}(\mu)C_{RN-2}(\mu)\cdots C_1(\mu)$, has no zeros in the domain $\mathrm{Im}\,\mu>0$ and also $R(\mu)(I-A(\mu))-I\in V_{N+}$, $\det\{R(\mu)(I-A(\mu))\}-1\in V_{1+}$, $\det\{R(\mu)(I-A(\mu))c\}\to 1$ as $|\mu|\to\infty$, $\mathrm{Im}\,\mu>0$. By Lemma 1 the matrices $C_n(\mu)$ can be constructed such that $R(\mu)\times(I-A(\mu))c\to I$ as $\mu\to i\infty$. Theorem 2 is proved.

3. ON EXPLICIT FORMULAS FOR A DISTRIBUTION OF THE STATIONARY EXPECTATION TIME IN A QUEUEING SYSTEM

Consider now the random walk $\{w_n\}_{n\geq 0}$ defined by the recurrence formula

$$w_n = \max\left(0,\, w_{n-1} + \xi^n_{\varkappa_{n-1}\varkappa_n}\right), \qquad n \geqslant 1,$$

satisfying the arbitrary initial conditions κ_0 and w_0, where $\xi^n_{kj}=\tau^{s,e}_{kj}-\tau^{e,n}_{kj}$, $\tau^{s,n}_{kj}$ and $\tau^{e,n}_{kj}$ are independent non-negative random variables. The variable w_n can be interpreted as the waiting time of the nth call which has entered the system <SM,1,SM,1> with a semi-Markov flow of calls and the semi-Markov rate of service, from the moment of its entry until the moment of service. Here $\tau_{kj}^{s,n}$ denotes the service time of the nth call, $\tau_{kj}^{e,n}$ denotes the time between the arrivals of the $(n-1)^{th}$ call and of the nth call provided $\kappa_{n-1}=k$, $\kappa_n=j$.

It is not hard to show that there exists

$$\lim_{n\to\infty} \sum_{k\in\mathcal{N}} \pi_k P_k (w_n \leqslant x) = \sum_{k\in\mathcal{N}} \pi_k P(\tilde{\bar{Y}} \leqslant x \mid \tilde{\varkappa}_0 = k),$$

where $\tilde{\bar{Y}}=\sup_{n\geq 0} \tilde{Y}_n$ is the supremum of partial sums of the process $\{\tilde{Y}_n, \tilde{\kappa}_n\}_{n\geq 0}$ "converted" to $\{Y_n, \kappa_n\}$, with evolution matrix

$$\tilde{A}(\mu) = \operatorname{diag}(\pi_1^{-1}, \ldots, \pi_N^{-1}) A^T(\mu) \operatorname{diag}(\pi_1, \ldots, \pi_N)$$

(see [4], compare with [2 , Chapter 1]). Hence it is of interest to find explicit formulas for the distribution $\bar{Y}$.

Let

$$\tilde{A}(\mu) = \| \tilde{f}_{kj}(\mu) \|;$$

$$\tilde{f}_{kj}(\mu) = \mathrm{M}\left(e^{i\mu\xi_{jk}^n}, \tilde{\varkappa}_1 = j \mid \tilde{\varkappa}_0 = k\right);$$

$$\tilde{f}_{kj+}(\mu) = \mathrm{M}\left(e^{i\mu\tau_{jk}^{s,n}}, \tilde{\varkappa}_1 = j \mid \tilde{\varkappa}_0 = k\right);$$

$$\tilde{f}_{kj-}(\mu) = \mathrm{M}\left(e^{-i\mu\tau_{jk}^{e,n}}, \tilde{\varkappa}_1 = j \mid \tilde{\varkappa}_0 = k\right); \qquad k, j \in \mathcal{N}.$$

Since $\tau_{kj}^{s,n}$ and $\tau_{kj}^{e,n}$ are independent, we have

$$\tilde{f}_{kj}(\mu) = \tilde{f}_{kj+}(\mu) \tilde{f}_{kj-}(\mu), \ \tilde{f}_{kj\pm}(\mu) \in V_{1\pm}; \qquad k, j \in \mathcal{N}.$$

If $\tilde{f}_{kj+}(\mu)$ is a rational function, there exists a purely imaginary pole μ_{kj}^{Γ} (principal pole) having the property that $\operatorname{Im} \mu_{kj,m} \leq \operatorname{Im} \mu_{kj}^{\Gamma}$, $m=1,2,\ldots,r$, where $\mu_{k,j},\ldots,\mu_{kj,r}$ are all the poles of $\tilde{f}_{kj+}$.

In our case Lemma 1 from Section 19 in the monograph [2] has the following form.

LEMMA 2. If $\tilde{f}_{kj+}(\mu)=\frac{q_{kj}(\mu)}{r_{kj+}(\mu)}$ is the irreducible relation of polynomials, $k,j \in N$, then $\tilde{A}^{+}(\mu)$ is a rational matrix

$$\tilde{A}^{+}(\mu) = \left\| \frac{p_{kj}(\mu)}{r^{+}_{kj}(\mu)} \right\|,$$

where

$$r^{+}_{kj}(\mu) = r_{kj+}(\mu) \left\{ \prod_{m=1}^{l_{kj}} (\mu - \mu_{kj,m}) \right\}^{-1}; (\mu_{kj,1}, \ldots, \mu_{kj,l_{kj}})$$

is the intersection of the set of zeros of the function $r_{kj+}(\mu)$ and the set of zeros $\tilde{f}_{kj-}(\mu)$ (taking into account the multiplicity); the point μ^{Γ}_{kj} does not belong to this intersection.

REMARK. A modification of the example cited on page 148 in the monograph [2] shows that the rationality of $\tilde{A}^{(+)}(\mu)$ does not imply in general the rationality of all the functions $\tilde{f}_{kj+}(\mu)$. The rationality of $\tilde{A}^{+}(\mu)$ implies the rationality of $\tilde{f}_{kj+}(\mu)$, if $\tilde{f}_{kj-}(\mu)$ has a finite number of zeros in the lower half-plane, $k,j \in N$.

As a corollary of Lemma 2 and that of Theorem 1, we obtain the following result.

THEOREM 3. Let conditions A_2, A_3 be satisfied. For $\tilde{A}_{+}(\mu)$ to be a rational matrix, it is sufficient that the functions $\tilde{f}_{kj}(\mu)$, $k,j \in N$, be rational. If $\tilde{f}_{kj+}(\mu)$, $\frac{q_{kj}(\mu)}{r_{kj+}(\mu)}$ is an irreducible relation of polynomials,

$$r^{+}_{kj}(\mu) = r_{kj+}(\mu) \left\{ \prod_{m=1}^{l_{kj}} (\mu - \mu_{kj,m}) \right\}^{-1}; (\mu_{kj,1}, \ldots, \mu_{kj,l_{kj}})$$

is the intersection of the set of zeros $r_{kj+}(\mu)$ and of the set of zeros $\tilde{f}_{kj-}(\mu)$ (taking into account the multiplicity); $k,j \in N$; $r(\mu)$ is the least common multiple of all $r^{+}_{kj}(\mu)$; R is the power of the polynomial $r(\mu)$, then the function $\det\{(I-A(\mu))r(\mu)\}$ has in the domain $\operatorname{Im}\mu<0$ exactly RN zeros and $\tilde{A}_{+}(\mu)=(r(\mu))^{-1}B(\mu)$, where $B(\mu)$ is the matrix of the polynomials to find which is possible according to the method contained in the proof of Theorem 1.

The rationality of the vector $\|M((e^{i\mu\bar{\bar{Y}}}|\tilde{\kappa}_0=k)\|$, does not, in general, imply the rationality of all $\tilde{f}_{kj+}(\mu)$, as is seen from Assertion 2 of Theorem 1 and Remark following Lemma 2.

From Theorem 2 and an analog of Lemma 2 we have

THEOREM 4. Let conditions A_2, A_3 be satisfied. In order that $\tilde{A}_{+}(\mu)$ be representable as $\tilde{A}_{+}(\mu)=R(\mu)(I-\tilde{A}(\mu))c$, where $R(\mu)$ is a rational matrix, c is a constant, it is sufficient that the functions $\tilde{f}_{kj-}(\mu)$, $k,j \in N$, be rational.

The author is grateful to A.A. Borovkov for the statement of the problem and his valuable comments.

REFERENCES

[1] K. Arndt. "On Properties of Boundary Functionals of a Random Walk on a Markov Chain." *Math. Operations forsch. und Statist., ser. Statistics,* 12, 1 (1981): 85-100.

[2] A.A. Borovkov. *Stochastic Processes in Queueing Theory.* New York: Springer-Verlag, 1976.

[3] A.A. Borovkov. "Stability Theorems and Estimates of the Rate of Convergence of the Components of Factorizations for Walks Defined on Markov Chains." *Theory Prob. Applications*, 25 (1980): 325-334.

[4] E. Pressman. "Metody faktorizatsii i granichnaya zadacha dlya summ sluchainykh velichin, zadannykh na tsepi Markova" (Methods of Factorization and Boundary Value Problems For Sums of Random Variables Given on a Markov Chain). *Izv. AN SSSR, ser. Matem.*, 33, 4 (1969): 861-900. (In Russian.)

ON A PROBLEM OF MARTINGALES ON A PLANE

V.M. Borodikhin

In [1], Strook and Varadhan formulated and solved the so-called problem of martingales, equivalent to the problem of existence and weak uniqueness of a solution of the stochastic differential equation

$$x(t) = x + \int_s^t b(u, x(u))\, du + \int_s^t \sigma(u, x(u))\, d\beta(u), \tag{1}$$

where β is the Brownian motion process on the line, the coefficient b is bounded and measurable, and σ is bounded and continuous.

In [2], I.I. Gikhman and Pyasetskaya prove the existence of a solution of a stochastic differential equation which is an analog of Eq. (1) for two-parameter random processes. In this article we introduce an analog of a solution of the problem of martingales for two-parameter processes, prove the analogs of several assertions given in [1], in particular, establish the equivalence between specifying the measure as a solution of the martingale problem and that of the stochastic differential equation considered in [2]. The results of [2] imply therefore the existence of a solution of the two-parameter problem of martingales in the case of a diffusion coefficient which is continuous and separated from zero and infinity. Another formulation of the martingale problem is handled by Tudor in [3].

1. TWO-DIMENSIONAL MARKOV TIMES

We denote by (Ω, F) a measurable space where $\Omega=C(R_+^2, R^d)$ is the space of continuous functions with values in R^d, defined on the set $R_+^2=[0,\infty)\times[0,\infty)$; F denotes the least σ-algebra with measurable mappings $x_z:\Omega\to R^d$, $z\in R_+^2$, $x_z(\omega)=\omega(z)$. In the sequel we denote the value of the trajectory $\omega\in\Omega$ at the point z by $x_z(\omega)$ or $x(z,\omega)$, omitting sometimes the index ω; Ω is regarded as a complete separable metric space with a topology of uniform convergence on the compacts of R_+^2.

We introduce on R_+^2 the following relations, one of which specifies the array of R_+^2. Let $z=(s,t)$, $z'=(s',t')\in R_+^2$. Then, by definition, $z\le z'$ if $s\le s'$ and $t\le t'$; $z\prec z'$ if $s\le s'$ or $t\le t'$. Let z_0, $z_1\in R_+^2$, $z_0\le z_1$. Then,

$F_{z_1}^{z_0}=\sigma\{x_z(\cdot);\ z_0\le z\le z_1\}$ denotes the least σ-algebra with respect to which the functions $x_z(\cdot)\,z_0\le z\le z_1$ are measurable. If $z_1=(\infty,\infty)$ then the σ-algebra is written as F^{z_0}.

Similarly,

$$\mathcal{G}_{z_1}^{z_0}=\sigma\{x_z(\cdot);\ z_0\leqslant z<z_1\}.$$

Also, let

$$\mathcal{F}_{s\infty}^{z_0}=\sigma\left\{\bigcup_{t_0\leqslant t<\infty}\mathcal{F}_{(s,t)}^{z_0}\right\};\quad \mathcal{F}_{\infty t}^{z_0}=\sigma\left\{\bigcup_{s_0\leqslant s<\infty}\mathcal{F}_{(s,t)}^{z_0}\right\}$$

be σ-algebras generated by families of sets given in the braces. For some $z_0\in R_+^2$ we write $I_{z_0}=\{z\in R_+^2;\ z_0\le z\}$. We call the function $\zeta:\Omega\to R_+^2$ a Markov z_0-time if $z_0\le\zeta(\omega)$ for all $\omega\in\Omega$ and for any $z\in I_{z_0}$ $\{\omega\in\Omega:\ \zeta(\omega)\le z\}\in F_z^{z_0}$. For each Markov z_0-time we can define the following algebras:

$$\mathcal{F}_\zeta^{z_0}=\left\{A\in\mathcal{F}^{z_0}:A\cap(\zeta\leqslant z)\in\mathcal{F}_z^{z_0}\quad\text{for each}\quad z\in I_{z_0}\right\};$$

$$\mathcal{G}_\zeta^{z_0}=\left\{A\in\mathcal{F}^{z_0}:A\cap(\zeta\leqslant z)\in\mathcal{G}_z^{z_0}\quad\text{for each}\quad z\in I_{z_0}\right\}.$$

We shall prove next two theorems to generalize the result obtained by Strook and Varadhan [Theorem 0.1 in [1]], to the case of two-parameter sample trajectories.

<u>THEOREM 1</u>. Let P be a probability measure on (Ω,F^{z_0}), $z_0\in R_+^2$, and let ζ be a Markov z_0-time. Then there exists a function $Q(\omega,A)$ such that

I. $Q(\omega,A)$ is a probability measure on (Ω,F^{z_0}) for each $\omega\in\Omega$;

II. $Q(\omega,A)$ is $G_\zeta^{z_0}$-measurable for each $A\in F^{z_0}$;

III. $Q(\omega,A_\omega)=1$, where

$$A_\omega=\{\omega':x(z,\omega')=x(z,\omega)\quad\text{for}\quad z_0\leqslant z<\zeta(\omega)\};$$

IV. $Q(\omega, A)=P(A/G_{\zeta}^{z_0})$ P-almost surely (a.s.).

We call the function $Q(\omega, A)$, as usual, a conditional regular distribution with respect to measure P, corresponding to the σ-algebra $G_{\zeta}^{z_0}$ (see [4]).

THEOREM 2. This theorem is formulated as Theorem 1 with the only exception that instead of $G_{\zeta}^{z_0}$ we take $F_{\zeta}^{z_0}$ and instead of A_{ω} we take $B_{\omega}=\{\omega': x(z,\omega')=x(z,\omega) \text{ for } z_0\le z\le\zeta(\omega)\}$.

Proof. To prove the Theorem, we introduce a certain type of stopping, preserving the continuity of the trajectory:

$$x_z^{z'} = \begin{cases} x_z(\omega) \text{ if } z \nless z', \\ x_{(s',t)}+x_{(s,t')}-x_{(s',t')} \quad \text{otherwise;} \end{cases}$$

or, in other words,

$$x_z^{z'} = \begin{cases} x_z(\omega), \text{ if } z'\not\le z, \\ x_{(s',t)}+x_{(s,t')}-x_{(s',t')}, \text{ if } z'\le z. \end{cases}$$

It is easy to see that for fixed $z'\in R_+^2$ and $\omega\in\Omega$ the function $x_z^{z'}(\omega)$ is continuous in z, and for fixed z and ω continuous in z'.

LEMMA 1. Let ζ be a Markov z_0-time. Then for any $z\in I_{z_0}$ the mapping $\omega\to x_z^{\zeta(\omega)}(\omega)$ is measurable.

Proof. Let $U\subset R^d$ be an arbitrary open set. It suffices to show that

$$\{\omega: x_z^{\zeta(\omega)}(\omega)\in U\}\cap(\zeta\leqslant z')\in \mathcal{G}_{z'}^{z_0} \tag{2}$$

for each $z'\in I_{z_0}$. By rectangles we mean subsets R_+^2 of the type $(z_1,z_2]=(s_1,s_2]\times(t_1,t_2]$ or $[z_1,z_2]\neq[s_1,s_2]\times[t_1,t_2]$, where $z_1=(s_1,t_1)\le z_2=(s_2,t_2)$. It is not hard to see that if

$z_1, z_2 \in I_{z_0}$, $z_1 \le z_2$, then

$$\{\omega: \zeta \in (z_1, z_2]\} \in G^{z_0}_{z_2} ; \qquad \{\omega: \zeta \in [z_1, z_2]\} \in G^{z_0}_{z_2} .$$

We denote by $\{D_{m,n}\}^{\infty}_{m,n=1}$ the countable family of rectangles with the following properties

a) $D_{m,n} \cap D_{m,k} = \varnothing$, $n \neq k$;

b) $\bigcup\limits_{n=1}^{\infty} D_{m,n} = \{z : z_0 \leqslant z \leqslant z'\}$;

c) $\operatorname{diam}(D_{m,n}) \leqslant 1/m$.

Then, obviously, $(\zeta \in D_{m,n}) \in G^{z_0}_{z'}$ for each m, n. By the continuity of $x^{z'}_z \omega$ in z' we have the relation

$$\{x^{\zeta}_z \in U\} \cap (\zeta \leqslant z') = \bigcup_{k=1}^{\infty} \bigcap_{m=k}^{\infty} \bigcup_{n=1}^{\infty} \{x^{\zeta}_z \in U\} \cap (\zeta \in D_{m,n}) =$$

$$= \bigcup_{k=1}^{\infty} \bigcap_{m=k}^{\infty} \bigcup_{n=1}^{\infty} \left\{x^{z_{m,n}}_z \in U\right\} \cap (\zeta \in D_{m,n}), \tag{3}$$

where $z_{m,n}$ is one of the vertices of $D_{m,n}$. Since

$$\{x^{z_1}_z \in U\} = \{x_z \in U;\ z < z_1\} \cup \{x_{(s_1,t)} + x_{(s,t_1)} - x_{(s_1,t_1)} \in U\};$$

$$\{z_1 \leqslant z\};\ \{x_z \in U\} \in \mathcal{F}^{z_0}_z \subset \mathcal{G}^{z_0}_{z'} \quad \text{for} \quad z < z_1,\ z_1 \leqslant z';$$

$$\{x_{(s_1,t)} + x_{(s,t_1)} - x_{(s_1,t_1)} \in U\} \in \mathcal{G}^{z_0}_{z_1} \subset \mathcal{G}^{z_0}_{z'} \quad \text{for} \quad z_1 \leqslant z,\ z_1 \leqslant z',$$

then $\{x^{z_{m,n}}_z \in U\} \in G^{z_0}_{z'}$ for all $m,n \ge 1$. The relation (3) implies now (2) and the assertion of Lemma 1.

The next lemma is similar to Lemma 0.1 of [1].

<u>LEMMA 2</u>. Let $z_0 \in R^2_+$ and let ζ be a Markov z_0-time. Let

$$\mathfrak{B}^{z_0}_{\zeta} = \{C : C = \{\omega \in \Omega : x^{\zeta}_{z_1}(\omega) \in \Gamma_1,\ x^{\zeta}_{z_2}(\omega) \in \Gamma_2, \ldots, x^{\zeta}_{z_k}(\omega) \in \Gamma_k\};\ z_i \in I_{z_0},\ \Gamma_i \in \mathcal{B}_{R^d},\ i = \overline{1, k},\ k = 1, 2, \ldots\}.$$

Let $\mathcal{B}_E$ be the σ-algebra of Borel subsets of E. Then

$$\mathcal{G}^{z_0}_{\zeta} = \sigma(\mathfrak{B}^{z_0}_{\zeta}).$$

Proof. The proof of Lemma 2 differs only in minor details from that of Lemma 0.1 of [1].

The proof of Theorem 1 can be obtained now through an obvious modification of the considerations in the proof of Lemma 0.2 and Theorem 0.2 of [1], if we replace Ω^s, ϕ^s, τ^s, $\Omega^s(\tau^s)$, ϕ^s_τ in the notations (1) respectively by:
$\Omega^{z_0}=G(I_{z_0},R^d)$; $\varphi^{z_0}:\Omega\to\Omega^{z_0}$ is such that if $\omega^{z_0}=\phi^{z_0}(\omega)$, then $x(z,\omega^{z_0})=x(z,\omega)$ $\forall z\in I_{z_0}$; if ζ is a Markov z_0-time, then $\zeta^{z_0}:\Omega\to R^2_+$ is such that

$$\zeta^{z_0}(\varphi^{z_0}(\omega))=\zeta(\omega)\ \forall\omega\in\Omega;\ \Omega^{z_0}(\zeta)=$$
$$=\{\omega^{z_0}\in\Omega^{z_0}: x(z,\omega^{z_0})=$$
$$=x_z^{\zeta^{z_0}(\omega^{z_0})}\ \forall z\in I_{z_0}\};$$

$\phi^{z_0}_\zeta:\Omega\to\Omega^{z_0}(\zeta)$ is such that $x(z,\phi^{z_0}_\zeta(\omega))=x^\zeta_z(\omega)$ $\forall z\in I_{z_0}$.

REMARK 1. Theorem 1 still holds if the Markov z_0-time is defined by the following: $\forall z\in I_{z_0}$ $\{\omega:\ \zeta(\omega)\leq z\}\in G^{z_0}_z$ and $\{\omega:\ \zeta(\omega)< z\}\in G^{z_0}_z$.

REMARK 2. The proof of Theorem 2 is similar to that of Theorem 1, if for the stopping instead of $x^{z'}_z(\omega)$ we take $x_{z\wedge z'}(\omega)$, where $z\wedge z'=(\min(s,s'),\min(t,t'))$.

2. PROPERTIES OF EXPONENTIAL MARTINGALES

For the function $\theta(z,\omega)$ defined on $I_{z_0}\times\Omega$ with values in R^d or $R^d\otimes R^d$, we introduce the following condition:

(z_0). The mapping $(z,\omega)\to\theta(z,\omega)$ is measurable with respect to $\mathscr{B}_{I_{z_0}}\times\mathscr{F}^{z_0}$ and for the fixed $z\in I_{z_0}$ the mapping

$\omega \to \theta(z,\omega)$ is $F_z^{z_0}$-measurable.

Let the function $a: I_{z_0} \times \Omega \to R^d \otimes R^d$ satisfy the condition (z_0) and let it be bounded: $\exists A>0$ such that

$$0 \le \langle a\lambda,\lambda\rangle \le A|\lambda^2| \quad \text{for all} \quad \lambda \in R^d . \tag{4}$$

Let $z=(s,t) \le z'=(s',t')$, $B=(z,z']$. For the function $\theta(z)$ given on R_+^2, let

$$\theta(B) = \theta(z, z'] = \theta(s', t') - \theta(s, t') - \theta(s', t) + \theta(s, t).$$

The function $\theta(B)$ defined on rectangles of the form $(z,z']$, is additive; we extend this function with respect to the additivity to the set of finite combinations of rectangles of this particular form.

The random process $\{\theta(z,\omega),\ z \in I_{z_0}\}$ with values in R^d, given on a probability space (Ω, F^{z_0}, P) and adapted to the family of σ-algebras $(F_z^{z_0},\ z \in I_{z_0})$, is said to be a strong exponential (or e-) martingale associated with the function $a(\cdot)$, if this process satisfies the condition

$$\mathbf{E}\{X_\lambda^z(z')/\mathcal{G}_z^{z_0}\} = 1 \ \mathbf{P}\text{-a.s.} \tag{5}$$

for all $\lambda \in R^d$; $z, z' \in I_{z_0}$, $z \le z'$, where

$$X_\lambda^z(z') = \exp\left\{\langle\lambda, \theta(z,z']\rangle - \frac{1}{2}\left\langle\lambda, \int_{(z,z']} a(u)\lambda du\right\rangle\right\}. \tag{6}$$

<u>LEMMA 3</u>. Let the function $a(z,\omega)$ satisfy the conditions (z_0) and (4); let $\{\theta(z,\omega),\ z \in I_{z_0}\}$ be a separable e-martingale associated with $a(\cdot)$. Then for any $z_1, z_2 \in I_{z_0}$, $z_1 \le z_2$, and for any $r>0$

$$\mathbf{P}\left\{\sup_{z\in(z_1,z_2]}|\theta(z_1,z]|\geq r\right\}\leq\frac{4de}{e-1}\exp\left\{-\frac{r^2}{2d(1+c)A|(z_1,z_2]|}\right\},\tag{7}$$

where

$$c=1/e+1/e^2,\ |(z_1,z_2]|=(s_2-s_1)(t_2-t_1).$$

Proof. We note that for fixed $s\geq s_1$ $\{X_\lambda^{z_1}(s,t),\ F_{\infty t}^{z_0},\ t\geq t_0\}$ is a P-martingale, and for fixed $t\geq t_1$ $\{X_\lambda^{z_1}(s,t),\ F_{s\infty}^{z_0},\ s\geq s_1\}$ is a P-martingale. Indeed, let $t'\geq t\geq t_1$; then P-a.s.

$$\mathbf{E}\{X_\lambda^{z_1}(s,t')/\mathscr{F}_{\infty t}^{z_0}\}=\mathbf{E}\Big\{\exp\Big\{\langle\lambda,\theta(z_1,z]+\theta((s_1,t),(s,t')]\rangle-$$
$$-\frac{1}{2}\Big\langle\lambda,\int_{(z_1,z]}a(u)\cdot\lambda du+\int_{((s_1,t),(s,t')]}a(u)\cdot\lambda du\Big\rangle\Big\}\Big/\mathscr{F}_{\infty t}^{z_0}\Big\}=$$
$$=X_\lambda^{z_1}(s,t)\,\mathbf{E}\{X_\lambda^{(s_1,t)}(s,t')/\mathscr{F}_{\infty t}^{z_0}\}=$$
$$=X_\lambda^{z_1}(s,t)\cdot\mathbf{E}\Big\{\mathbf{E}\Big\{X_\lambda^{(s_1,t)}(s,t')/\mathscr{G}_{(s_1,t)}^{z_0}\Big\}\Big/\mathscr{F}_{\infty t}^{z_0}\Big\}=X_\lambda^{z_1}(s,t).$$

Applying now the Doob inequality [Chapter VII of [5]] in the way it was done, for example, in [6], we can show that

$$r\mathbf{P}\left\{\sup_{(s,t)\in(z_1,z_2]}X_\lambda^{z_1}(s,t)\geq r\right\}\leq$$
$$\leq\frac{e}{e-1}+\frac{e}{e-1}\sup_{(s,t)\in(z_1,z_2]}\mathbf{E}\{X_\lambda^{z_1}(s,t)\ln^+X_\lambda^{z_1}(s,t)\}.\tag{8}$$

We note that $\inf\{\alpha>0:\ e^{\alpha x}\geq x^+\ \forall x\in R\}=1/e$. This implies that $\ln^+x\leq x^{1/e}\ \forall_{x>0}$. Hence (8) implies

$$r\cdot\mathbf{P}\left\{\sup_{(s,t)\in(z_1,z_2]}X_\lambda^{z_1}(s,t)\geq r\right\}\leq\frac{e}{e-1}\left\{1+\sup_{(s,t)\in(z_1,z_2]}\mathbf{E}\{X_\lambda^{z_1}(s,t)\}^{1+1/e}\right\}.\tag{9}$$

Next,

$$\sup_{(s,t)\in(z_1,z_2]} \mathbf{E}\{X_\lambda^{z_1}(s,t)\}^{1+1/e} = \sup_{(s,t)\in(z_1,z_2]} \mathbf{E}\Big\{\exp\Big[\langle\lambda(1+1/e),\,\theta(z_1,z)]\rangle -$$

$$-\frac{1}{2}\Big\langle\lambda(1+1/e),\ \int_{(z_1,z]} a(u)\lambda(1+1/e)\,du\Big\rangle + \frac{1}{2}(1/e+1/e^2)\Big\langle\lambda,$$

$$\int_{(z_1,z]} a(u)\lambda du\Big\rangle\Big]\Big\} \leqslant \exp\Big\{\frac{1}{2}cA|\lambda|^2|(z_1,z_2]|\Big\}. \qquad (10)$$

Noting (9) and (10), we have

$$\mathbf{P}\Big\{\sup_{z\in(z_1,z_2]}\Big\langle\frac{\lambda}{|\lambda|},\ \theta(z_1,z]\Big\rangle \geqslant r\Big\} \leqslant$$

$$\leqslant \mathbf{P}\Big\{\sup_{z\in(z_1,z_2]} X_\lambda^{z_1}(z) \geqslant \exp\Big[r|\lambda| - \frac{|\lambda|^2}{2}A|(z_1,z_2]|\Big]\Big\} \leqslant$$

$$\leqslant \frac{e}{e-1}\Big(1+\exp\Big[\frac{1}{2}cA|\lambda|^2|(z_1,z_2]|\Big]\Big)\cdot\exp\Big[-r|\lambda| +$$

$$+\frac{1}{2}A|\lambda|^2\cdot|(z_1,z_2]|\Big] = \frac{e}{e-1}\Big\{\exp\Big[-r|\lambda| + \frac{1}{2}A|\lambda|^2\cdot|(z_1,z_2]|\Big] +$$

$$+\exp\Big[-r|\lambda| + \frac{1}{2}(1+c)A|\lambda|^2|z_1,z_2]|\Big]\Big\}. \qquad (11)$$

Let $|\lambda| = \frac{r}{(1+c)A|(z_1,z_2]|}$. Then (11) implies

$$\mathbf{P}\Big\{\sup_{z\in(z_1,z_2]}\Big\langle\frac{\lambda}{|\lambda|},\ \theta(z_1,z]\Big\rangle \geqslant r\Big\} \leqslant \frac{2e}{e-1}\exp\Big\{-\frac{r^2}{2(1+c)A|(z_1,z_2]|}\Big\}.$$

This, in turn, implies (7). //

<u>LEMMA 4</u>. Under the conditions of Lemma 3 for any $\lambda_0>0$, $k\geq 0$ the family of random variables

$$\{|\theta(z_1,z]|^k X_\lambda^{z_1}(z)\}_{|\lambda|\leqslant\lambda_0,\ z\in(z_1,z_2]}$$

is uniformly integrable.

<u>Proof</u>. The proof of Lemma 4 follows from Lemma 2 of [3] and from the following inequalities:

$$\mathbf{P}\left\{\sup_{z\in(z_1,z_2]} |\theta(z_1,z_2]|^h X_\lambda^{z_1}(z) \geq r^h \exp(\lambda_0 r)\right\} \leq$$

$$\leq \mathbf{P}\left\{\sup_{z\in(z_1,z_2]} |\theta(z_1,z]|^h \exp[\langle\lambda,\theta(z_1,z]\rangle] \geq r^h \exp(\lambda_0 r)\right\} \leq$$

$$\leq \mathbf{P}\left\{\sup_{z\in(z_1,z_2]} |\theta(z_1,z_2]| \geq r\right\} \leq \frac{4de}{e-1}\exp\left\{-\frac{r^2}{4d(1+c)A|(z_1,z_2]|}\right\}.$$

COROLLARY. Under the conditions of Lemma 3, $\forall\ \lambda\in R^d$:

I) $\mathbf{E}\left\{\langle\lambda,\theta(z_1,z_2]\rangle/\mathcal{G}_{z_1}^{z_0}\right\} = 0$ $\mathbf{P}$-a.s. ;

II) $\mathbf{E}\left\{\langle\lambda,\theta(z_1,z_2]\rangle^2 - \left\langle\lambda, \int_{(z_1,z_2]} a(u)\cdot\lambda du\right\rangle \Big/ \mathcal{G}_{z_1}^{z_0}\right\} = 0$ $\mathbf{P}$-a.s. ;

III) $\mathrm{E}|\theta(z_1,z_2]|^4 \leq C|(z_1,z_2]|^2$, where C depends only on A.

LEMMA 5. Let $\{\theta(z,\omega),\ z\in I_{z_0}\}$ be P-a.s. a continuous strong e-martingale associated with $a(\cdot)$. Then, for any two Markov z_0-times ζ_1, ζ_2 such that for some constant M

$$z_0 \leq \zeta_1 \leq \zeta_2 \leq (M-1,\ M-1),$$

we have

$$\mathbf{E}\left\{X_\lambda^{\zeta_1}(\zeta_2)/\mathcal{G}_{\zeta_2}^{z_0}\right\} = 1 \qquad \mathbf{P}\text{-a.s.}$$

Proof. We prove Lemma 5 following along the lines of Theorem 3.1 of [1]. For $z=(s,t)$, $n\in N$ we let $z^{(n)}=(s^{(n)},t^{(n)})$, where

$$s^{(n)} = ([ns]+1)/n, \qquad t^{(n)} = ([nt]+1)/n .$$

Then, $\zeta_1^{(n)} \leq \zeta_2^{(n)} \leq (M,M)$, and, using an assertion similar to Theorem 34 in Chapter IV of [7], it is possible to show that $\zeta_i^{(n)}$ are Markov z_0-times.

Let $A\in G_{\zeta_1}^{z_0}$. Also, let

$$A^{k,l} = A \cap \{\zeta_1^{(n)} = (k/n,\, l/n)\} \in \mathcal{G}^{z_0}_{(k/n,l/n)};$$

$$A^{k,l}_{p,q} = A^{k,l} \cap \{\zeta_2^{(n)} = (p/n,\, q/n)\} \in \sigma\left\{\mathcal{G}^{z_0}_{(k/n,l/n)} \cup \mathcal{F}^{z_0}_{(p/n,q/n)}\right\};$$

$$B^{k,l}_{p,q} = A^{k,l} \cap \{(p/n,\, q/n) \leqslant \zeta_2^{(n)}\} \in \mathcal{G}^{z_0}_{(p/n,q/n)}.$$

Here for each $n \in N$ $(k/n, \ell/n) = z_{k,\ell}$, $(p/n, q/n) = z_{p,q}$ run through the rectangle $(z_0, (M,M)]$, $z_{k,\ell} \le z_{p,q}$. If A is a rectangle of the form $(z, z']$ or the sum of finitely many such rectangles, we write

$$X_\lambda(A) = \exp\left\{\langle \lambda, \theta(A)\rangle - \frac{1}{2}\left\langle \lambda, \int_A a(u)\lambda\, du\right\rangle\right\}.$$

For each (k, ℓ) we have

$$\int_{B^{k,l}_{k,l}} 1\, d\mathbf{P} = \int_{B^{k,l}_{k,l}} X_\lambda^{z_{k,l}}(z_{k+1,l+1})\, d\mathbf{P} = \int_{A^{k,l}_{k+1,l+1}} X_\lambda^{\zeta_1^{(n)}}(\zeta_2^{(n)})\, d\mathbf{P} +$$

$$+ \int_{B^{k,l}_{k,l} \setminus A^{k,l}_{k+1,l+1}} X_\lambda^{z_{k,l}}(z_{k+1,l+1})\, d\mathbf{P} = \int_{A^{k,l}_{k+1,l+1}} X_\lambda^{\zeta_1^{(n)}}(\zeta_2^{(n)})\, d\mathbf{P} +$$

$$+ \int_{B^{k,l}_{k,l} \setminus A^{k,l}_{k+1,l+1}} X_\lambda^{z_{k,l}}(z_{k+1,l+2})\, d\mathbf{P} = \int_{A^{k,l}_{k+1,l+1} \cup A^{k,l}_{k+1,l+2}} X_\lambda^{\zeta_1^{(n)}}(\zeta_2^{(n)})\, d\mathbf{P} +$$

$$+ \int_{B^{k,l}_{k,l} \setminus \left(A^{k,l}_{k+1,l+1} \cup A^{k,l}_{k+1,l+2}\right)} X_\lambda^{z_{k,l}}(z_{k+1,l+2})\, d\mathbf{P} = \ldots = \int_{\bigcup_{j \geqslant 1} A^{k,l}_{k+1,l+j}} X_\lambda^{\zeta_1^{(n)}}(\zeta_2^{(n)})\, d\mathbf{P} +$$

$$+ \int_{B^{k,l}_{k+1,l}} X_\lambda^{z_{k,l}}\left(\frac{k+1}{n},\, M\right) d\mathbf{P} = \int_{\bigcup_{j \geqslant 1} A^{k,l}_{k+1,l+j}} X_\lambda^{\zeta_1^{(n)}}(\zeta_2^{(n)})\, d\mathbf{P} +$$

$$+ \int_{B^{k,l}_{k+1,l}} X_\lambda\left(\left(z_{k,l}, \left(\frac{k+1}{n},\, M\right)\right] \cup (z_{k+1,l},\, z_{k+2,l+1}]\right) d\mathbf{P} =$$

$$= \int_{\bigcup_{j \geqslant 1} A^{k,l}_{k+1,l+j}} X_\lambda^{\zeta_1^{(n)}}(\zeta_2^{(n)})\, d\mathbf{P} + \int_{A^{k,l}_{k+2,l+1}} X_\lambda^{z_{k,l}}(z_{k+2,l+1})\, d\mathbf{P} +$$

$$+\int_{B^{k,l}_{k+1,l}\setminus A^{k,l}_{k+2,l+1}} X_\lambda\left(\left(z_{k,l},\left(\frac{k+1}{n},M\right)\right]\cup(z_{k+1,l},z_{k+2,l+1}]\right)d\mathbf{P}=$$

$$=\ldots=\int_{\bigcup\limits_{j\geq 1,i\geq 1}A^{k,l}_{k+i,l+j}} X_\lambda^{\zeta_1^{(n)}}(\zeta_2^{(n)})\,d\mathbf{P}=\int_{A^{k,l}} X_\lambda^{\zeta_1^{(n)}}(\zeta_2^{(n)})\,d\mathbf{P}.$$

Summing up in (k,ℓ), we obtain

$$\int_A 1d\mathbf{P}=\int_A X_\lambda^{\zeta_1^{(n)}}(\zeta_2^{(n)})\,d\mathbf{P}, \tag{12}$$

and therefore

$$\mathbf{E}\left\{X_\lambda^{\zeta_1^{(n)}}(\zeta_2^{(n)})/\mathcal{G}^{z_0}_{\zeta_1}\right\}=1 \qquad \text{P-a.s.}$$

Since

$$\mathbf{P}\left\{\sup_n|\theta(\zeta_1^{(n)},\zeta_2^{(n)}]|\geq r\right\}\leq\mathbf{P}\left\{\sup_{z_0\leq z\leq z'\leq(M,M)}|\theta(z,z']|\geq r\right\}\leq$$

$$\leq 4\mathbf{P}\left\{\sup_{z_0\leq z\leq(M,M)}|\theta(z_0,z]|\geq r/4\right\}\leq\frac{16de}{e-1}\exp\left\{-\frac{r^2}{32d(1+c)A|(z_0,(M,M)]|}\right\},$$

then $\left\{X_\lambda^{\zeta_1^{(n)}}(\zeta_2^{(n)})\right\}_{n=1}^\infty$ are uniformly integrable; hence in (12) we can pass to the limit in n, which implies

$$\mathbf{E}\left\{X_\lambda^{\zeta_1}(\zeta_2)/\mathcal{G}^{z_0}_{\zeta_1}\right\}=1 \qquad \text{P-a.s.} \qquad //$$

COROLLARY. Let ζ be a bounded Markov z_0-time, let $\{\theta(z,\omega),\ z\in I_{z_0}\}$ be a P-a.s. continuous strong e-martingale with $a(\cdot)$, and let Q_ω be a regular conditional distribution with respect to P, corresponding to $G_\zeta^{z_0}$. Then there exists a set $N\in G_\zeta^{z_0}$, $P(N)=0$, such that for $\omega\notin N$ and $\lambda\in R^d$, $\{\theta(\zeta(\omega),z),\ z\in I_{\zeta(\omega)}\}$ is a strong e-martingale associated with $a(\cdot)$, with respect to the measure of Q_ω.

Proof. Let $(0,0)\leq z_1\leq z_2$, $B\in G_{\zeta}^{z_0}$, $A\in G_{\zeta+z_1}^{z_0}$. Then, using Lemma 5, we have

$$\mathbf{E}^{\mathbf{P}}\left\{\mathbf{I}_B(\omega)\,\mathbf{E}^{Q_\omega}\left\{\mathbf{I}_A(\omega)\,X_\lambda^{\zeta+z_1}(\zeta+z_2)\right\}\right\}=\mathbf{E}^{\mathbf{P}}\left\{\mathbf{I}_{A\cap B}(\omega)\,X_\lambda^{\zeta+z_1}(\zeta+z_2)\right\}=$$
$$=\mathbf{P}(A\cap B)=\mathbf{E}^{\mathbf{P}}\left\{\mathbf{I}_B(\omega)\,Q_\omega(A)\right\}.$$

Thus, for any $z_1\leq z_2$, $\lambda\in R^d$, $A\in G_{\zeta+z_1}^{z_0}$ we can find

$$N_{z_1,z_2}^{A,\lambda}\in\mathcal{G}_{\zeta}^{z_0},\ \mathbf{P}\left(N_{z_1,z_2}^{A,\lambda}\right)=0,\quad\text{such that for}\quad\omega\notin N_{z_1,z_2}^{A,\lambda}$$
$$\mathbf{E}^{Q_\omega}\left\{\mathbf{I}_A(\omega')\,X_\lambda^{\zeta+z_1}(\zeta+z_2)\right\}=Q_\omega(A).$$

The end of the proof is the same as that in Theorem 3.1 of [1].

Now we define the strong e-martingale more precisely, specifying conditions on the boundary of the domain I_{z_0}.

Let $x\in R^d$ and let the functions $a_{s\cdot}:R_+\times\Omega\to R^d\otimes R^d$, $a_{\cdot t}:R_+\times\Omega\to R^d\otimes R^d$ satisfy the following conditions:

(s_0). $a_{\cdot t}(s,\omega)$ is $\mathcal{B}_{R_+}\times F^{z_0}$-measurable and for each fixed $s\geq s_0$ it is $F_{(s,t)}^{z_0}$-measurable.

(t_0). $a_{s\cdot}(t,\omega)$ is $\mathcal{B}_{R_+}\times F^{z_0}$-measurable and for each fixed $t\geq t_0$ it is $F_{(s,t)}^{z_0}$-measurable. Let also $a:R_+^2\times\Omega\to R^d\otimes R^d$ satisfy the condition (z_0) and let all these functions satisfy the condition (4). The random process $\{\theta(z,\omega),\ z\in I_{z_0}\}$ is said to be a strong e-martingale with respect to $(z_0,\ x,\ a_{s_0\cdot},\ a_{\cdot t_0},\ a)$ and the measure P, if it satisfies the condition (z_0) (see (5)) as well as the following conditions:

(5°) $\mathbf{P}\{\theta(z_0)=x\}=1$;

(5′) $\left\{Y^{(t)}_{\lambda,s_0\cdot}(t_0),\mathscr{F}^{z_0}_{\infty t},\ t\geqslant t_0\right\}\ \forall\lambda\in R^d - \mathbf{P}$- martingale;

(5″) $\left\{Y^{(s)}_{\lambda,\cdot t_0}(s_0),\mathscr{F}^{z_0}_{s\infty},\ s\geqslant s_0\right\}\ \forall\lambda\in R^d - \mathbf{P}$- martingale,

where

$$Y^{(t')}_{\lambda,s\cdot}(t)=\exp\left\{\langle\lambda,\theta(s,t')-\theta(s,t)\rangle-\frac{1}{2}\left\langle\lambda,\int_t^{t'}a_{s\cdot}(u)\lambda du\right\rangle\right\};$$

$$Y^{(s')}_{\lambda,\cdot t}(s)=\exp\left\{\langle\lambda,\theta(s',t)-\theta(s,t)\rangle-\frac{1}{2}\left\langle\lambda,\int_s^{s'}a_{\cdot t}(v)\lambda dv\right\rangle\right\};$$

$$t'\geqslant t,\ s'\geqslant s.$$

<u>LEMMA 6</u>. If $\{\theta(z,\omega),\ z\in I_{z_0}\}$ is a strong e-martingale with respect to $(z_0,\ x,\ a_{s_0\cdot},\ a_{\cdot t_0},\ a)$ and the measure P, then it is a marginal e-martingale, i.e., for each fixed $s\geq s_0$ $\{Y^{(t)}_{\lambda,s\cdot}(t_0),\ F^{z_0}_{\infty t},\ t\geq t_0\}$ is a P-martingale of $\forall\lambda\in R^d$ and for each fixed $t\geq t_0$ $\{Y^{(s)}_{\lambda,\cdot t}(s_0),F^{z_0}_{s\infty},\ s\geq s_0\}$ is a P-martingale $\forall\lambda\in R^d$. In this case, the functions $a_{s\cdot}$, $a_{\cdot t}$ in the definition of $Y^{(t')}_{\lambda,s\cdot}(t)$, $Y^{(s')}_{\lambda,\cdot t}(s)$, are

$$a_{s\cdot}(t)=a_{s_0\cdot}(t)+\int_{s_0}^{s}a(u,t)\,du,\quad a_{\cdot t}(s)=a_{\cdot t_0}(s)+\int_{t_0}^{t}a(s,v)\,dv.$$

<u>Proof</u>. Let $z_0=(s_0,t_0)$, $z_1=(s_0,t)$, $z_2=(s,t')$, where $s\geq s_0$, $t'\geq t\geq t_0$. Then P-a.s.

$$\mathbf{E}\left\{Y^{(t')}_{\lambda,s\cdot}(t_0)/\mathscr{F}^{z_0}_{\infty t}\right\}=$$

$$=\mathbf{E}\left\{\exp\left[\langle\lambda,\theta(s,t)+\theta(s_0,t')-\theta(s_0,t)-\theta(s,t_0)\rangle-\right.\right.$$

$$\left.-\frac{1}{2}\left\langle\lambda,\int_{t_0}^{t'}a_{s_0}\ (v)\lambda dv+\int_{t_0}^{t}\left(\int_{s_0}^{s}a(u,v)\lambda du\right)dv\right\rangle\right]\times$$

$$\times \mathbf{E}\left\{\exp\left[\langle\lambda, \theta(z_1 z_2]\rangle - \right.\right.$$

$$\left.\left.\left.- \frac{1}{2}\left\langle\lambda, \int_{(z_1, z_2]} a(u, v)\lambda du dv\right\rangle\right] \Big/ \mathcal{G}_{z_1}^{z_0}\right\} \Big/ \mathcal{F}_{\infty t}^{z_0}\right\} =$$

$$= \exp\left[\langle\lambda, \theta(s, t) - \theta(s, t_0)\rangle - \frac{1}{2}\left\langle\lambda, \int_{t_0}^{t} a_{s\cdot}(v)\lambda dv\right\rangle\right] \times$$

$$\times \mathbf{E}\left\{\exp\left[\langle\lambda, \theta(s_0, t') - \theta(s_0, t)\rangle - \right.\right.$$

$$\left.\left.- \frac{1}{2}\left\langle\lambda, \int_{t}^{t'} a_{s_0\cdot}(v)\lambda dv\right\rangle\right] \Big/ \mathcal{F}_{\infty t}^{z_0}\right\} = Y_{\lambda, s\cdot}^{(t)}(t_0).$$

Following the discussion in [1], we say that the probability measure P on (Ω, F^{z_0}) is the solution of the martingale problem for $(z_0, x, a_{s_0\cdot}, a_{\cdot t_0}, a)$ if the process $\theta(z,\omega) = x_z(\omega)$, $z \in I_{z_0}$, is a strong e-martingale with respect to $(z_0, x, a_{s_0\cdot}, a_{\cdot t_0}, a)$ and the measure P. The corollaries from Lemma 4 together with Lemma 6 enable us to assert that the solution of the martingale problem is one also in the sense of Definition 4 in [3].

In order to describe the solution of the martingale problem in terms of stochastic differential equations, we need to introduce a stochastic integral with respect to the strong e-martingale. This can be done in the usual way, as in [1] or [8]. The mapping $\phi: I_{z_0} \times \Omega \to R^d$ is said to be F^{z_0}-predictable if it is measurable with respect to the σ-algebra generated by the sets $(z, z'] \times A$, where $z_0 \le z \le z'$, $A \in F_z^{z_0}$.

Let $\{\theta, (z,\omega), z \in I_{z_0}\}$ be a P-a.s. continuous strong e-martingale associated with the function $a(\cdot)$ satisfying

the conditions (z_0), (4). We denote by $L^2(a)$ the set of all F^{z_0}-predictable functions ϕ for which
$\mathbf{E}\int_{(z_0,z]} \langle\varphi(u), a(u)\varphi(u)\rangle du < \infty$ for all z, $z_0 \leq z$.

For the simple function

$$\varphi(z,\omega) = \sum_{i=1}^{n} \alpha_i(\omega) I_{(z_i,z_i')}(z) \in \mathscr{L}^2(a)$$

(here α_i is measurable with respect to F^{z_0}) the stochastic integral is defined by the equality

$$\langle\varphi, d\theta\rangle(z) = \int_{z_0}^{z} \langle\varphi(u), d\theta(u)\rangle = \sum_{i=1}^{n} \langle\alpha_i, \theta((z_i, z_i'] \cap (z_0, z])\rangle.$$

The integral thus defined is linear and has the following properties: for any $z, z' \in I_{z_0}$, $z \leq z'$

I) $\mathbf{E}\{\langle\varphi, d\theta\rangle(z,z')]/G_z^{z_0}\} = 0$ P-a.s.;

II) $\mathbf{E}(\langle\varphi, d\theta\rangle(z,z'])^2 = \mathbf{E}\int_{(z,z']} \langle\varphi(u), a(u)\varphi(u)\rangle du.$

Hence it is possible to extend the integral $\langle\phi, d\theta\rangle$ to all $L^2(a)$ preserving these properties at the same time. Using Lemma 5 and its corollaries, as well as Lemmas 3, 4, we can prove the following property:

III. for any bounded $\phi \in L^2(a)$ and any $z, z' \in I_{z_0}$, $z \leq z'$,

$$\mathbf{E}\left\{\exp\left\{\langle\varphi, d\theta\rangle(z,z'] - \frac{1}{2}\int_{(z,z']} \langle\varphi(u), a(u)\varphi(u)\rangle du\right\} \Big/ \mathscr{G}_z^{z_0}\right\} = 1 \quad \text{P-a.s.}$$

In the same way as was done in [1], we introduce here the vector stochastic integral $\int_{z_0}^{z} \sigma(u)d\theta(u)$ of the matrix F^{z_0}-predictable function $\sigma: I_{z_0} \times \Omega \to R^d \otimes R^d$ for which

$$\mathbf{E} \int_{(z_0, z]} \langle \lambda, \sigma^* (u) a (u) \sigma (u) \lambda \rangle du < \infty \ \forall_\lambda \in R^d \forall z \in I_{z_0},$$

where the matrix σ^* is transposed relative to σ. If σ satisfies the condition $\exists B>0$

$$\left| \sum_{i,j=1}^{n} \sigma_{ij} \lambda_i \lambda_j \right| \leqslant B |\lambda|^2 \ \forall \lambda \in R^d, \tag{13}$$

then

IV. the integral $\eta(z)=\int_{z_0}^{z} \sigma(u)d\theta(u)$ is also a strong e-martingale associated with the function $\sigma^* a\sigma$, and has the property $\langle \phi, d\eta \rangle = \langle \sigma^* \phi, d\theta \rangle$.

Lastly, we introduce one- and two-parameter processes of Brownian motion on a probability space (Ω, F^{z_0}, P). By the two-parameter z_0-process of Brownian motion we mean the function $\beta(z,\omega)$: $I_{z_0} \times \Omega \to R^d$ satisfying the condition (z_0), P-a.s. continuous and such that

$$\mathbf{P}\left\{\beta(z_2) \in \Gamma / \mathscr{G}_{z_1}^{z_0}\right\} = \frac{1}{(2\pi |(z_1, z_2]|)^{d/2}} \times$$

$$\times \int_{\Gamma} \exp\left\{- \frac{|y - \beta(s_1, t_2) - \beta(s_2, t_1) + \beta(s_1, t_1)|^2}{2 |(z_1, z_2]|}\right\} dy;$$

$$\mathbf{P}\{\beta(z) = 0;\ z \in [s_0, \infty) \times \{t_0\} \cup \{s_0\} \times [t_0, \infty)\} = 1$$

for any $z_1, z_2 \in I_{z_0}$, $z_1 \le z_2$, $\Gamma \in \mathcal{B}_{R^d}$.

By the z_0-process of Brownian motion along the line $s=s_1$ we mean the function $\beta_{s_1 \cdot}(z,\omega)$: $I_{z_0} \times \Omega \to R^d$ on the line $s=s_1$, P-a.s. continuous and satisfying the condition (z_0), and such that

$$\mathbf{P}\left\{\beta_{s_1\cdot}(s_1, t_2) \in \Gamma / F^{z_0}_{\infty t_1}\right\} = \frac{1}{[2\pi(t_2 - t_1)]^{d/2}} \int_\Gamma \exp\left\{-\frac{|y - \beta_{s_1\cdot}(s_1, t_1)|^2}{2|t_2 - t_1|}\right\} dy;$$

$$\mathbf{P}\{\beta_{s_1\cdot}(s_1, t_0) = 0\} = 1$$

for any $t_0 \le t_1 \le t_2$, $\Gamma \in \mathscr{B}_{R^d}$.

THEOREM 3. Let P be a probability measure on (Ω, F^{z_0}); and let the functions a, $a_{s_0\cdot}$, $a_{\cdot t_0}$ be symmetric, satisfy the conditions (z_0), (t_0), (s_0), respectively, as well as the following condition:

$$A'|\lambda|^2 \leqslant \langle\lambda, a\lambda\rangle \leqslant A|\lambda|^2 \quad \forall\lambda \in R^d,$$

where $0<A'<A<\infty$. The (P-a.s.) continuous process $\{\theta(z,\omega), z\in I_{z_0}\}$ is a strong e-martingale with respect to $(z_0, x, a_{s_0\cdot}, a_{\cdot t_0}, a)$ and the measure P if and only if there exist z_0-processes of Brownian motion β, $\beta_{s_0\cdot}$, $\beta_{\cdot t_0}$ with respect to P, such that:

a) $\mathbf{P}\{\theta(s_0, t_0) = x\} = 1$;

b) $\theta(s_0, t) - \theta(s_0, t_0) = \int_{t_0}^{t} \sigma_{s_0\cdot}(v)\, d\beta_{s_0\cdot}(v),\ t \geqslant t_0$ P-a.s.;

$$\theta(s, t_0) - \theta(s_0, t_0) = \int_{s_0}^{s} \sigma_{\cdot t_0}(u)\, d\beta_{\cdot t_0}(u), \quad s \geqslant s_0;$$

c) $\theta(z_0, z] = \int_{z_0}^{z} \sigma(u)\, d\beta(u),\ z \in I_{z_0}$ P-a.s.,

where σ, $\sigma_{s_0\cdot}$, $\sigma_{\cdot t_0}$ are positive definite symmetric square roots of a, $a_{s_0\cdot}$, $a_{\cdot t_0}$.

Proof. By virtue of the results obtained in [1], it suffices to verify that the conditions "c" or (5) hold.

Let θ be a strong e-martingale. Then let

$\beta(z)=\int_{z_0}^{z}\sigma^{-1}(u)d\theta(u)$. It is not hard to verify that $\{\beta(z), z\in I_{z_0}\}$ has the following properties: $\beta(z)$ is P-a.s. continuous with probability 1, vanishes on the boundary of I_{z_0}. In addition, for any $z_1, z_2\in I_{z_0}$, $z_1\leq z_2$, $\lambda\in R^d$

$$\mathbf{E}\left\{\exp\left\{\langle\lambda, \beta(z_1, z_2]\rangle - \frac{1}{2}|\lambda|^2\cdot|(z_1, z_2]|\right\}\Big/ G_{z_1}^{z_0}\right\} = \mathbf{E}\left\{\exp\left\{\langle\lambda, \beta(z_1, z_2]\rangle - \right.\right.$$

$$\left.\left. - \frac{1}{2}\left\langle\lambda, \int_{(z_1,z_2]} (\sigma^{-1}(u))^* a(u)\sigma(u)\cdot\lambda du\right\rangle\right\}\Big/ G_{z_1}^{z_0}\right\} = 1 \quad \text{P-a.s.}$$

This in turn implies that β is a z_0-process of Brownian motion. The relation "c" can be proved in the same way as was done in Theorem 3.3 in [1], using property IV.

Conversely, if θ satisfies the relations "a", "b", and "c", θ is a strong e-martingale by property IV and the fact that β is a strong e-martingale associated with the identity matrix ($a\equiv E$).

REFERENCES

[1] D.W. Strook and S.R.S. Varadhan. "Diffusion Processes with Continuous Coefficients (I)." *Comm. on Pure and Applied Math.*, 22, 3 (1969).

[2] I.I. Gikhman and T.E. Pyasetskaya. "Ob odnom klasse stokhasticheskikh differentsial'nykh uravnenij s chastnymi proizvodnymi, soderzhashchikh dvuparametricheskij belyj shum" (On a Class of Stochastic Partial Differential Equations Containing Two-parameter White Noise).

In *Predel'nye theoremy dlya sluchainykh protsessov*, 71-92. Kiev: Naukova Dumka, 1977.

[3] C. Tudor. "A Theorem Concerning the Existence of the Weak Solution of the Stochastic Equation with Continuous Coefficients in the Plane." *Revue Roumain de mathématiques pure et appliquées*, 23, 9 (1977).

[4] I.I. Gikhman and A.V. Skorokhod. *Introduction to the Theory of Random Processes*. Philadelphia: Scripta Technica, W.B. Saunders Co., 1969.

[5] J.L. Doob. *Stochastic Processes*. New York: Wiley & Sons, 1953.

[6] R. Cairoli. "Une inégalité pour martingales à indices multiples et ses applicationes." *Lecture Notes in Mathematics*, Vol. 124 (1970): 1-27.

[7] P.A. Meyer. *Probability and Potentials*. Waltham, Mass.: Blaisdell Pub. Co., 1966.

[8] R. Cairoli and J.B. Walsh. "Stochastic Integral in the Plane." *Acta Math.*, Vol. 134 (1975): 111-183.

QUADRATIC VARIATION OF RANDOM SEQUENCES

G.P. Karev

Let $(\Omega, \mathfrak{F}, P)$ be a probability space, let $\{\mathfrak{F}_n\}$, $n=0,1,\ldots$, $\mathfrak{F}_n \subseteq \mathfrak{F}$ be an increasing sequence of σ-algebras, and let for any n, X_n be an $\mathfrak{F}_n$-measurable random variable such that $E|X_n|<\infty$. Let

$$S_n^2(X) = X_0^2 + \sum_{k=1}^{n} (X_k - X_{k-1})^2,\ S^2(X) = X_0^2 + \sum_{k=1}^{\infty} (X_k - X_{k-1})^2.$$

We call $S(X)$ the quadratic variation of the sequence $X=\{X_n\}$.

Properties of a quadratic variation of martingales have been investigated in many works (see [1-7]). Thus, Austin [1] proved that if X is an L_1-bounded martingale, then

$$S(X) < \infty \quad \text{almost everywhere}\ , \tag{1}$$

and Burkhölder [2,3] established that for any $\lambda>0$

$$\mathbf{P}(S(X) \geqslant \lambda) \leqslant \frac{c}{\lambda} \sup \mathbf{E}|X_n|,\ c = \text{const} \leqslant 3. \tag{2}$$

In this paper we give the conditions under which similar relations are satisfied for arbitrary random sequences.

Let $X_n^* = \max_{k \leqslant n} |X_k|;\ \mu = \inf(n : X_n^* > \lambda);\ \varepsilon_k = \mathbf{E}(X_{k+1} - X_k | \mathfrak{F}_k).$

LEMMA. For any $n \geq 1$

$$\mathbf{E}S^2_{\mu\wedge n-1} \leqslant 2\lambda \mathbf{E}|X_{\mu\wedge n-1}| - 2\mathbf{E} \sum_{k<\mu\wedge n} X_k \varepsilon_k. \tag{3}$$

Proof. Using the Doob identity

$$S_{n-1}^2 + X_{n-1}^2 = 2X_n X_{n-1} - 2\sum_{k=1}^{n} X_{k-1}(X_k - X_{k-1}),$$

we obtain

$$\mathbf{E}S^2_{\mu\wedge n-1} = 2\mathbf{E}X_{\mu\wedge n}X_{\mu\wedge n-1} - \mathbf{E}X^2_{\mu\wedge n-1} - 2\mathbf{E}\sum_{k<\mu\wedge n} X_k \varepsilon_k. \tag{4}$$

Now we estimate $2X_{\mu\wedge n}X_{\mu\wedge n-1} - X^2_{\mu\wedge n-1}$. We put $(\mu=k)=\chi\{\omega: \mu(\omega)=k\}$. Then

$$2X_{\mu\wedge n}X_{\mu\wedge n-1} - X^2_{\mu\wedge n-1} = \sum_{k=1}^{n} (\mu = k) X_{k-1}(2X_k - X_{k-1}) +$$
$$+ (\mu > n) X_{n-1}(2X_n - X_{n-1}) \leqslant \lambda \cdot 2|X_\mu|(\mu \leqslant n) -$$
$$- X^2_{\mu-1}(\mu \leqslant n) + X_n^2(\mu > n,)$$

since $|X_\mu|>\lambda>0$; $2X_{n-1}X_n-X_{n-1}^2\le X_n^2$. Then

$$2X_{\mu\wedge n}X_{\mu\wedge n-1}-X^2_{\mu\wedge n-1}\leqslant 2\lambda|X_\mu|(\mu\leqslant n)+\lambda|X_n|(\mu>n)\leqslant 2\lambda|X_{\mu\wedge n}|,$$

which plus (4) yields (3).

THEOREM 1. For all $\lambda>0$

$$\lambda\mathbf{P}(S_{n-1}>\lambda)\leqslant 3\mathbf{E}|X_{\mu\wedge n}|-\frac{2}{\lambda}\mathbf{E}\sum_{k<\mu\wedge n}X_k\varepsilon_k.$$

Proof. $P(S_{n-1}>\lambda)\le P(S_{n-1}>\lambda,\ X^*_{n-1}\le\lambda)+P(X^*_{n-1}>\lambda)$. Obviously,

$$\mathbf{P}(X^*_{n-1}>\lambda)\leqslant\mathbf{E}(X_\mu(\mu<n)).$$

Next,

$$(S_{n-1}>\lambda,\ X^*_{n-1}\leqslant\lambda)\subseteq(S_{\mu\wedge n-1}>\lambda).$$

Hence

$$\lambda\mathbf{P}(S_{n-1}>\lambda,\ X^*_{n-1}\leqslant\lambda)\leqslant\frac{1}{\lambda}\mathbf{E}(S^2_{\mu\wedge n-1})\leqslant 2\mathbf{E}|X_{\mu\wedge n}|-\frac{2}{\lambda}\mathbf{E}\sum_{k<\mu\wedge n}X_k\varepsilon_k$$

by the Lemma. Thus,

$$\lambda\mathbf{P}(S_{n-1}>\lambda)\leqslant 2\mathbf{E}|X_{\mu\wedge n}|-\frac{2}{\lambda}\mathbf{E}\sum_{k<\mu\wedge n}X_k\varepsilon_k+\mathbf{E}(X_\mu(\mu<n))\leqslant$$
$$\leqslant 3\mathbf{E}|X_{\mu\wedge n}|-\frac{2}{\lambda}\mathbf{E}\sum_{k<\mu\wedge n}X_k\varepsilon_k,$$

which was to be proved.

REMARK 1. Theorem 1 leads to the Burkhölder inequality (2) if X is a martingale.

THEOREM 2. Let $\sup(X_{n+1}-X_n)\in L_1$. Then $S^2(X)$ is integrable on any set where $\sup|X_n|$ is bounded and the series $\sum X_k\varepsilon_k$ is lower bounded. In particular, $S(X)<\infty$ is amost everywhere on the set

$$\{\sup|X_n|<\infty\}\cap\left\{\inf_n\sum_{k=1}^n X_k\varepsilon_k>-\infty\right\}.$$

Proof. Let $M = E\sup(X_{n+1}-X_n)$. Then

$$2EX_{\mu\wedge n}X_{\mu\wedge n-1} - EX^2_{\mu\wedge n-1} \leqslant 2\lambda E((\mu\leqslant n)(X_\mu - X_{\mu-1} + X_{\mu-1}) + \\ + \lambda EX_n(\mu > n) \leqslant 2\lambda(M + \lambda E(\mu\leqslant n)) + \lambda^2 E(\mu > n) \leqslant 2(\lambda M + \lambda^2).$$

By the equality (4)

$$\mathbf{E}S^2_{\mu\wedge n-1} \leqslant 2(\lambda M + \lambda^2) - 2\sum_{k<\mu\wedge n} X_k\varepsilon_k. \tag{5}$$

Let

$$t = \inf\left(n : \sum_{k=0}^{n} X_k\varepsilon_k < -K = \text{const}\right).$$

Then $ES^2_{\mu\wedge t-1} \leq 2(\lambda M + \lambda^2 + K)$, which, in turn, proves the Theorem.

REMARK 2. The condition $\sup(X_{n+1}-X_n)\in L_1$ can be replaced by the condition $\sup X_n \in L_1$. In this case the second half of the Theorem follows from Theorem 1.

The result obtained by Austin [1] is that the quadratic variation $S(X)$ of a L_1-bounded martingale is square integrable on any set where $\sup|X_n|$ is bounded. In [5, Vol. III, 1.4], a stronger assertion is proved: on the same set $S^2(X)$ is exponentially integrable.

Let $X_n \geq 0$ for all n. Let also $X^* = \sup X_n$; $Q_n = X_n/X^*_n$; $\delta_n = E(Q_{n+1}-Q_n)|\mathfrak{F}_n$. We show that $S^2(X)$ is exponentially integrable on any set where X^* and $\sup_n \left|\sum_{k=0}^{n} Q_k\delta_k\right|$ are bounded. More precisely, the following theorem holds, in proving which we follow along the lines of [5].

THEOREM 3. Let

$$\tau = \inf\left(n : \left|\sum_{k=0}^{n} Q_k\delta_k\right| > \frac{M}{2}\right).$$

Then

$$\mathbf{E}\left[(X^* < \lambda)\exp\left(\frac{1}{8(1+M)\lambda^2}\sum_{k=1}^{\tau}(X_k - X_{k-1})^2\right)\right] \leqslant 2e^{\frac{1}{4(1+M)}}.$$

Proof. In [5, Vol. III, 1.2] it is proved that if A_n is a nondecreasing random sequence such that

$$\mathbf{E}(A_n - A_{k-1}|\mathfrak{F}_k) \leqslant B = \text{const} < 1 \qquad n \geqslant k \geqslant 1, \tag{6}$$

then

$$\mathbf{E}(\exp(A_n - A_{k-1})\,|\,\mathfrak{F}_k) \leqslant \frac{1}{1-B}. \tag{7}$$

Let

$$A_n = \sum_{k=1}^{n}(Q_k - Q_{k-1})^2.$$

We have

$$A_n - A_{k-1} = Q_n^2 - Q_{k-1}^2 - 2\sum_{s=k}^{n} Q_{s-1}(Q_s - Q_{s-1}) \leqslant Q_n^2 + Q_{k-1}^2 - 2\sum_{s=k+1}^{n} Q_{s-1}(Q_s - Q_{s-1}),$$

yielding

$$\mathbf{E}(A_n - A_{k-1}|\mathfrak{F}_k) \leq 2 - 2\mathbf{E}\left(\sum_{s=k}^{n-1} Q_s\delta_s \middle| \mathfrak{F}_k\right). \tag{8}$$

Next, $X_n - X_{n-1} = X_n^* - X_{n-1}^* + X_{n+1}^*(Q_n - Q_{n-1})$, hence

$$(X_n - X_{n-1})^2 \leq 2X_n^*(X_n^* - X_{n-1}^*) + 2X_n X_n^*(Q_n - Q_{n-1})^2,$$

and

$$\frac{1}{X^*}\sum_{k=1}^{n}\frac{(X_k - X_{k-1})^2}{X_k^*} \leqslant 2 + 2\sum_{k=1}^{n}(Q_k - Q_{k-1})^2. \tag{9}$$

Note that $\sup_{k<\tau}\left|\sum_{s=k}^{\tau-1} Q_s\delta_s\right| \leqslant M$. Hence it follows from (8) and (9) that for $\alpha < \frac{1}{4}(1+M)$ the sequence

$$Y_n = \frac{\alpha}{X^*} \sum_{k=1}^{\tau\wedge n} \frac{(X_k - X_{k-1})^2}{X_k^*}$$

satisfies the conditions (6); by (7) we have

$$E\left(\exp \frac{\alpha}{X^*} \sum_{k=1}^{\tau} \frac{(X_k - X_{k-1})^2}{X_n^*}\right) \leqslant \frac{e^{2\alpha}}{1 - 4\alpha(1+M)},$$

and for $\alpha = \frac{1}{8}(1+M)$ we obtain

$$E\left[(X^* < \lambda) \exp\left(\frac{1}{8(1+M)\lambda^2} \sum_{k=1}^{\tau} (X_{k+1} - X_k)^2\right)\right] \leqslant 2e^{1/4(1+M)}.$$

This implies the boundedness of

$$\mathrm{E}\left[(X^* < \lambda) \exp\left(\frac{1}{8(1+M)\lambda^2} S_\tau^2\right)\right].$$

Since $X_0^2 < \lambda^2$ on the set $\{X^* < \lambda\}$.

REFERENCES

[1] D.G. Austin. "A Sample Function Property of Martingales." *Ann. Math. Stat.*, 37, 5 (1966): 1369-1397.

[2] D.L. Burkhölder. "Martingale Transforms." *Ann. Math. Stat.*, 37, 6 (1966): 1494-1504.

[3] D.L. Burkhölder. "Distribution Function Inequalities for Martingales." *Ann. Prob.*, 1, 1 (1973): 19-42.

[4] B.I. Davis. "On the Integrability of Martingale Square Functions." *Israel J. Math.*, 8, 2 (1970): 187-190.

[5] A.M. Garsia. "Martingale Inequalities." *Seminar Note on on Recent Progress*, 1973.

[6] P.-A. Meyer. "Martingales and Stochastic Integrals. I." In *Lecture Notes in Mathematics*, Vol. 284 (1972).

[7] A.A. Novikov. "Martingal'nye neravenstva" (Martingale Inequalities). In *Trudy shkoly-seminara po teorii sluchainykh protsessov. II, Druskininkaj, 1974*, 89-126. Vilnius, 1975. (In Russian.)

PROBABILITY INEQUALITIES FOR SUMS OF INDEPENDENT RANDOM VARIABLES WITH VALUES IN A BANACH SPACE

S.V. Nagaev

Let $X_1, X_2, \ldots, X_n$ be independent random variables (r.v.) assuming values in a separable Banach space B, and let $S_n = \sum_1^n X_i$, $S_{kj} = \sum_{k+1}^{j} X_i$. We denote by $|\cdot|$ the norm in the space B.

The aim of this article is to obtain estimates for the probabilities $P(|S_n| \geq u)$ (Theorems 1, 2, 3, Corollary 3). These estimates are formulated here in somewhat different terms than those in [1, 2]. Instead of $E|S_n|$ we use the variable u_0, which is defined by the relation $P(|S_n| \geq u_0) \leq \alpha$, where α is a fixed constant. It is seen that u_0 can be chosen such that $u_0 \leq E|S_n|/\alpha$.

It is not so clear, however, that $E|S_n|$ is estimable from above in terms of u_0 and $\sum_1^n E|X_j|^2$ (Corollary 5, also see [10]).

We note that our method for proving is different from those used in [1, 2]. Our method can be regarded as a modification of the method used in [3, Chapter VI, Section 3].

We recall that the random variable X with values in B is called symmetric if X and $-X$ are identically distributed. The random variable X^s is called the symmetrization of X if $X^s=X'-X''$, where X' and X'' are independent and $L(X')=L(X'')$.

Let

$$M_n = \max_{1\le k\le n} |S_k|, \qquad \bar{M}_n = \max_{1\le k\le j\le n} |S_{kj}| .$$

PROPOSITION 1. If $X_1, X_2, \ldots, X_n$ are symmetric then

$$P(M_n \ge u) \le 2P(|S_n| \ge u) . \tag{1}$$

This proposition is a generalization of the well-known Lévy inequality. The proof of this proposition can be found, for example, in [4, pp. 27-28], and [5].

PROPOSITION 2. Let X be any random variable with values in B. Then

$$\tfrac{1}{2}P(|X^s| \ge 2u) \le P(|X| \ge u) \le P(|X^s| \ge u-v)/P(|X| \le v) . \tag{2}$$

Proof. The left-hand side of the inequality (2) can be proved exactly as in the one-dimensional case (see, for instance, [6, p. 259]).

We prove the right-hand side of (2). Obviously, $|X^s| \ge |X'|-|X''|$, where X' and X'' are independent and $L(x')=L(X'')$.

On the other hand,

$$\{|X'|\geq u,\ |X''|\leq v\} \leq \{|X'|-|X''|\geq u-v\} .$$

Therefore,

$$P(|X^S|\geq u-v) \geq P(|X|\geq u)P(|X|\leq v) .$$

//

We note that the right-hand side of (2) can easily be derived from a more general assertion given in [7, p. 81, Lemma 4.1].

PROPOSITION 3. For any nonnegative s, t, u

$$\mathbf{P}(M_n\geqslant t+s+u)\leqslant P(M_n\geqslant t)\,P(\overline{M}_n\geqslant s)+\sum_1^n \mathbf{P}(|X_j|>u). \qquad (3)$$

Proof. We can assume without loss of generality that $P(M_n\geq t)>0$. Let $\tau=\inf\{k:\ |S_k|\geq t\}$. It is easy to see that

$$\mathbf{P}(M_n\geqslant t+s+u)\leqslant \sum_1^n \int\limits_{t\leqslant|x|\leqslant t+u} \mathbf{P}\Big(\max_{k\leqslant j\leqslant n}|S_{kj}+x|\geqslant t+s+u\Big)\times$$

$$\times\mathbf{P}(\tau=k,\ S_k\in dx)+\sum_1^n \mathbf{P}(\tau=k,\ |S_k|>t+u)\equiv A+B. \quad (4)$$

Obviously, $|S_{kj}+x|\leq|S_{kj}|+|x|$. Hence

$$\{|S_{kj}+x|\geq t+s+u\} \subseteq \{|S_{kj}|\geq t+s+u-|x|\} .$$

Therefore,

$$\{|S_{kj}+x|\geq t+s+u,\ |x|\leq t+u\} \subseteq \{|S_{kj}|\geq s\} .$$

We thus conclude that

$$A\leqslant\sum_1^n \mathbf{P}\Big(\max_{k\leqslant j\leqslant n}|S_{kj}|\geqslant s\Big)\,\mathbf{P}(\tau=k)\leqslant\mathbf{P}(\overline{M}_n\geqslant s)\,\mathbf{P}(M_n\geqslant t), \qquad (5)$$

since

$$\mathbf{P}\Big(\max_{k\leqslant j\leqslant n}|S_{kj}|\geqslant s\Big)\leqslant\mathbf{P}(\overline{M}_n\geqslant s)$$

and

$$\sum_1^n P(\tau = k) = P(M_n \geq t).$$

Next we estimate B. It is not hard to see that

$$\{\tau=k, \ |S_k| > t+u\} \subseteq \{|X_k| > u\} \ . \qquad (6)$$

Comparing the inequalities (4)-(6), we obtain (3). //

We note that the proof of Proposition 3 coincides, in essence, with the proof of Lemma 4.4 in [7].

COROLLARY 1. For any nonnegative s, t, u

$$P(M_n \geq t+s+u) \leq P(M_n \geq t)\, P(M_n \geq s/2) + \sum_1^n P(|X_j| > u). \qquad (7)$$

Proof. Obviously, $|S_{jk}| \leq |S_j| + |S_k|$. Therefore, $\bar{M}_n \leq 2M_n$. This implics that

$$P(\bar{M}_n \geq s) \leq P(M_n \geq s/2) \ .$$

It remains only to substitute this estimate into the right-hand side of (3).

PROPOSITION 4. For any integer $m>1$

$$\mathbf{P}(M_n \geq \lambda) \leq \mathbf{P}^m(M_n \geq \lambda/4m) + \mathbf{P}(M_n < \lambda/4m)^{-1} \sum_1^n \mathbf{P}(|X_j| > \lambda/2m).$$

A similar inequality has been obtained in [3, p. 454] for bounded random variables with values in a Hilbert space.

Proof. Letting in (7)

$$t = (k-1)\lambda/m \ , \qquad s = \lambda/2m \ , \qquad u = \lambda/2m \ ,$$

we have

$$\mathbf{P}(M_n \geq k\lambda/m) \leq \mathbf{P}(M_n \geq (k-1)\lambda/m)\, \mathbf{P}(M_n \geq \lambda/4m) +$$

$$+ \sum_1^n \mathbf{P}(|X_j^{\xi}| > \lambda/2m).$$

Let

$$a = \mathbf{P}(M_n \geqslant \lambda/4m); \quad b = \sum_1^n \mathbf{P}(|X_j| > \lambda/2m);$$

$$a_k = \mathbf{P}(M_n \geqslant k\lambda/m).$$

Then $a_k \le aa_{k-1}+b$, $k \geqslant 2$. Whence

$$a_m \leqslant a_1 a^{m-1} + b \sum_0^{m-1} a^k \leqslant a^m + b/(1-a).$$

On the other hand, $a_m = P(M_n \ge \lambda)$. //

<u>COROLLARY 2</u>. If X_j, $j=\overline{1,n}$, are symmetric, then for any integer $m>1$ such that $P(|S_n| \ge \lambda/4m)\} < 1/2$,

$$\mathbf{P}(|S_n| \geqslant \lambda) \leqslant 2^m \mathbf{P}^m(|S_n| \geqslant \lambda/4m) +$$
$$+ (1 - 2\mathbf{P}(|S_n| \geqslant \lambda/4m))^{-1} \sum_1^n \mathbf{P}(|X_j| > \lambda/2m).$$

<u>Proof</u>. If suffices to note that $P(|S_n| \ge \lambda) \le P(M_n \ge \lambda)$ and apply Propositions 1 and 4.

<u>THEOREM 1</u>. If $u_0 > 0$ and $t>0$ are such that

$$P(M_n \ge u_0) \le 1/6 , \qquad A_t/u_0^t \le 1/36 ,$$

then $\forall\, u > u_0$

$$\mathbf{P}(M_n \geqslant u) < \frac{1}{2} \exp\left\{- c_0 (u/u_0)^{\frac{\ln 2}{\ln 3}}\right\} + 11 \cdot 3^{t+1} A_t/u^t,$$

where $c_0 = (\ln 3)/2$, $A_t = \sum_1^n \mathbf{E}|X_i|^t$.

<u>Proof</u>. By Proposition 3,

$$P(M_n \ge 3u) \le P^2(M_n \ge u) + A_t/u^t .$$

Let

$$u_k = 3^k u_0 , \quad k \geqslant 1 , \quad \varepsilon = A_t/u_0^t , \quad a_k = P(M_n \geqslant u_k) .$$

Then

$$a_k \le a_{k-1}^2 + \varepsilon/3^{(k-1)t} . \qquad (8)$$

Let

$$k_0 = \max \{n :\ 3^{2^k}\ \varepsilon/3^{(k-1)t} \le 1/4\}\ .$$

We note that the function $3^{2^u - ut}$ is convex. Hence

$$\max_{1 \leqslant h \leqslant h_0} 3^{2^h}\varepsilon/3^{(h-1)t} = \max \left[9\varepsilon,\ 3^{2^{h_0}}\varepsilon/3^{(h_0-1)t}\right] \leqslant 1/4. \tag{9}$$

Let $\lambda_k - 3^{2^k}$, $b_k - a_k\lambda_k$. Then

$$b_k \le b_{k-1}^2 + \varepsilon 3^{2^k-(k-1)t}\ ,$$

since $\lambda_k = \lambda_{k-1}^2$.

Noting (9), we obtain $b_k \le b_{k-1}^2 + 1/4$, $1 \le k \le k_0$. Obviously, $b_0 = \lambda_0 a_0 \le 1/2$, and therefore $b_k \le 1/2$ for any $k \le k_0$. This implies that for $k \le k_0$

$$a_k \le 3^{-2^k}/2\ . \tag{10}$$

Let $u_{k-1} \le u \le u_k$. Then

$$P(M_n \geqslant u) \le P(M_n \geqslant u_{k-1}) \le 3^{-2^{k-1}}/2\ .$$

Obviously, $2^{k-1} = 3^{k(\ln 2)/\ln 3}/2 = (u_k/u_0)^{(\ln 2)/\ln 3}/2.$

Therefore, $\forall\ u \le u_{k_0}$

$$\mathbf{P}(M_n \geqslant u) \leqslant \frac{1}{2}\exp\left\{-c_0 (u/u_0)^{\frac{\ln 2}{\ln 3}}\right\}, \tag{11}$$

where $c_0 = (\ln 3)/2$.

We consider now $u > u_{k_0}$. Obviously, $\forall\ n \ge 1$ holds $a_n \le a_{n-1}$. Therefore, $\forall\ k > k_0$

$$a_k \le a_{k_0} a_{k-1} + \varepsilon/3^{(k-1)t}\ .$$

Whence

$$a_k \leqslant a_{k_0}^{k-k_0+1} + \varepsilon \sum_{k_0+1}^{k} a_{k_0}^{k-j}/3^{(j-1)t}. \tag{12}$$

Obviously,

$$\sum = \sum_{k_0+1}^{k} a_{k_0}^{k-j}/3^{(j-1)} = \sum_{k_0+1}^{k} (3^t a_{k_0})^{k-j}/3^{(k-1)t}. \tag{13}$$

By Definition of k_0, $3^{2^{k_0+1}-k_0 t}\varepsilon > 1/4$ is satisfied. Whence

$$3^{2^{k_0}-k_0 t/2} > \varepsilon^{-1/2}/2 . \tag{14}$$

In particular,

$$3^{k_0 t/2-2^{k_0}} < 1/3 . \tag{15}$$

It follows from (10) and (15) that

$$3^t a_{k_0} < 1/6 \tag{16}$$

if $k_0 \geq 2$.

Combining (13) and (16), we conclude that for $k_0 \geq 2$

$$\sum \leq 6/5 \cdot 3^{(k-1)t} . \tag{17}$$

By (10) and (14) $a_{k_0} \leq (\varepsilon/3^{k_0 t})^{1/2}$. Hence

$$a_{k_0}^{k-k_0+1} \leq \varepsilon^{(k-k_0+1)/2} / 3^{k_0(k-k_0+1)t/2} . \tag{18}$$

It is easily seen that $\forall\ k>k_0$

$$k_0(k-k_0+1) \geq 2(k-1) , \tag{19}$$

if $k_0 \geq 2$. On the other hand, $\forall\ k>k_0$

$$\varepsilon^{(k-k_0+1)/2} \leq \varepsilon . \tag{20}$$

Comparing the inequalities (18)-(20), we obtain

$$a_{k_0}^{k-k_0+1} \leq \varepsilon/3^{(k-1)t} , \qquad k > k_0 , \quad k_0 \geq 2 . \tag{21}$$

From (12), (17) and (21) it follows that

$$a_k \le 3\varepsilon/3^{(k-1)t} , \quad k > k_0 , \quad k_0 \ge 2 . \tag{22}$$

We assume now that $k_0=1$. By (8),

$$a_1 \le a_0^2 + \varepsilon \le 1/18 . \tag{23}$$

On the other hand, by the definition of k_0

$$3^{4-t}\varepsilon > 1/4 . \tag{24}$$

Therefore,

$$3^t < 4\cdot 3^4\varepsilon \le 9 . \tag{25}$$

Comparing (23) and (25), we conclude that

$$3^t a_1 < 1/2 . \tag{26}$$

It follows from (13) and (25) that

$$\sum < 2/3^{(k-1)t} . \tag{27}$$

On the other hand, by (24),

$$3^{-t}\varepsilon > 1/4\cdot 3^4 . \tag{28}$$

Comparing (23) and (28), we obtain $a_1<18\varepsilon/3^t$. Since $18\varepsilon\le 1/2$, then $\forall\ k>1$ we have the inequality

$$a_1^k < 9\varepsilon/3^{tk} . \tag{29}$$

Combining (12), (27) and (29), we obtain

$$a_k < 11\varepsilon/3^{(k-1)t} , \quad k_0 = 1 , \quad k > 1 . \tag{30}$$

By (22) and (30) we conclude that $\forall\ k_0$ we have $a_k<11\varepsilon/3^{(k-1)t}$, if $k>k_0$. This implies that

$$P(M_n \ge u_k) < 11\cdot 3^t A_t/u_k^t , \quad k > k_0 .$$

If $u_{k-1}<u<u_k$, then

$$P(M_n \geq u) \leq P(M_n \geq u_{k-1}) < 11\cdot 3^{t+1} A_t/u^t . \qquad (31)$$

The inequalities (11) and (31) prove the assertion of Theorem 1.

COROLLARY 3. Let X_j, $j=\overline{1,n}$, be symmetric and let $u_0>0$, $t>0$ be such that

$$P(|S_n| \geq u_0) \leq 1/12 , \qquad A_t/u_0^t \leq 1/36 .$$

Then $\forall\ u>u_0$

$$\mathbf{P}(|S_n| \geqslant u) \leqslant \frac{1}{2} \exp\left\{- c_0 (u/u_0)^{\frac{\ln 2}{\ln 3}}\right\} + 11\cdot 3^{t+1} A_t/u^t,$$

where $c_0=(\ln 2)/\ln 3$.

Proof. By the inequality (1),

$$P(M_n \geq u_0) \geq 2P(|S_n| \geq u_0) \leqslant 1/6 .$$

Thus, the conditions of Theorem 1 are satisfied. It remains only to note that

$$P(|S_n| \geq u) \leq P(M_n \geq u) .$$

THEOREM 2. If u_0 and t satisfy the conditions

$$P(|S_n| \geq u_0) \leq 1/24 , \qquad A_t/u_0^t \leq 1/36 , \qquad (32)$$

then $\forall\ u>u_0$

$$\mathbf{P}(|S_n| \geqslant u) < \exp\left\{- c_0 ((u-u_0)/2u_0)^{\frac{\ln 2}{\ln 3}}\right\} + 6^{t+2} A^t/2 (u-u_0)^t,$$

where $c_0=(\ln 3)/\ln 2$.

Proof. By the first inequality in (2)

$$P(|S_n^s| \geq 2u_0) \leq 2P(|S_n| \geq u_0) \leq 1/12 . \qquad (33)$$

It is easily seen that $\forall\ t>0$

$$E|X_j^s|^t \leq 2^t E|X_j|^t .$$

Hence

$$A_t^s \equiv \sum_1^n E|X_j^s|^t \le 2^t A_t . \qquad (34)$$

Therefore,

$$A_t^s/(2u_0)^t \le 1/36 . \qquad (35)$$

Using Corollary 3, we infer that $\forall\ u \ge 2u_0$

$$\mathbf{P}(|S_n^s| \geqslant u) < \frac{1}{2}\exp\left\{-c_0\left(\frac{u}{2u_0}\right)^{\frac{\ln 2}{\ln 3}}\right\} + 11\cdot 3^{t+1}A_t^s/u^t,$$

which, by the second inequality in (2) and the inequality (34), yields

$$\mathbf{P}(|S_n| \leqslant u_0)\,\mathbf{P}(|S_n| \geqslant u) < \frac{1}{2}\exp\{-c_0((u-u_0)/2u_0)^{\ln 2/\ln 3}\} + \\ + 33\cdot 6^t A_t/(u-u_0)^t,$$

where $u \ge 3u_0$.

It remains only to use the estimate

$$P(|S_n| \le u_0) \ge 23/24 > 1/2 . \qquad (36)$$

<u>THEOREM 3</u>. Let u_0 and t satisfy the conditions (32). Then for each integer $m>1$ there exist constants $c_0(m)$, $c_1(m)$, $c_2(t,m)$ such that $\forall\ u \ge (8m+1)u_0$:

$$\mathbf{P}(|S_n| \geqslant u) \leqslant c_1(m)\exp\{-c_0(m)((u-u_0)/u_0)^{\ln 2/\ln 3}\} + \\ + c_2(t,m)\left(A_t/(u-u_0)^t\right)^m + \frac{6}{5}\sum_1^n \mathbf{P}(|X_j| > (u-u_0)/2m). \qquad (37)$$

We can put

$$c_0(m) = 2^{-1}m(8m)^{-(\ln 2)/\ln 3}\ln 3 ; \qquad c_1(m) = 2^m ;$$

$$c_2(t,m) = (11\cdot 3^{t+1}\cdot 2^{3t+2}m^t)^m .$$

<u>Proof</u>. By the second inequality in (2),

$$P(|S_n| \geq u) \leq P(|S_n^s| \geq u-u_0)/P(|S_n| \leq u_0) \ . \tag{38}$$

On the other hand, by (33) $\forall\ u \geq (8m+1)u_0$

$$P(|S_n^s| \geq (u-u_0)/4m) \leq P(|S_n^s| \geq 2u_0) \leq 1/12 \ . \tag{39}$$

Applying now Corollary 2, we have

$$\mathbf{P}(|S_n^s| \geqslant u-u_0) \leqslant 2^m \mathbf{P}^m(|S_n^s| \geqslant (u-u_0)/4m) + \\ + \frac{6}{5}\sum_1^n \mathbf{P}(|X_j| > (u-u_0)/2m).$$

By (35) and (39), for $2u_0$ the conditions of Corollary 3 are satisfied. Hence $\forall\ u \geq (8m+1)u_0$

$$\mathbf{P}(|S_n^s| \geqslant (u-u_0)/4m) \leqslant \frac{1}{2}\exp\{-c_0((u-u_0)/8u_0m)^{(\ln 2)/\ln 3}\} + \\ + 11\cdot 3^{t+1}(4m)^t A_t^s/(u-u_0)^t.$$

Comparing the last two inequalities, we infer that

$$\mathbf{P}(|S_n^s| \geqslant u-u_0) \leqslant 2^m\Big(\frac{1}{2}\exp\{-c_0((u-u_0)/8u_0m)^{(\ln 2)/\ln 3}\} + \\ + 11\cdot 3^{t+1}(4m)^t A_t^s/(u-u_0)^t\Big)^m + \frac{6}{5}\sum_1^n \mathbf{P}(|X_j| > (u-u_0)/2m). \tag{40}$$

The inequalities (34), (36), (38), (40) and $(a+b)^m \leqslant 2^{m-1}(a^m+b^m)$ yield (37).

<u>REMARK 1</u>. If $P(|S_n| \geq u_0) \leq 1/24$, but $A_t/u_0^t > 1/36$, it is not permissible to use immediately the estimate (37). In this case it is more convenient to put $u_0' = (36A_t)^{1/t}$. Since u_0' satisfies the conditions (32), $\forall\ u > (8m+1)u_0'$

$$\mathbf{P}(|S_n| \geqslant u) < c_3(t,m)(A_t/u^t)^m + \frac{6}{5}\sum_1^n \mathbf{P}(|X_j| > 4u_0/(8m+1)).$$

Besides (37), we have used also the inequality $u-u_0' > 8mu/(8m+1)$, which holds for $\forall\ u > (8m+1)u_0'$.

On the other hand, if $u \le u_0'$, then

$$P(|S_n| \ge u) \le 1 \le (u_0'/u)^{tm} \le 36^m (A_t/u^t)^m .$$

Let $B_n^2 = \sum_1^n E|X_j|^2$.

REMARK 2. Let B be of type 2, i.e., there exists a constant C such that $E|S_n|^2 \le CB_n^2$ for any $X_1, X_2, \ldots, X_n$ with zero mathematical expectations. Then $u_0 = (24C)^{1/2} B_n$ satisfies the condition $P(|S_n| \ge u_0) \le 1/24$.

Applying now the inequality (37), we arrive at an infinite-dimensional analog of the estimate (47) from [8] with the difference that $(u/B_n)^{(\ln 2)/\ln 3}$ appears instead of $(u/B)^2$. The inequalities obtained in [1, 2] are free from this shortcoming.

THEOREM 4. If $P(|S_n| \ge u_0) \le 1/24$, then $\forall\ t > 1$ there exists a constant $a(t)$ such that†

$$E|S_n|^t < a(t) \max [u_0^t, 36A_t] . \qquad (41)$$

As $a(t)$ we can take

$$a(t) = 4t\,(17/16c_0(2))^{(t\ln 3)/\ln 2}\,\Gamma\left(\frac{t\ln 3}{\ln 2}\right) + \\ + (17/16)^{t-1}\,16^{-t}\,36^{-2}c_2(t, 2) + (17/4)^t\,36^{-1},$$

where the constants $c_0(2)$ and $c_2(t,2)$ are the same as in Theorem 3.

Proof. We assume first that both conditions of (32) are satisfied. Putting $m=2$ in (37), we have $\forall\ u > 17u_0$

† For symmetric X_i a somewhat more general result is obtained in [9, p. 8, Inequality 3.17].

$$\mathbf{P}(|S_n| \geqslant u) \leqslant 4\exp\left\{-c_0(2)\left((u-u_0)/u_0\right)^{\alpha_0}\right\} +$$

$$+ c_2(t, 2)\left(A_t/(u-u_0)^t\right)^2 + \frac{6}{5}\sum_1^n \mathbf{P}(|X_j| > (u-u_0)/4), \quad (42)$$

where $\alpha_0 = (\ln 2)/\ln 3$.

Obviously,

$$\mathbf{E}|S_n|^t = t\int_0^\infty u^{t-1}\mathbf{P}(|S_n| \geqslant u)\,du \equiv t\int_0^{17u_0} + t\int_{17u_0}^\infty . \quad (43)$$

It is not hard to see that

$$t\int_0^{17u_0} \leqslant 17^t u_0^t. \quad (44)$$

Now we estimate $\int_{17u_0}^\infty$. Obviously, $u > 17u_0 \Rightarrow u - u_0 > 16u_0$. Therefore, $\forall\ u < 17u_0$

$$u = u - u_0 + u_0 < \frac{17}{16}(u-u_0) .$$

Whence

$$\int_{17u_0}^\infty \frac{u^{t-1}}{(u-u_0)^{2t}}\,dt < (17/16)^{t-1}\,t^{-1}(16u_0)^{-t}.$$

Since

$$A_t < u_0^t/36 , \quad (45)$$

then

$$A_t^2\int_{17u_0}^\infty \frac{u^{t-1}}{(u-u_0)^{2t}}\,dt < (17/16)^{t-1}\,16^{-t}36^{-2}t^{-1}u_0^t. \quad (46)$$

Consider next the integral

$$I = \int_0^\infty u^{t-1}\exp\left\{-c_0(2)\left((u-u_0)/u_0\right)^{\alpha_0}\right\}du.$$

Obviously,

$$u > 17u_0 \Rightarrow u-u_0 > (16/17)u . \tag{47}$$

Therefore, $\forall\ u>17u_0$

$$(u-u_0)^{\alpha_0} > (16/17)u^{\alpha_0} .$$

Using this estimate, we easily obtain

$$I < u_0^t \int_0^\infty u^{t-1} \exp\left\{-(16/17)\, c_0(2)\, u^{\alpha_0}\right\} du = u_0^t (17/16c_0(2))^{t/\alpha_0}\, \Gamma(t/\alpha_0). \tag{48}$$

Finally, by (47)

$$\int_{17u_0}^\infty u^{t-1}\mathbf{P}(|X_j| > (u-u_0)/4)\, du < (17/4)^t \int_0^\infty u^{t-1}\mathbf{P}(|X_j| > u)\, du = (17/4)^t\, t^{-1}\mathbf{E}|X_j|^t. \tag{49}$$

By (45) and (49)

$$\sum_1^n \int_{17u_0}^\infty u^{t-1}\mathbf{P}(|X_j| > (u-u_0)/4)\, du < (17/4)^t\, 36^{-1}u_0^t. \tag{50}$$

From (42), (46), (48) and (50) it follows that

$$t\int_{17u_0}^\infty < u_0^t\big(4t\,(17/16c_0(2))^{t/\alpha_0}\,\Gamma(t/\alpha_0) + (17/16)^{t-1}\,16^{-t}36^{-2}c_2(t,2) + (17/4)^t\,36^{-1}\big). \tag{51}$$

Combining (43), (44) and (51), we obtain

$$E|S_n|^t < a(t)u_0^t , \tag{52}$$

if the conditions (32) are satisfied.

Assume now that $A_t/u_t^0 > 1/36$. Then $u_0' = (36A_t)^{1/t}$ satisfies the conditions (32) and by (52)

$$E|S_n|^t < 36a(t)A_t . \tag{53}$$

The assertion of the Theorem follows from (52) and (53).

COROLLARY 4. For any $t>1$

$$E|S_n|^t < 12a(t) \max [2(E|S_n|)^t, 3A_t] .$$

Proof. Obviously, $P(|S_n|\geq u)\leq E|S_n|/u$. Hence it is possible to put in (41) $u_0=24E|S_n|$, which completes the proof. Corollary 4 generalizes Theorem 2 of [2] and makes it more precise.

COROLLARY 5. If $P(|S_n|\geq u_0)\leq 1/24$, then

$$E|S_n|^2 < a(2) \max [u_0^2, 36B_n^2] ; \tag{54}$$

$$E|S_n| < a^{1/2}(2) \max [u_0, 6B_n] . \tag{55}$$

Proof. The inequality (54) is obtainable from (41) for $t=2$. The inequality (55) follows from (54) and the inequality $E|S_n|\leq(E|S_n|^2)^{1/2}$.

The mathematical expectation $E|S_n|$ is frequently used in the investigations concerning distributions in Banach spaces (see, for example, [1, 2, 7]).

Corollary 5 shows that instead of $E|S_n|$ one can use u_0 in many cases.

REFERENCES

[1] V.V. Yurinskij. "Exponential Bounds for Large Deviations," *Theory Prob. Applications*, 1 (1974): 154-155.

[2] I.F. Pinelis. "On the Distribution of Sums of Independent Random Variables with Values in a Banach Space." *Theory Prob. Applications*, 3 (1978): 608-615.

[3] I.I. Gikhman and A.V. Skorokhod. *Introduction to the Theory of Random Processes.* Philadelphia: Scripta Technica, W.B. Saunders Co., 1969.

[4] J.P. Kahane. *Some Random Series of Functions.* Lexington, Massachusetts: Heath, 1968.

[5] V.V. Buldygin. "On the Lévy Inequality for Random Variables in a Banach Space." *Theory Prob. Applications,* 1 (1974): 156-159.

[6] M. Loève. *Theory of Probability.* Princeton, New Jersey: Van Nostrand, 1963.

[7] I. Hoffman-Jorgensen. *Probability in B-spaces.* Aarhus University, 1977. (Lecture Notes ser., no. 48.)

[8] D.Kh. Fuk and S.V. Nagaev. "Probability Inequalities for Sums of Independent Random Variables." *Theory Prob. Applications,* 4 (1971): 643-660.

[9] N.C. Jain and M.B. Marcus. "Integrability of Infinite Sums of Independent Vector-valued Random Variables." *Trans. Amer. Math. Soc.,* Vol. 212 (1975): 1-35.

[10] V.V. Yurinskij. "Exponential Inequalities for Sums of Independent Random Vectors." *J. Multivar. Anal.,* Vol. 6, No. 4 (1976): 473-499.

INDUCTIVE LIMITS OF DIRECTED SYSTEMS OF CONTINUOUS MEASURES

G.V. Nedogibchenko and L.Ya. Savel'ev

By a continuous measure we mean the continuous additive mapping of a topological Boolean algebra into a topological Abelian semigroup [1]. In this article we define the inductive limit of the directed systems of continuous measures and describe some of its properties. The inductive limits of sequences of continuous measures are discussed in [2].

1. INDUCTIVE MEASURES OF TOPOLOGICAL BOOLEAN ALGEBRAS AND RINGS

The topological algebra is defined as the algebra with topology, for which each operation is continuous in each variable.

1.1. Topological Boolean Algebras and Rings

By a Boolean algebra $(A,\vee,\wedge,\setminus)$ we call the set A for which the order $\leq$, the upperboundary $\vee$, the lower boundary $\wedge$ and the relative complement $\setminus$ are defined. The Boolean algebra $(A,\vee,\wedge,\setminus)$ defines the Boolean ring $(A,+,\cdot)$ with the sum $+$ and the product $\cdot$. The sum $+$ in the Boolean ring $(A,+,\cdot)$ is equal to the difference $-$, and the product $\cdot$ is idempotent: $x+x=0$, $x\cdot x=x$ $(x\in A)$.

The operations $\vee$, $\wedge$, $\setminus$ and $+$, $\cdot$ are expressible through the equalities

$$x+y = (x\backslash y) \vee (y\backslash x) , \qquad x\cdot y = x\wedge y ;$$

$$x\vee y = x+y+xy , \quad x\wedge y = x\cdot y , \quad x\backslash y = x+x\cdot y ;$$

the relative complement $\backslash$ is not commutative.

For the Boolean algebra $(A,\vee,\wedge,\backslash)$ there exists the smallest element 0 being the zero of the ring $(A,+,\cdot)$. If for the Boolean algebra $(A,\vee,\wedge,\backslash)$ there exists the largest element 1, the latter is the unity of the Boolean algebra $(A,+,\cdot)$. For the Boolean algebra with unity the complement $'$ is defined: $x'=1\backslash x=1+x$, $x\backslash y=x\cdot y'$.

We consider the Boolean algebra $(A,\vee,\wedge,\backslash)$, the Boolean ring $(A,+,\cdot)$ and the topology T for the set A. We adopt that the topology T is compatible with the operation $\circ$ for A if and only if the transformations $x\to a\circ x$ and $x\to x\circ a$ of the set A are continuous for each $a\in A$. The Boolean algebra $(A,\vee,\wedge,\backslash)$ and the topology T compatible with each of the operations $\vee$, $\wedge$, $\backslash$ for the set A, form the topological Boolean algebra $(A,\vee,\wedge,\backslash,T)$. The Boolean ring $(A,+,\cdot)$ and the topology T compatible with each of the operations $+$, $\cdot$ for the set A, form the topological Boolean ring $(A,+,\cdot,T)$.

We say that the topology T is (sequentially) compatible with the order $\leq$ for set A if and only if for each upper bounded increasing (sequence) directed system in A there exists an upper boundary of the set of values and the (sequence) directed system converges to it, and for each lower bounded decreasing (sequence) directed system in A there exists a lower boundary of the set of values $(B,\vee,\wedge,\backslash)$ and the (sequence) directed system converges to it (compare with [3]).

Now we consider the subalgebra $(B,\vee,\wedge,\setminus)$ of the Boolean algebra $(A,\vee,\wedge,\setminus)$ and the Boolean ring $(B,+,\cdot)$ defined by this subalgebra. We adopt that $(B,\vee,\wedge,\setminus)$ is the ideal for $(A,\vee,\wedge,\setminus)$ if and only if the ring $(B,+,\cdot)$ is the ideal for the ring $(A,+,\cdot)$.

Some elementary properties of topological Boolean algebras and rings are described in [1].

1.2. Inductive Limits of Topological Boolean Algebras and Rings

We consider the increasing directed system of Boolean algebras $(C_i,\vee_i,\wedge_i,\setminus_i)$ and its inductive limit $(C,\vee,\wedge,\setminus)$:

$$C_i \subseteq C_k,\ \vee_i \subseteq \vee_k,\ \wedge_i \subseteq \wedge_k,\ \setminus_i \subseteq \setminus_k\ (i \leqslant k);$$
$$C = \cup C_i,\ \vee = \cup \vee_i,\ \wedge = \cup \wedge_i,\ \setminus = \cup \setminus_i. \tag{1}$$

<u>REMARK</u>. Operations are parts of Cartesian products of sets. Hence for them the inclusion and the union are defined.

For the set C_i we take the topology $\mathcal{T}_i$ compatible with each of the operations $\vee_i$, $\wedge_i$, $\setminus_i$. We assume that the topology $\mathcal{T}_i$ for C_i is induced by the topology $\mathcal{T}_k$ for C_k for each $i \leq k$:

$$\mathcal{T}_i = C_i \cap \mathcal{T}_k . \tag{2}$$

The class

$$\mathcal{T} = \{T \mid C_i \cap T \in \mathcal{T}_i \quad \forall i\} \tag{3}$$

of the parts T of the set C, the traces $C_i\ T$ of which on each set C_i belong to the topologies $\mathcal{T}_i$, is the largest among the topologies for C, under which each identical embedding $e_i : C_i \to C$ is continuous [4, Chapter 4, Section 2.4]. The topology $\mathcal{T}$ is the inductive limit of the directed system of topologies $\mathcal{T}_i$ and is denoted by $\lim\operatorname{ind} \mathcal{T}_i$.

Now we consider the mapping $f : C \to H$ of the space $(C, \mathcal{T})$ into an arbitrary topological space $(H, \mathcal{U})$. The definition of the inductive limit implies [4, Section 1.2, Proposition 6] the following.

CONTINUITY CRITERION. The mapping f is continuous if and only if each composition $f \circ e_i$ is continuous.

Using this criterion, it is easy to prove

PROPOSITION 1. The topology $\mathcal{T}$ is compatible with each of the operations $\vee$, $\wedge$, $\setminus$ for the set C.

Proof. We consider $a \in C$, the operation $\circ$ for C and the transformations s, t of the set C, defined by the equalities $s(x) = a \circ x$, $t(x) = x \circ a$ $(x \in C)$.

Let $a \in C_i$. For each $j \in J$ there exists $k \in J$ such that $i \le k$, $j \le k$ and $C_i \subseteq C_k$, $C_j \subseteq C_k$. The compositions $s \circ e_j$, $t \circ e_j$ are narrowings by C_j of the transformations of the set C_k and they all are continuous. Therefore s, t are also continuous.

Thus, $(C, \vee, \wedge, \setminus, \mathcal{T})$ is a topological Boolean algebra. It is called the inductive limit of the directed system of topological Boolean algebras $(C_i, \vee_i, \wedge_i, \setminus_i, \mathcal{T}_i)$ and is denoted by $\lim\operatorname{ind} (C_i, \vee_i, \wedge_i, \setminus_i, \mathcal{T}_i)$.

1.3. Properties of Inductive Limits of Directed Systems

The assumptions of this subsection describe some elementary properties of the topology $T = \lim\operatorname{ind} T_i$.

CLOSURE CRITERION. The set T' is closed under the topology T, if and only if $C_i \cap T'$ is closed under the topology T_i for each i.

This criterion follows immediately from the definitions.

We adopt to say that topological spaces in which each point is separated from any other point by a certain neighborhood is separable. This property is equivalent to the closure of one-point sets of the space (T_1).

COROLLARY. The space (C, T) is separable if and only if the space (C_i, T_i) is separable for each i.

Proof. Indeed, the intersection $C_i \cap \{c\}$ is zero or $\{c\}$ for each index i and each point $c \in C$.

We denote by T_i' and T' the classes of closed parts of sets C_i and C under the topologies T_i and T. The equalities (2), (3) are equivalent to the equalities

$$\mathcal{T}'_i = C_i \cap \mathcal{T}'_k \qquad (i \leqslant k), \tag{2'}$$

$$\mathcal{T}' = \{T' \mid C_i \cap T' \in \mathcal{T}'_i \qquad (\forall i)\}. \tag{3'}$$

PROPOSITION 2. If C_i is closed in the space (C_k, T_k) for each $i \leq k$, the C_i is closed in the space (C, T) for each i.

Proof. Indeed, if Proposition 2 is satisfied, $C_i \in T_k'$ and $C_i \cap C_j \in C_j \cap T_k' = T_j'$ for each $i \leq k$, $j \leq k$. Therefore, $C_i \in T'$.

PROPOSITION 3. If C_i is closed in the space (C_k, T_k) for each $i \leq k$, then $T_i = C_i \cap T$ for each i.

Proof. The continuity of the identical embedding $e_i: C_i \to C$ implies that $C_i \cap \mathcal{T} \subseteq \mathcal{T}_i$ for each i. We prove the inverse inclusion. We take an arbitrary $T_i \in \mathcal{T}_i$ and consider $T = T_i \cup (C \setminus C_i)$. It is clear that $T_i = C_i \cap T$. We show that $T \in \mathcal{T}$. By the equality (2) for each $i \le k$, $j \le k$ and some $T_k \in \mathcal{T}_k$ we have

$$C_j \cap T = C_j \cap C_k \cap T = C_j \cap (T_i \cup (C_k \setminus C_i)) =$$
$$= C_j \cap ((C_i \cap T_k) \cup (C_k \setminus C_i)) = C_j \cap (T_k \cup (C_k \setminus C_i)) \in \mathcal{T}_j.$$

Therefore, $T \in \mathcal{T}$, $T_i \in C_i \cap \mathcal{T}$ and $\mathcal{T}_i \subseteq C_i \cap \mathcal{T}$.

PROPOSITION 4. If $(C_i, \vee_i, \wedge_i, \setminus_i)$ is the ideal for $(C_k, \vee_k, \wedge_k, \setminus_k)$ for each $i < k$, and the topology $\mathcal{T}_i$ is (sequentially) compatible with the order $\le_i$ for C_i for each i, the topology $\mathcal{T}$ is (sequentially) compatible with the order $\le$ for C.

Proof. It follows from this proposition that the boundedness of the (sequence) directed system x in C implies its boundedness in some C_i. Indeed, let $x_\alpha \le a$ for each index α, $a \in C_i$ and $x_\alpha, a \in C_j$. Then $x_\alpha = a \cdot x_\alpha \in C_i$.

Since the topology $\mathcal{T}_i$ is (sequentially) compatible with the order $\le_i$ for the set C_i, the bounded increasing (sequence) directed system x converges in the space $(C_i, \mathcal{T}_i)$ to the upper boundary u of the set of its values. Therefore, x converges to u in the space $(C, \mathcal{T})$. The same situation is observed with decreasing (sequences) directed systems.

1.4. Properties of Inductive Limits of Sequences

We consider now the inductive limit $(C, \vee, \wedge, \setminus, \mathcal{T})$ of the increasing sequence of topological Boolean algebras $(C_n, \vee_n, \wedge_n, \setminus_n, \mathcal{T}_n)$:

$$C_n \subseteq C_{n+1},\ \vee_n \subseteq \vee_{n+1},\ \wedge_n \subseteq \wedge_{n+1},\ \setminus_n \subseteq \setminus_{n+1};$$
$$C = \cup C_n,\ \vee = \cup \vee_n,\ \wedge = \cap \wedge_n,\ \setminus = \cup \setminus_n; \tag{4}$$
$$\mathcal{T}_n = C_n \cap \mathcal{T}_{n+1}. \tag{5}$$

It is not hard to show that $\mathcal{T} = \lim\operatorname{ind} \mathcal{T}_n$ possesses some other properties in addition to those listed in Subsection 1.3 (compare with [5]).

PROPOSITION 5. $\mathcal{T}_n = C_n \cap \mathcal{T}$ for each n.

Proof. The continuity of the identical embedding $e_n : C_n \to C$ implies that $C_n \cap \mathcal{T} \subseteq \mathcal{T}_n$ for each n. We prove the inverse inclusion. Let $T_n \in \mathcal{T}_n$. According to the induction principle the equality (5) implies the existence of the sequence of sets $T_{n+p} \in \mathcal{T}_{n+p}$ such that

$$T_{n+p} = C_{n+p} \cap T_{n+p+q} \qquad (p,q \geq 0).$$

Let $T = \cup T_{n+q}$ $(q \geq 0)$. Since the sequence of sets T_{n+p} $(p \geq 0)$ increases, $T = \cup T_{n+p+q}$ $(q \geq 0)$ for each p. Hence

$$C_{n+p} \cap T = \cup (C_{n+p} \cap T_{n+p+q}) = T_{n+p} \qquad (q \geq 0).$$

In particular, $C_n \cap T = T_n$. At the same time

$$C_m \cap T = C_m \cap (C_n \cap T) \in C_m \cap \mathcal{T}_n = \mathcal{T}_m \qquad (m \leq n).$$

Therefore, $T \in \mathcal{T}$ and $\mathcal{T}_n \subseteq C_n \cap \mathcal{T}$.

Now we consider the sequence x of points C and the point $a \in C$. Proposition 5 implies the following.

COROLLARY. If each space $(C_n, \mathcal{T}_n)$ is separable, $x \to a$ in the space $(C, \mathcal{T})$ if and only if $x \to a$ in one of the spaces

(C_n, T_n).

Proof. We assume that the set of values of the sequence x is not contained in either set C_n and $a \in C_m$. In this case there exists a strictly increasing sequence of numbers $k(n)$, such that $x(k(n)) \notin C_{m+n}$ for each n. We consider the set X of all these values of the sequence x and its complement $T = C \setminus X$.

Since $C_m \cap X = 0$, $a \in T$. For each p the set $C_p \cap X = \{x(k(n)) \mid n < p-m\}$ is finite. Since the space (C_p, T_p) is separable, it is closed under the topology T_p. Therefore X is closed under the topology T and $T \in T$. Therefore T is an open neighborhood of the point a in the space (C, T). This implies that the sequence x together with the subsequence chosen does not converge to the point a in the space (C, T).

We assume that the set of values of the sequence x is contained in the set C_m and $a \in C_m$ for some m. It follows from Proposition 5 that in this case the convergence of the sequence x to the point a in the space (C, T) is equivalent to the convergence of x to a in the space (C_m, T_m).

1.5. Sequential Topologies

Topologies under which each sequential closed set is closed will be called sequential. Among sequential topologies for the set C, which contain the topology T for C, there is the smallest topology. We call it a sequential topology generated by T; we denote it by I. The class of closed sets under the topology I is equal to the class of sequentially closed sets under the topology T. The sequential topology I is the largest among topologics for the set C, under which the sequential limit contains the sequential limit under the topology T [6, Chapter 1, Section 7]. Sequential limits under the topologies T and I are equal. If $(C,\vee,\wedge,\setminus,T)$ is a topological Boolean algebra, $(C,\vee,\wedge,\setminus,I)$ also is a topological Boolean algebra [1, Subsection 5.3].

Next, we consider the inductive limit $(C,\vee,\wedge,\setminus,T)$ of the sequence of topological Boolean algebras $(C_n,\vee_n,\wedge_n,\setminus_n,T_n)$; the sequential topology I generated by the topology T, and the topological Boolean algebra $(C,\vee,\wedge,\setminus,I)$; the sequence of sequential topologies I_n generated by the topologies T_n, and the sequence of topological Boolean algebras $(C_n,\vee_n,\wedge_n,\setminus_n,I_n)$. We assume that for each n the set C_n is closed in the space (C_{n+1},T_{n+1}) and the space (C_n,T_n) is separable. The closure of the C_n in (C_{n+1},T_{n+1}) implies the closure of C_n in (C_{n+1},I_{n+1}), the separability of (C_n,T_n) implies the separability of (C_n,I_n).

<u>LEMMA</u>. $I_n=C_n\cap I_{n+1}$ for each n.

<u>Proof</u>. The trace Y on C_n of each sequentially open for T_{n+1} part Z of the set C_{n+1} is sequentially open under the topology T_n. At the same time, $T_n=C_n\cap T_{n+1}$. Hence $C_n\cap I_{n+1}\subseteq I_n$. At the same time each sequentially open under the topology T_n part Y of the set C_n is the trace on C_n of the set $Z=Y\cup(C_{n+1}\setminus C_n)$. It follows from the closure of the set C_n in the space (C_{n+1},T_{n+1}) that Z is sequentially open under the topology T_{n+1}. Hence $I_n\subseteq C_n\cap T_{n+1}$.

Therefore, $\mathcal{I}_n = C_n \cap \mathcal{I}_{n+1}$.

The Lemma proved plus the equalities (4) enable us to examine the inductive limit $\lim\operatorname{ind} \mathcal{I}_n$ of the sequence of topologies $\mathcal{I}_n$.

PROPOSITION 6. $\mathcal{I} = \lim\operatorname{ind} \mathcal{I}_n$.

Proof. We consider the arbitrary sequence x and the point a in C. It follows from Proposition 5 that the convergence $x \to a$ in the space $(C, \mathcal{T})$ implies the convergence $x \to a$ in some space $(C_n, \mathcal{T}_n)$. Since the sequential limits under the topologies $\mathcal{T}_n$ and $\mathcal{I}_n$ are equal, $x \to a$ in the space $(C, \lim\operatorname{ind} \mathcal{I}_n)$. Therefore, $x \to a$ in the space $(C, \lim\operatorname{ind} \mathcal{I}_n)$.

The sequential topology $\mathcal{I}$ is the largest among the topologies $\mathcal{R}$ for the set C, under which the convergence $x \to a$ in the space $(C, \mathcal{T})$ implies the convergence $x \to a$ in the space $(C, \mathcal{R})$ for each sequence x and each point a in C. Hence it follows that $\lim\operatorname{ind} \mathcal{I}_n \subseteq \mathcal{I}$.

Now we consider the arbitrary number n, the sequence x and the point a in C_n. Since the convergence $x \to a$ in the space $(C_n, \mathcal{T}_n)$ implies the convergence of the sequence $e_n(x) = x$ to the point a in the space $(C, \mathcal{T})$, the identical embedding $e_n : C_n \to C$ is continuous under the topologies $\mathcal{I}_n$ and $\mathcal{I}$ [7, Proposition 5.2, Corollary 2]. The topology $\lim\operatorname{ind} \mathcal{I}_n$ is the largest among the topologies $\mathcal{Q}$ for the set C, such that each identical embedding $e_n : C_n \to C$ is continuous under the topologies $\mathcal{I}_n$ and $\mathcal{Q}$. Hence it follows that $\mathcal{I} \subseteq \lim\operatorname{ind} \mathcal{I}_n$. //

1.6. The Inductive Limit of Closures

We consider again the increasing directed system of topological Boolean algebras $(C_i, \vee_i, \wedge_i, \setminus_i, T_i)$ and its inductive limit $(C, \vee, \wedge, \setminus, T)$. Let $A_i \subseteq C_i$ be such that $A_i = A_k \cap C_i$ for each $i \leq k$; we denote by A the inductive limit of the directed system of the sets A_i. This limit is equal to the union: $A = \cup A_i$. We note that

$$A_i \cap C_j = A_k \cap C_i \cap C_j = A_k \cap C_j \cap C_i = A_j \cap C_i$$

for each $i \leq k$, $j \leq k$. Using the distributivity of the union plus these equalities, we convince ourselves that under the assumptions made concerning A_j the following lemma holds.

LEMMA. $A \cap C_j = A_j$ for each j.

We consider the closure $\bar{A}_i$ of the set A_i in the space (C_i, T_i) and the closure $\bar{A}$ of the set A in the space (C, T).

PROPOSITION 7. If $\bar{A}_i = \bar{A}_k \cap C_i$ for each $i \leq k$, then $\bar{A} = \cup \bar{A}_i$.

Proof. It is seen that $A = \cup A_i \subseteq \bar{A}_i = B$. At the same time $A_i = A_i \cap C_i \subseteq A \cap C_i \subseteq \bar{A} \cap C_i$ for each i. Since the set $\bar{A}$ is closed in the space (C, T), the set $\bar{A} \cap C_i$ is closed in the space (C_i, T_i). Hence from $A_i \subseteq \bar{A} \cap C_i$ it follows that $\bar{A}_i \subseteq \bar{A} \cap C_i$. This implies that $B \subseteq \bar{A}$. Therefore, $A \subseteq B \subseteq \bar{A}$. Therefore, to prove Proposition 7 it suffices to show that the set B is closed in the space (C, T). By the Lemma, from the condition $\bar{A}_i = \bar{A}_k \cap C_i$ $(i \leq k)$ it follows that $B \cap C_j = \bar{A}_j$, and hence $B \cap C_j$ is closed in (C_j, T_j) for each j. Therefore, B is closed in (C, T). //

Let us say that the set A_i bounds the set C_i iff for

each element c of C_i there exists an element $a\geq c$ of A_i. We assume in addition that $(A_i, \vee_i, \wedge_i, \backslash_i)$ is the subalgebra of the algebra $(C_i, \vee_i, \wedge_i, \backslash_i)$ for each i. Proposition 7 implies the following.

<u>COROLLARY</u>. If $(A_i, \vee_i, \wedge_i, \backslash_i)$ is the ideal for $(A_k, \vee_k, \wedge_k, \backslash_k)$ and A_i bounds C_i for each $i\leq k$, then $\bar{A}=\cup\bar{A}_i$.

<u>Proof</u>. Indeed, $A_i=A_k\cap C_i\subseteq\bar{A}_k\cap C_i$, the set $\bar{A}_k\cap C_i$ is closed in the space $(C_i, \mathcal{T}_i)$ and, therefore, $\bar{A}_i\subseteq\bar{A}_k\cap C_i$ for each $i\leq k$. Let $c\in\bar{A}_k\cap C_i$. We take the directed system x in A_k converging to c in the space $(C_k, \mathcal{T}_k)$. If A_i bounds C_i, there exists $a\in A_i$ such that $a\geq c$. In this case the directed system ax converges to $ac=c$ in the space $(C_k, \mathcal{T}_k)$. If $(A_i, \vee_i, \wedge_i, \backslash_i)$ is the ideal for $(A_k, \vee_k, \wedge_k, \backslash_k)$, all of the values of ax belong to the set A_i, and ax converges also to c in the space $(C_i, \mathcal{T}_i)$. Therefore, $c\in\bar{A}_i$ and $\bar{A}_k\cap C_i\subseteq\bar{A}_i$. Thus, $\bar{A}_i=\bar{A}_k\cap C_i$ for each $i\leq k$. By Proposition 7 it follows that $\bar{A}=\cup\bar{A}_i$.

Next we consider an example illustrating that the closure $\bar{A}$ in the space $(C, \mathcal{T})$ of the inductive limit A of the directed system of sets A_i need not be equal to the inductive limit $\cup\bar{A}_i$ of the directed system of their closures $\bar{A}_i$ in the space $(C_i, \mathcal{T}_i)$.

We take the inductive limit $(\mathcal{C}, \cup, \cap, \backslash, T)$ of the increasing sequence of topological Boolean algebras $(\mathcal{C}_n, \cup_n, \cap_n, \backslash_n, T)$ in which $\mathcal{C}_n$ is the class of all segments of the interval $[0, n+1]$ of the real line, and T_n is the sequential order topology for this class [7, Section 1.7.2]. Let

$$J_m = [1/(m+1),\ 1/m[\ , \qquad Q_m = Q\cap[m,m+1[\ ;$$

$$J_K = \cup J_m\ , \qquad Q_K = \cup Q_m (m\in K)\ , \qquad U_K = J_K + Q_K$$

for each number m and each finite set of numbers K. In particular, $U_k=0$ for $K=0$. We denote by A_n the class of all unions $U_k\cup F$ of the sets U_k with $K\subseteq\{1,\ldots,n\}$ and of all finite sets $F\subseteq Q\cap[1,n+1[$. It is not hard to verify that $(A_n, \cup_n, \cap_n, \setminus_n)$ is the subalgebra of the algebra $(C_n, \cup_n, \cap_n, \setminus_n)$.

For each $K\subseteq\{1,\ldots,n\}$ the set Q_K is the union of some increasing sequence of finite sets $F_K\in A_n$. Hence the set $J_K=U_K\setminus Q_K$ is the intersection of the decreasing sequence of sets $G_K=U_K\setminus F_K\in A_n$. Therefore, the set J_K belongs to the closure $\bar{A}_n$ of the class A_n in the space (C_n,T_n). In particular, $H_n=J_{\{1,\ldots,n\}}\in\bar{A}_n$. At the same time, $H_n\in C_0$ for each n. Hence the sequence of the set H_n converges to its union $]0,1[$ in the space (C,T). But the interval $]0,1[$ does not belong to the union $\cup\bar{A}_n$ of the classes $\bar{A}_n$, since all of the sets of the class $\bar{A}_n$ together with sets of the class A_n are contained in the interval $[1/(n+1),\ n+1]$. Therefore, the union $\cup\bar{A}_n$ is not closed in the space (C,T) and need not be equal to the closure $\bar{A}$ in this space A of the classes A_n.

1.7. Adjunction of an Identity Element

Each topological Boolean algebra is isomorphic to the subalgebra of some topological Boolean algebra with an identity element. This can easily be proved using the general definition of the inductive limit of the family of topologies as well as Stone's theorem on the representation of Boolean algebras. Suppose

$(A, \vee, \wedge, \setminus, T)$ is the topological Boolean algebra;

$(D, \vee, \wedge, \setminus)$ is the Boolean algebra with a set of elements $D=\{0,1\}$;

H is the set of all homomorphisms of A in D;

P is the class of all parts of the set H;

$(P, \cup, \cap, \setminus)$ is the Boolean algebra with elements given by the sets of the class P and with operations given by the union, intersection, and a relative complement:

$$\varphi : A \to \mathcal{P},\ \varphi(a) = \{h \in H \mid h(a) = 1\} \qquad (a \in A);$$
$$\varphi' : A \to \mathcal{P},\ \varphi'(a) = \{h \in H \mid h(a) = 0\} \qquad (a \in A);$$
$$\mathcal{A} = \varphi(A),\ \mathcal{A}' = \varphi'(A),\ \mathcal{B} = \mathcal{A} \cup \mathcal{A}';$$

U is the largest topology for B, under which the mappings ϕ and ϕ' are continuous;

$T=A\cap U$ is the topology for A, induced by the topology U.

PROPOSITION 8. $(B, \cup, \cap, \setminus, U)$ is a topological Boolean algebra with an identity element; $(A, \vee, \wedge, \setminus, T)$ is isomorphic to the subalgebra $(A, \cup, \cap, \setminus, T)$.

Proof. By Stone's theorem ϕ is the isomorphism of the Boolean algebra $(A, \vee, \wedge, \setminus)$ onto the subalgebra $(A, \cup, \cap, \setminus)$ of the Boolean algebra $(P, \cup, \cap, \setminus)$ ([8, Chapter VIII, Section 40]; [9, Chapter I, Section 2]). It is not hard to verify that $(B, \cup, \cap, \setminus)$ is the subalgebra of the Boolean algebra $(P, \cup, \cap, \setminus)$. The identity H belongs to B since $H=\phi'(0)$. Therefore, $(A, \cup, \cap, \setminus)$ is the subalgebra of the Boolean algebra $(B, \cup, \cap, \setminus)$

with an identity.

We prove now that $(\mathcal{B}, \cup, \cap, \setminus, U)$ is a topological Boolean algebra. Let $B \in \mathcal{B}$ and let $\cup_B$, $\cap_B$, C_N, D_N, denote the transfers of the set $\mathcal{B}$ with the coefficient B:

$$\cup_B(X) = B \cup X , \quad \cap_B(X) = B \cap X , \quad C_B(X) = B \setminus X , \quad D_B(X) = X \setminus B.$$

Instead of C_H we agree to write also C: $C(X) = H \setminus X = X'$ $(X \in \mathcal{B})$.

All the transfers considered are some compositions of $\cap_B$ and C. Hence, to prove that the transfers are continuous, it suffices to show that $\cap_B$ and C are continuous. It follows from the definition of the topology U that the continuity of these mappings is equivalent to the continuity of the compositions $\cap_B \phi$, $\cap_B \phi'$ and $C\phi$, $C\phi'$ [4, Chapter 1, Section 2, Proposition 6]. Since $C\phi = \phi'$ and $C\phi' = \phi$, the compositions $C\phi$ and $C\phi'$ are continuous together with ϕ and ϕ'.

For $\cap_B \phi$ and $\cap_B \phi'$ we consider two cases: $B = \phi(A)$ and $B = \phi'(A)$, $(a \in A)$.

Let $B = \phi(a)$ $(a \in A)$. Then

$$\cap_B(\phi(x)) = \phi(a) \cap \phi(x) = \phi(a \wedge x) \qquad (x \in A) ;$$

and the continuity of $\cap \phi$ follows from the continuity of the transfer $x \to a \wedge x$ of the set A and the mapping ϕ. If $B = \phi'(a)$ $(a \in A)$, then

$$\cap_B(\phi(x)) = \phi'(a) \cap \phi(x) = \phi(x \setminus a) \qquad (x \in A)$$

and the continuity of $\cap_B \phi$ follows from the continuity of the transfer $x \to x \setminus a$ of the set A and the mapping ϕ. The continuity of the composition $\cap_B \phi'$ can be proved in a similar way.

To prove the second part of Proposition 8 it suffices to show that $\phi(T)=T$. This follows from the definition of the topology U, one-to-one correspondence of the mappings ϕ, ϕ' and the equality $A \cap A'=0$ [4, Chapter 1, Section 2, Proposition 8]. //

2. INDUCTIVE LIMITS OF CONTINUOUS MEASURES

By a continuous measure we mean a continuous additive mapping of the topological Boolean algebra onto a topological Boolean subgroup. The inductive limit of the directed system of continuous measures is a continuous measure for which a continuous extension exists.

2.1. The Definition of the Inductive Limit of Continuous Measures

We consider the Boolean algebra $(A, \vee, \wedge, \setminus)$ and the Abelian semigroup $(H,+)$ with zero. The mapping $m: A \to H$ for which $m(0)=0$ and $m(x \vee y)=m(x)+m(y)$ for each disjunct x, y of A, is said to be the additive mapping of A into H, or the measure on A with values in H. Let us take the topology T for the set A compatible with the operations $\vee$, $\wedge$, $\setminus$, and the topology U for the set H with the operation $+$ (the transform $t \to h+t$ of the set H is continuous for each $h \in H$). We consider the topological Boolean algebra $(A, \vee, \wedge, \setminus, T)$ and the topological Abelian semigroup $(H, +, U)$ with zero. A continuous additive mapping of A into

H is said to be the continuous measure on A with values in H.

Assume that there exists a closed part F and a part G of the set H, satisfying the following conditions:

1. $0\in F$;
2. $F+F\subseteq G$ $(F=0+F\subseteq G)$;
3. the induced topology $G\cap\mathcal{U}$ is generated by some uniformity $\mathcal{W}$ for G;
4. the restriction of the sum $+$ for H to the set $F\times F$ is uniformly continuous.

In this case we say that $(H,+,\mathcal{U})$ and F, G, $\mathcal{W}$ form a uniform Abelian semigroup $(H,+,\mathcal{U},F,G,\mathcal{W})$.

When $F=G=H$, a uniform Abelian semigroup $(H,+,\mathcal{U},\mathcal{W})$ is formed. The sum $+$ for H is then uniformly continuous. If the uniform space $(G,\mathcal{W})$ is separable and (sequentially) complete, we say that the uniform Abelian semigroup $(H,+,\mathcal{U},F,G,\mathcal{W})$ is separable and (sequentially) complete.

If the set of values $m(A)$ of the measure $m:A\to H$ is contained in the closed set F, we say that the measure m assumes on values in the uniform Abelian semigroup $(H,+,\mathcal{U},F,G,\mathcal{W})$.

We consider now the increasing directed system of topological Boolean algebras $(A_i,\vee_i,\wedge_i,\setminus_i,\mathcal{T}_i)$ and its inductive limit $(A,\vee,\wedge,\setminus,\mathcal{T})$, the topological Abelian semigroup $(H,+,\mathcal{U})$ with zero, the increasing directed system of continuous measures $m_i:A_i\to H$ and its inductive limit $m:A\to H$ equal to the union of these measures: $m=\lim\operatorname{ind} m_i=\cup m_i$.

REMARK. The measures $m_i : A_i \to H$ are parts of the Cartesian product of the sets A and H. Hence, for them the union and the order in inclusion are defined. The increase of the directed system of the measures m_i implies that $m_i \subseteq m_k$ (m_i is the restriction to m_k) for $i \le k$.

PROPOSITION 9. The inductive limit m of the increasing directed system of the continuous measures $m_i : A_i \to H$ is a continuous measure on $A = \cup A_i$ with values in H.

Proof. Since for each i, j there exists $k \ge i$, $k \ge j$ and $m_i \subseteq m_k$, $m_j \subseteq m_k$, the correspondence $m = \cup m_i$ is one-to-one. It follows from the equality $A = \cup A_i$ that m is defined on A. The additivity of each measure m_i implies the additivity of m; and the continuity of each measure m_i implies the continuity of m by the continuity criterion plus the equality $me_i = m_i$ for the composition of the identical embedding $e_i : A_i \to A$ with m.

Proposition 9 implies the following.

COROLLARY. Each continuous measure m on $A = \cup A_i$ with values in H is the inductive limit of the increasing directed system of its restrictions $m_i : A_i \to H$.

2.2. A Topological Extension of the Inductive Limit of Continuous Measures

We consider the increasing directed systems of topological Boolean algebras $(C_i, \vee_i, \wedge_i, \setminus_i, T_i)$ and the inductive limit $(C, \vee, \wedge, \setminus, T)$. Let $A_i \subseteq C_i$ such that A_i equals $A_k \cap C_i$ for each $i \le k$ and is the set of elements of a subalgebra of the Boolean algebra $(C_i, \vee_i, \wedge_i, \setminus_i)$ for each i.

Let $\bar{A}_i$ be the closure of the set A_i in the space $(C_i, \mathcal{T}_i)$, $B=\cup\bar{A}_i$, $A=\cup A_i$, $\bar{A}$ be the closure of the set A in the space $(C, \mathcal{T})$.

We consider next the topological Abelian semigroup $(H,+,\mathcal{U})$ with zero, the increasing directed system of continuous measures $m_i: A_i \to H$ and the inductive limit $m: A \to H$. We assume that for each i there exists a closed set $F_i \subseteq H$, a set $G_i \subseteq H$ and a uniformity $\mathcal{W}_i$ for G_i, which satisfy conditions 1-4 of Subsection 2.1 for $F=F_i$, $G=G_i$ and $\mathcal{W}=\mathcal{W}_i$. We also assume that for each i the measure m_i assumes values in the uniform Abelian semigroup $(H,+,\mathcal{U},F_i,G_i,\mathcal{W}_i)$ and that this semigroup is Hausdorff and complete.

Under the above assumptions we have the following theorem.

<u>THEOREM 1</u>. There exists a unique continuous measure $n: B \to H$, extending the continuous measure $m: A \to H$.

<u>Proof</u>. For each $i \le k$ we denote by $\bar{A}_i^{(k)}$ the closure of the set A_i in the space $(C_k, \mathcal{T}_k)$. Since $A_i = C_i \cap A_k$ and $\mathcal{T}_i = C_i \cap \mathcal{T}_k$, then $\bar{A}_i = \bar{A}_i^{(i)} \subseteq \bar{A}_i^{(k)} \subseteq \bar{A}_k^{(k)} = \bar{A}_k$. By the theorem on extension of continuous measures [1, Theorem 1], there exists a unique, continuous under the topologies $\mathcal{T}_k$ and $\mathcal{U}$, measure $\bar{m}_i^{(k)}: \bar{A}_i^{(k)} \to H$, extending the measure $m_i: A_i \to H$, $\bar{m}_i = \bar{m}_i^{(i)} \subseteq \bar{m}_i^{(k)} \subseteq \bar{m}_k^{(k)} = \bar{m}_k$.

We take now the inductive limit n of the increasing directed system of the measures $\bar{m}_i: \bar{A}_i \to H$. By Proposition 9 this limit is a measure on $B=\cup\bar{A}_i$, continuous under the topologies $\mathcal{T}$ and $\mathcal{U}$ and having values in H. Since $\bar{m}_i$ extends

m_i for each i, $n=\cup\bar{m}_i$ extends $m=\cup m_i$. Then $B\subseteq\bar{A}$ and the uniqueness of this extension follows from its continuity. //

We assume in addition that $\bar{A}_i=C_i\cap\bar{A}_k$ for each $i\leq k$. In this case Proposition 7 and Theorem 1 imply the following.

COROLLARY 1. There exists a unique continuous measure $\bar{m}:\bar{A}\to H$, extending the continuous measure $m:A\to H$.

This corollary will hold if instead of $\bar{A}_i=C_i\cap\bar{A}_k$ we assume that $(A_i,\vee_i,\wedge_i,\backslash_i)$ is the ideal for $(A_k,\vee_k,\wedge_k,\backslash_k)$ for each $i\leq k$ and A_i restrict C_i for each i.

Next we consider the increasing directed system of Boolean algebras $(C_i,\vee_i,\wedge_i,\backslash_i)$, the inductive limit $(C,\vee,\wedge,\backslash)$ the topology $\mathcal{I}$ for C compatible with the operations $\vee$, $\wedge$, $\backslash$, the increasing directed system of topological Boolean algebras $(C_i,\vee_i,\wedge_i,\backslash_i,\mathcal{T}_i)$ with induced topologies $\mathcal{T}_i=C_i\cap\mathcal{I}$ for C_i, the inductive limit $(C,\vee,\wedge,\backslash,\mathcal{T})$. From the definitions it follows that each identical embedding $e_1:C_i\to C$ is continuous under the topologies $\mathcal{T}_i$ and $\mathcal{I}$. Therefore, $\mathcal{I}\subseteq\mathcal{T}$.

Let us take the subalgebra $(A,\vee,\wedge,\backslash)$ of the algebra $(C,\vee,\wedge,\backslash)$; the topological Abelian semigroup $(H,+,\mathcal{U})$ with zero; the measure $m:A\to H$, continuous under the topologies $\mathcal{I}$ and $\mathcal{U}$; the increasing directed system of the measures $m_i:A_i\to H$, continuous under the topologies $\mathcal{T}_i$ and $\mathcal{U}$ and being the restrictions of m to the sets $A_i=C_i\cap A$. We assume that for m_i all the conditions under which Theorem 1 was proved, are satisfied. Then we have the following.

COROLLARY 2. There exists a unique, continuous under the topologies $\mathcal{T}$ and $\mathcal{U}$, measure $n:B\to H$, extending the measure

$m: A \to H$, continuous under the topologies $\mathcal{I}$ and $\mathcal{U}$.

<u>REMARK</u>. Since $\mathcal{I} \subseteq \mathcal{T}$ and $B \subseteq \bar{A}$, Corollary 2 describes as well the unique, continuous under the topologies $\mathcal{I}$ and $\mathcal{U}$, extension of the measure m to the set B, if such an extension exists.

2.3. The Sequential Extension of the Inductive Limit of Continuous Measures

Again, we consider the increasing directed system of topological Boolean algebras $(C_i, \vee_i, \wedge_i, \setminus_i, \mathcal{T}_i)$ and the inductive limit $(C, \vee, \wedge, \setminus, \mathcal{T})$. Let $A_i \subseteq C_i$ such that A_i equals $A_k \cap C_i$ for each $i \le k$ and is the set of elements of the subalgebra of the Boolean algebra $(C_i, \vee_i, \wedge_i, \setminus_i)$ for each i.

Let $\mathcal{I}_i$ be the sequential topology generated by the topology $\mathcal{T}_i$ for C_i; let $\tilde{A}_i$ be the closure of the set A_i in the space $(C_i, \mathcal{I}_i)$; let $D = \cup \tilde{A}_i$; $A = \cup A_i$; and let $\bar{A}$ be the closure of the set A in the space $(C, \mathcal{T})$.

We consider the topological Abelian semigroup $(H, +, \mathcal{U})$ with zero, the increasing directed system of continuous, under the topologies $\mathcal{T}_i$ and $\mathcal{U}$, measures $m_i : A_i \to H$ and the inductive limit $m: A \to H$. We assume that for m_i all the conditions under which Theorem 1 was proved, are satisfied, except the condition for completeness of the uniform Abelian semigroup $(H, +, \mathcal{U}, F_i, G_i, \mathcal{W}_i)$. We assume that this semigroup is sequentially complete.

Under the foregoing assumptions we have the following theorem.

THEOREM 2. There exists a unique continuous measure $s:D\to H$, extending the continuous measure $m:A\to H$.

Proof. By the theorem on sequential extension of a continuous measure [1, Theorem 2] for each i there exists a continuous, under the topologies $\mathcal{T}_i$ and $\mathcal{U}$, measure $\tilde{m}_i:\tilde{A}_i\to H$, extending the measure $m_i:A_i\to H$. We denote by $\tilde{A}_i^{(k)}$ the closure of the set A_i in the space $(C_k,\mathcal{I}_k)$ for each $i\leq k$. Since $A_i=C_i\cap A_k$ and $\mathcal{T}_i=C_i\cap\mathcal{T}_k$, then $\mathcal{I}_i\supseteq C_i\cap\mathcal{I}_k$ (see the proof of the lemma in Subsection 1.5) and $\tilde{A}_i=\tilde{A}_i^{(i)}\subseteq\tilde{A}_i^{(k)}\subseteq\tilde{A}_k^{(k)}=\tilde{A}_k$ for $i\leq k$. From $\tilde{A}_i\subseteq\tilde{A}_k$ and $m_i\subseteq m_k$ it follows that $\tilde{m}_i\subseteq\tilde{m}_k$ for $i\leq k$.

Now we take the inductive limit s of the increasing directed system of continuous measures $\tilde{m}_i:\tilde{A}_i\to H$. By Proposition 9 this limit is a continuous, under the topologies $\mathcal{T}$ and $\mathcal{U}$, measure on $D=\cup\tilde{A}_i$ with values in H. Since $\tilde{m}_i$ extends m_i for each i, then $s=\cup\tilde{m}_i$ extends $m=\cup m_i$. In this case $D\subseteq\bar{A}$ and the uniqueness of this extension follows from its continuity. //

We assume in addition that C_i is closed in the space $(C_k,\mathcal{T}_k)$ and $\tilde{A}_i=C_i\cap\tilde{A}_k$ for each $i\leq k$. The argument similar to that used in proving the lemma in Subsection 1.5, shows that $\mathcal{I}_i=C_i\cap\mathcal{I}_k$ for each $i\leq k$. Therefore, it is possible to consider the topology $\mathcal{I}=\lim\operatorname{ind}\mathcal{I}_i$ for the set C. We denote by $\tilde{A}$ the closure of the set $A=\cup A_i$ in the space $(C,\mathcal{I})$. Since $\mathcal{I}_i\supseteq\mathcal{T}_i$ for each i, then $\mathcal{I}\supseteq\mathcal{T}$ and $\tilde{A}\subseteq\bar{A}$.

Under the additional assumptions, Proposition 7 and Theorem 2 imply the following.

COROLLARY. There exists a unique continuous measure $\tilde{m}:\tilde{A}\to H$, extending the continuous measure $m:A\to H$.

This corollary still holds if instead of $\tilde{A}_i = C_i \cap \tilde{A}_k$ we assume that $(A_i, \vee_i, \wedge_i, \setminus_i)$ is the ideal for $(A_k, \vee_k, \wedge_k, \setminus_k)$ for each $i \le k$ and A_i restricts C_i for each i.

REMARK. In a particular case, when the directed system in question is the sequence of separable topological Boolean algebras, the topology S is the sequential topology generated by the topology $\mathcal{T}$ (Proposition 6).

2.4. Example

We give here an example illustrating an extension of the inductive limit of the directed system of continuous measures. Let us consider the set E, the class $\mathcal{P}$ of all the parts of E and the topological Boolean algebra $(\mathcal{P}, \cup, \cap, \setminus, S)$ with standard operations $\cup$, $\cap$, $\setminus$ and the sequential order topology S ([7, 1.7.2]; [10, Section 2]). We take the increasing directed system of sets $E_i \subseteq E$, which determines the increasing directed system of subalgebras $(\mathcal{C}_i, \cup_i, \cap_i, \setminus_i, T_i)$ of the topological Boolean algebra $(\mathcal{P}, \cup, \cap, \setminus, S)$ in which $\mathcal{C}_i$ is the class of all the parts of the set E_i, and the operations $\cup_i$, $\cap_i$, $\setminus_i$ and the topology T_i are induced. We consider the inductive limit $(\mathcal{C}, \cup, \cap, \setminus, T)$ of this system.

The set $\mathcal{C}_i$ is closed in the space $(\mathcal{P}, S)$ for each i. Therefore, $\mathcal{C}_i$ is closed in $(\mathcal{C}_k, T_k)$ for each $i \le k$ [4, Chapter 1, Section 3]. The space $(\mathcal{P}, S)$ is Hausdorff [10, Section 2]. Therefore, the space $(\mathcal{C}_i, T_i)$ is Hausdorff. It follows from Propositions 2 and 3 that for i the set $\mathcal{C}_i$

is closed in the space $(\mathcal{C},T)$ and is a subspace of $(\mathcal{C},T)$. Since for each $i\leq k$ the Boolean algebra $(\mathcal{C}_i,\cup_i,\cap_i,\setminus_i)$ is the ideal for $(\mathcal{C}_k,\cup_k,\cap_k,\setminus_k)$ and the topology T_i is sequentially adapted to the order $\subseteq_i$ for $\mathcal{C}_i$, the topology T is sequentially adapted to the order $\subseteq$ for $\mathcal{C}$ (Proposition 4).

<u>REMARK</u>. The Boolean algebra $(\mathcal{C},\cup,\cap,\setminus)$ is a subalgebra of the Boolean algebra $(\mathcal{P},\cup,\cap,\setminus)$. But the topology T need not coincide with the topology $\mathcal{C}\cap S$ for $\mathcal{C}$, induced by the topology S for $\mathcal{P}$, i.e., $(\mathcal{C},T)$ need not be a subspace of the topological space $(\mathcal{P},S)$. To make oneself sure that this is the case, it suffices to consider the strictly increasing sequence of sets E_n and note that the sequence $X_n=E_{n+1}\setminus E_n$ converges to 0 under the topology $\mathcal{C}\cap S$ and diverges under the topology T (Corollary of Proposition 5).

From the equalities $T_i=\mathcal{C}_i \cup S=\mathcal{C}_i \cup T$ it follows that $\mathcal{C}\cup S\subseteq T$. At the same time, T_i is the sequential order topology S_i for the set $\mathcal{C}_i$. This is confirmed by arguing in the similar way as in proving the lemma in Subsection 1.5.

Let $\mathcal{A}_i\subseteq\mathcal{C}_i$ be such that $\mathcal{A}_i=\mathcal{C}_i\cap\mathcal{A}_k$ for each $i\leq k$, $E_i\in\mathcal{A}_i$ and $\mathcal{A}_i$ is the set of elements of a subalgebra of the Boolean algebra $(\mathcal{C}_i,\cup,\cap,\setminus)$ (and, therefore, $(\mathcal{P},\cup,\cap,\setminus)$) for each i. Let $\bar{\mathcal{A}}_i$ be the closure of $\mathcal{A}_i$ in the space $(\mathcal{C}_i,T_i)$; $\mathcal{A}=\cup\mathcal{A}_i$; and let $\bar{\mathcal{A}}$ be the closure of $\mathcal{A}$ in the space $(\mathcal{C},T)$. The closure $\bar{\mathcal{A}}_i$ is the class $\sigma(\mathcal{A}_i)$ of the parts of the set E_i being elements of the σ-ring generated by the class $\mathcal{A}_i$ [7, Proposition 10.2].

The conditions $E_i\in\mathcal{A}_i$ and $\mathcal{A}_i=\mathcal{C}_i\cap\mathcal{A}_k$ imply that

$(\mathcal{A}_i, \cup, \cap, \setminus)$ is the ideal for $(\mathcal{A}_k, \cup, \cap, \setminus)$ for each $i \leq k$ and $\mathcal{A}_i$ restricts $\mathcal{C}_i$ for each i. Indeed, since $\mathcal{A}_i = \mathcal{C}_i \cap \mathcal{A}_k \subseteq \mathcal{A}_k$ for $i \leq k$, then $A_i \cap A_k \subseteq A_i \subseteq E_i$, $A_i \cap A_k \in \mathcal{C}_i$ and $A_i \in \mathcal{A}_k$, $A_i \cap A_k \in \mathcal{A}_k$, $A_i \cap A_k \in \mathcal{C}_i \cap \mathcal{A}_k = \mathcal{A}_i$ for each $A_i \in \mathcal{A}_i$, $A_k \in \mathcal{A}_k$ for $i \leq k$. Furthermore, $E_i \in \mathcal{A}_i$ is the unity of the Boolean algebra $(\mathcal{C}_i, \cup, \cap, \setminus)$ for each i. Therefore, $\bar{\mathcal{A}} = \cup \bar{\mathcal{A}}_i$ (Corollary of Proposition 7). From the sequential coordination of the topology T to the order $\subseteq$ it follows that the closure $\bar{\mathcal{A}}$ contains the class $\delta(\mathcal{A})$ of parts of the set E, that is the elements of the δ-ring generated by the class $\mathcal{A}$ (by a δ-ring we mean the ring of sets which is closed with respect to countable intersections).

We consider now the Hausdorff commutative semigroup $(H, +, \mathcal{U})$ with zero. We assume in addition that the sum is continuous in both variables at the same time. We take the increasing directed system of continuous measures $m_i : \mathcal{A}_i \to H$ and the inductive limit $m : \mathcal{A} \to H$. We assume that for each i the set $m(\mathcal{A}_i)$ is relatively compact. Next we take the compact sets $F_i \supseteq m_i(\mathcal{A}_i)$ and $G_i \supseteq F_i + F_i$. (For example, $F_i = \overline{m_i(\mathcal{A}_i)}$ and $G_i = F_i + F_i$. The set $F_i + F_i$ is compact since it is the image of the compact $F_i \times F_i$ under the continuous mapping $(x,y) \to x+y$.) There exists a unique uniformity $\mathcal{W}_i$ generating the induced topology $G_i \cap \mathcal{U}$ for G_i. The restriction of the sum for H on the set $F_i \times F_i$ is uniformly continuous as any continuous mapping of the compact.

Thus, for $F = F_i$, $G = G_i$ and $\mathcal{W} = \mathcal{W}_i$, conditions 1-4 of Subsection 2.1 are satisfied. Therefore, Corollary 1 of Theorem 1 is applicable to the continuous measure $m:A \to H$; and there exists a unique continuous measure $\bar{m}:\bar{A} \to H$ extending m. This implies that there exists a unique continuous measure $n:B \to H$ on $B=\delta(A)$, extending m.

2.5. The Application to Vector Measures

The extension scheme described in Subsection 2.4 is applicable in particular to vector measures. Let H be the set of points of a Banach space having the Radon-Nikodym property, and let U be the strong topology for H. In this case the continuity of the measure m_i implies the countable additivity of m_i and boundedness of its variation ([11, Section 1, Proposition 1; Section 2, Lemma 4; Section 5, Corollary 1] and [1, 2.2, Proposition]). If the measure m_i is non-atomic, the relative compactness of $m_i(A_i)$ is necessary for the existence of a continuous measure $\bar{m}_i:\bar{A}_i \to H$, extending the measure m_i. This follows from Lyapunov's theorem for vector measures assuming values in a Banach space with the Radon-Nikodym property [12, Chapter IX, Section 1, Theorem 10].

Let H be the set of points of the weakly sequentially complete Banach space and let U be the weak topology for H. In this case the continuity of the measures implies their weak countable additivity, which, in turn, implies strong countable additivity ([12, Chapter 1, Section 4, Corollary 1 of Theorem 2]; [13, Theorem 2.3]). The relative weak compactness of $m_i(A_i)$ is necessary for the existence of the continuous measure $\bar{m}_i:\bar{A}_i \to H$, extending the measure m_i. This follows from the theorem on a countably additive extension for vector measures with values in a Banach space ([12, Chapter 1, Section 5, Corollary 7 of Theorem 1 and Theorem 2]; [13, Theorem 4.1]; compare with [14, Corollaries 1 and 2]; [15]).

REFERENCES

[1] L.Ya. Savel'ev. "Extension of Continuous Measures." *Siberian Mathematical J.*, 20, 5 (1979): 765-771.

[2] L.Ya. Savel'ev. "Inductive Limits of Sequences of Continuous Measures." *Soviet Math., Doklady*, 20, 4 (1979): 854-857.

[3] E.E. Floyd. "Boolean Algebras with Pathological Order Topologies." *Pacific J. of Mathematics*, Vol. 5 (1955): 687-689.

[4] N. Bourbaki. *General Topology. Basic Structures.* Addison-Wesley, 1966. Russian translation: Moscow: Nauka, 1968.

[5] D.J.H. Garling. "A Generalized Form of Inductive Limit Topology for Vector Spaces." *Proceedings of the London Mathematical Society*, 14, 53 (1964): 1-28.

[6] N. Bourbaki. *General Topology. Basic Structures.* Russian translation: Moscow: Nauka, 1958.

[7] L.Ya. Savel'ev. *Lektsii po matematicheskomu analizu.* Chast' 4. Prilozhenie. Novosibirsk: Novosibirskij Gosudarstvennyj Universitet, 1975. (In Russian.)

[8] P. Halmos. *Measure Theory*. New York: Van Nostrand, 1950.

[9] D.A. Vladimirov. *Boolevy algebry*. Moscow: Nauka, 1969.

[10] L.Ya. Savel'ev. "On Order Topologies and Continuous Measures." *Sibirskij matem. zh.*, 6, 6 (1965): 1357-1364. (In Russian.)

[11] L.Ya. Savel'ev. "Conditions for the Extendibility of Vector Measures." *Siberian Mathematical J.*, 9, 4 (1968): 697-707.

[12] I. Diestel and I.I. Uhl. *Vector Measures*. Providence, Rhode Island: Amer. Math. Soc., 1977.

[13] I. Kluvanek. "K teorii vektornykh mer" (On the Theory of Vector Measures). Chast' I. *Mat.-Fyz. Časopis SAV*, 11, 3 (1961): 173-191.

[14] I. Kluvanek. "K teorii vektornykh mer." Chast' II. *Mat.-Fyz. Časopis SAV*, 16, 1 (1966): 76-81.

[15] R.G. Bartle, N. Dunford, and I.T. Schwartz. "Weak Compactness and Vector Measures." *Canadian J. Math.*, Vol. 7 (1955): 289-305.

AN ESTIMATE FOR DENSITY OF THE DISTRIBUTION OF THE INTEGRAL TYPE

A.I. Sakhanenko

1. AN INTRODUCTION AND THE MAIN IDEA OF THE PROOF

In some works (see, for example, [1,2] and the bibliography therein), while studying the rate of convergence of distributions of functionals of the form

$$I(\xi_n) = \int_0^1 f(\xi_n(t), t)\, dt,$$

where $\{\xi_n\}$ is the sequence of random broken lines, and f is a function measurable in both variables, the requirements imposed on f include the existence of bounded density of the distribution $I(\xi)$, where ξ is the limit process. (In the works mentioned ξ is a usual Wiener process w. We shall assume in the sequel that either $\xi=w$ or ξ is a "Brownian bridge" w_0, i.e., the Wiener process w conditioned by $w(1)=0$.) In other words, it is required that

$$L(f, \xi) \equiv \sup_{x,h>0} h^{-1}\mathbf{P}(x \leqslant I(\xi) \leqslant x+h) < \infty. \qquad (1)$$

It is appropriate to pose the problem of finding simple sufficient conditions to have (1) satisfied. Until recently, the most essential result in this research has belonged, apparently, to S. Sawyer [1], who showed that $L(f,w)<\infty$ for

$$f(x, t) = a(t)x^2 + b(t)x + c(t), \quad 0 < \int_0^1 [a^2(t) + b^2(t)]\, dt < \infty.$$

However, some time ago (independently of the author who announced some of his results in [3]) I.S. Borisov [4] obtained sufficient conditions to have (1) satisfied for functions of the form

$$f(x,t) = f(x) \quad . \qquad (2)$$

This paper, as well as [3] and the respective results in [4], were written with the purpose to respond to the hypothesis stated by A.A. Borovkov, that in order to have (1) satisfied, it suffices to have $f'(x)$ separate from zero in some neighborhood of zero. In particular, when $f(x)$ has a derivative for $|x|\leq b$, $b>0$, which can change the sign only for $x=0$, it follows from [4] that to have (1) satisfied, it suffices for some $C>0$

$$|f'(x)| > C\exp(-\pi^2/2(40)^2x^2) \qquad \text{for} \quad |x| \leq b .$$

On the other hand, it is shown in [4] that

$$L(d(x),w_0) = \infty , \qquad \text{where} \quad d(x) = \exp(-\pi^2/8x^2) .$$

The objective of this paper is to obtain the estimate for $L(f,\xi)$, which will be given in Theorem of Section 2 and proved in Sections 3 and 4. This theorem implies, in particular, that $L(f,\xi)<\infty$, if for $|x|\leq b$, $b>0$, the sign of $f'(x,t)$ depends only on the sign of x, and for some $C>0$

$$|f'(x,t)| > C|x|^{-k}d(x)\theta(x,\xi) \qquad \text{for} \quad k > 11 , \tag{3}$$

where $\theta(x,w)=1$, $\theta(x,w_0)=|x|^{-1}$, and the derivatives here and after are taken by the first argument. The fact that this result is close to the unimprovable one is confirmed by the following:

$$L(f,\xi) = \infty , \quad \text{if} \quad |f'(x)|/(|x|^{-3}d(x)\theta(x,\xi)) \to 0 \quad \text{as} \quad x \to 0 .$$

The last statement follows from the results of [4] since

$$\mathbf{P}\left(\max_{0\leq t\leq 1}|\xi(t)|< x\right)\Big/f(x)\to\infty \quad \text{as} \quad x\to 0.$$

The results of [4] are based on the infinite divisibility of the distribution $\int_0^\tau f(w(t))dt$, where τ is the first moment at which a level b is reached; these results do not extend to the functions which do not satisfy (2). We have obtained estimates for $L(f,\xi)$ using the following obvious relation.

LEMMA 1. Let on a probability space $\Omega_1\times\Omega_2$ a random variable $X=X(\omega_1,\omega_2)$ be given. We assume that the conditional distribution of the random variable X for fixed ω_1 has density bounded by the constant $Y(\omega_1)$. Then the random variable X has density bounded by the constant $MY(\omega_1)$.

In particular, it readily follows (see Lemma 2) that if the sign of $f'(x,t)$ is constant and for some $C>0$

$$|f'(x,t)| \geq C \quad \forall\, x\ \forall\, t \in [0,1] ,$$

then

$$L(f,w) \leq \sqrt{3/2\pi}\ C^{-1} ; \qquad L(f,w_0) \leq \sqrt{6/\pi}\ C^{-1} .$$

Difficulties in proving the basic result arise because we allow the derivative to change the sign and impose restrictions only in some neighborhood of zero.

2. BASIC RESULTS

We fix $0<T\leq 1$ and consider continuous monotone nondecreasing functions of their arguments $0<\varepsilon(a)\leq a/2$ and $g(a,t)>0$, defined for $0<a\leq\sqrt{T}$ and $0<t\leq\varepsilon^2(a)$. Let $K(y,u,a)$ denote a rectangle on the (x,t)-plane with vertices at points $(y\pm\varepsilon(a),u)$ and $(y\pm\varepsilon(a),u+\varepsilon^2(a))$, and let $H(a)$ denote the set of all points (y,u), $|y|\leq a-\varepsilon(a)$, $0\leq u\leq T$, such that for all $(x,t)\in K(y,u,a)$ the function $f'(x,t)$ has the constant sign and $|f'(x,t)|\geq g(a,t-u)$.

By $\alpha(a)$ we denote the first entry of the set $H(a)$ and by $\kappa(a)$ the first entry to one of the levels $\pm a$. If $\alpha(a)$ is not defined, we assume $\alpha(a)=\infty$. Let $\theta(T,a,w)=1$, and $\theta(T,a,w_0)=\min\{a^{-1},(1-T)^{-\frac{1}{2}}\}$, $\theta(1,a,w_0)=a^{-1}$.

THEOREM. If for all $a\in(0,\sqrt{T})$ we have $P(\alpha(a)<\kappa(a)\,|\,\kappa(a)\leq T)=1$, then

$$L(f,\xi)\leqslant c\int_0^{\sqrt{T}}da\int_0^{\varepsilon^2(a)}\exp\left(-\frac{\varepsilon^2(a)}{2t}-\frac{\pi^2T}{8a^2}\right)\frac{\varepsilon(a)\theta(T,a,\xi)}{a^3t^3g(a,t/2)}\,dt.$$

A symbol c replaces here and below some positive absolute constants.

In particular, if in some neighborhood of zero $\{(x,t):|x|\leq b,\ 0<t\leq T\}$, $b>0$, the sign of $f'(x,t)$ depends only on the sign of x, we can put, having chosen $\varepsilon(a)$,

$$g(a,u) = \min\{|f'(x,t)|:\ b\geq|x|\geq a-2\varepsilon(a),\ u\leq t\leq T\}\qquad\text{for } a\leqslant b$$

and $g(a,u)=g(b,u)$ for $a>b$. Then the condition of the Theorem will be satisfied automatically for $H(a)=\{(x,t):|x|=a-\varepsilon(a)\}$. If we consider as an example the function

$$g(a,t)=C\exp\left(-C_0^2a^6/t\right)|a|^{-k}\exp\left(-\pi^2T/8a^2\right)\theta(T,a,\xi)$$

for $C>0$, $C_0>0$, $k>11$, $\varepsilon(a)=2C_0a^3$, we obtain (3) and all the results announced in [3].

3. AUXILIARY ASSERTIONS

LEMMA 2. Let $\xi(t)$ be a Gaussian process on $[A,B]$ with the covariance function

$$r(u,v)\equiv \mathbf{M}(\xi(u)-\mathbf{M}\xi(u))(\xi(v)-\mathbf{M}\xi(v))\geqslant 0.$$

We assume that the sign of the function $f'(x,t)$ does not depend on (x,t) and for some function $g(t)$

$$|f'(x,t)| \geq g(t) \geq 0 \quad \forall\ (x,t) .$$

Then the random variable $\int_A^B f(\xi(t),t)dt$ has density bounded by the constant $(2\pi\sigma^2)^{-\frac{1}{2}}$, where $\sigma^2=\int_A^B\int_A^B r(u,v)g(u)g(v)dudv$.

Proof. We note that the normally distributed random variable $\eta=\int_A^B \xi(t)g(t)dt$ and the Gaussian process $\zeta(t)=\xi(t)-\sigma^{-2}h(t)\eta$, where $h(t)=\int_A^B r(u,t)g(u)du$, are independent. For the fixed trajectory of $\zeta(t)$, the densities $p(\cdot)$ and $p_\psi(\cdot)$ of the random variables η and $\psi(\eta)\equiv\int_A^B f(\zeta(t)+\sigma^{-2}h(t)\eta,t)dt$, where $\psi(\cdot)$ is the monotone function, satisfy the relation $p_\psi(y)=p(\psi^{-1}(y))/|\psi'(\psi^{-1}(y))|$. Since

$$|\psi'(z)| = \left|\int_A^B \sigma^{-2}h(t)f'\big(\zeta(t)+\sigma^{-2}h(t)z,\ t\big)\,dt\right| \geqslant \int_A^B \sigma^{-2}h(t)\,g(t)\,dt = 1,$$

then

$$\sup_y p_\psi(y) \leqslant \sup_y p(y) = (2\pi D\eta)^{-1/2} = (2\pi\sigma^2)^{-1/2} \equiv Y(\zeta),$$

and the required estimate follows from Lemma 1.

For $0\leq t<T\leq 1$, $\delta>0$ let $\mu=\max_{t\leq u\leq T}|w(u)|$,

$$\lambda(a,x) = P(\mu<a,\ |w(1)|<\delta \mid w(t)=x) .$$

As before, $d(a)=\exp(-\pi^2/8a^2)$, and all the derivatives are taken by the first argument. Let

$$\varphi(x,t) = (2\pi t)^{-1/2}\exp(-x^2/2t),\ \Phi(x,t) = \int_{-\infty}^{x}\varphi(y,t)\,dy.$$

LEMMA 3. For $t\leq\varepsilon^2(a)\leq a^2/4$ and $a^2\leq T$

$$\lambda'(a,x) \leqslant \psi(a,T,\delta) \equiv ca^{-3}d(a/\sqrt{T})\min\{1,\ \delta a^{-1},\ \delta(1-T)^{-1/2}\}.$$

Proof. Let

$$q(a,x,z) = \frac{d}{dz}\mathbf{P}(\mu < a,\ w(T) < z \mid w(t) = x).$$

In this case q' as the function of a and z is the density of the joint distribution μ and w(T) by condition w(t)=x. From the formula (5.7) on page 404 in [5] it follows that

$$q(a,x,z) = \sum_{k=-\infty}^{\infty} [\varphi(z-x+4ka,\ T-t) - \varphi(z+x+2a+4ka,\ T-t)].$$

Applying now the Poisson summation formula [5, p. 710] for $d=d(a/\sqrt{T-t})$, we have

$$q(a,x,z) = \frac{1}{2a}\sum_{k=1}^{\infty} d^{k^2}\left[\cos\frac{(z-x)k\pi}{2a} - \cos\frac{(z+x+2a)k\pi}{2a}\right]. \qquad (4)$$

Using the inequalities $|x| \le a$ and $|z| \le a$, it is easy to obtain

$$q'(a,x,z) \leqslant c\sum_{k=1}^{\infty} d^{k^2}\left[a^{-4}k^2 + a^{-2}k + a^{-2}\right] \equiv q_1(a).$$

Since $a^2 \le T \le 1$ and $d<1$, it follows from the relations

$$d^{k^2} \leqslant d^k,\ \sum_{k=1}^{\infty} d^{k-1} = (1-d)^{-1},\ \sum_{k=1}^{\infty} kd^{k-1} = (1-d)^{-2},\ \sum_{k=1}^{\infty} k(k-1)d^{k-2} = 2(1-d)^{-3}$$

that $q_1(a) \le ca^{-4}d(1-d)^{-3} \equiv q_2(a)$. From the restrictions on t and a there follows

$$d = d(a/\sqrt{T})/d(a/\sqrt{t}) \le d(a/\sqrt{T})/d(2) \le d(1)/d(2) < 1,$$

and hence

$$q'(a,x,z) \le q_2(a) \le ca^{-4}d(a/\sqrt{T}). \qquad (5)$$

The formula of total probability for the values of $w(T)$ immediately yields

$$\lambda(a,x)=\int_{-a}^{a} q(a,x,z)\,\mathbf{P}(|w(1)|<\delta\,|\,w(T)=z)\,dz,$$

which, together with (5) and the equality $q(a,x,\pm a)=0$ following from (4), gives us

$$\lambda'(a,x)=\int_{-a}^{a} q'(a,x,z)\,\Phi_0(z)\,dz\leqslant ca^{-4}d\left(a/\sqrt{T}\right)\int_{-a}^{a}\Phi_0(z)\,dz, \tag{6}$$

where $\Phi_0(z)=\Phi(z+\delta,1-T)-\Phi(z-\delta,1-T)$. From the inequalities $\Phi_0(z)\leq 1$ and $\Phi_0(z)\leq 2\delta/\sqrt{2\pi(1-T)}$ we obtain

$$\int_{-a}^{a}\Phi_0(z)\,dz\leqslant\min\{2a,\,\delta a(1-T)^{-1/2}\}. \tag{7}$$

On the other hand,

$$\int_{-a}^{a}\Phi_0(z)\,dz\leqslant\int_{-\infty}^{\infty}\Phi_0(z)\,dz=2\delta,$$

which together with (6) and (7) imply the required assertion.

We note that when the conditions of Lemma 3 are satisfied for all $A\subset[0,\infty)$

$$\mathbf{P}(\mu\in A,\ |w(1)|<\delta/w(t)=x)\leqslant\mathbf{P}(|w(1)|<\delta/w(t)=x)=$$
$$=\Phi(\delta-x,\,1-t)-\Phi(-\delta-x,\,1-t)\leqslant\min\{1,\,\delta\}. \tag{8}$$

We fix $\varepsilon>0$, $t_0>0$ and $t_1=(3/4)t_0$, and let τ denote the first after t_1 moment at which the process $w(t)$ crosses one of the levels $\pm\varepsilon$.

LEMMA 4. For $B\subset[(7/8)t_0,\infty)$ and $\phi_0(t)=28\varepsilon t^{-1}\phi(\varepsilon,t)$ we have the inequality

$$\mathbf{P}(|w(t_1)|<\varepsilon,\ \tau\in B)\leqslant\int_B\varphi_0(t)\,dt.$$

Proof. Let τ^+ and τ^- be the first after t_1 moments at which the process $w(t)$ crosses the levels $+\varepsilon$ and $-\varepsilon$, respectively. From [5, p. 403] for $|x|\leq\varepsilon$ and $t>t_1$ we have

$$q(x,t)\equiv\frac{d}{dt}\mathbf{P}(\tau^+<t\mid w(t_1)=x)=(\varepsilon-x)(t-t_1)^{-1}\varphi(\varepsilon-x,t-t_1)\leqslant$$
$$\leqslant 2\varepsilon(t-t_1)\varphi(\varepsilon-x,t-t_1).$$

Because the process w is symmetric and Markov, we obtain

$$\mathbf{P}(|w(t_1)|<\varepsilon,\ \tau\in B)\leqslant 2\mathbf{P}(|w(t_1)|<\varepsilon,\ \tau^+\in B)=$$
$$=2\iint\limits_{\substack{|x|<\varepsilon\\ t\in B}}\varphi(x,t_1)\,q(x,t)\,dxdt\leqslant 2\iint\limits_{t\in B}2\varepsilon(t-t_1)^{-1}\varphi(x,t_1)\times$$
$$\times\varphi(\varepsilon-x,t-t_1)\,dxdt=4\varepsilon\int\limits_{B}(t-t_1)^{-1}\varphi(\varepsilon,t)\,dt\ ,$$

which together with the inequality $(t-t_1)^{-1}\leq 7t^{-1}$ holding for $t>(7/8)t_0$, imply the required assertion.

4. PROOF OF THE MAIN THEOREM

We fix y, $0<T\leq 1$, $h>0$, $a_0>0$, $\delta>0$, $0<t_0\leqslant\varepsilon^2=\varepsilon^2(a_0)$, $t_1=(3/4)t_0$ and the sets $A\subset[a_0,\infty)$ and $B\subset[(7/8)t_0,t_0]$. As before, we denote now by $\alpha=\alpha(a_0)$ the first moment at which the set $H(a_0)$ is reached. Also, let $\mu(t)=\max\limits_{t\leq u\leq T}|w(u)|$, $w_1(t)=w(\alpha+t)-w(\alpha)$, $\mu_1=\max\limits_{0\leq u\leq t_1}|w_1(u)|$, let β be the first moment at which the process $w_1(t)$ reaches the boundaries of the set $K=K(0,0,a_0)$ and we define γ as the first after t_1 moment at which the process $w_1(t)$ reaches the boundaries of the set K. Let $I(u,v)=\int_u^v f(w(t),t)dt$, $I_1=I(0,\alpha)$, $I_2=I(\alpha,\alpha+t_1)$, $I_3=I(\alpha+t_1,1)$, $I=I(0,1)$.

We estimate first $p(A,B)=h^{-1}P(\Omega)$, where

$$\Omega=\{I\in(y,y+h),\ \beta\in B,\ \mu(0)\in A,\ |w(1)|<\delta\}\ .$$

The Markov property of the process w implies

$$\Omega=\{I_1+I_2+I_3\in(y,\ y+h),\ \mu_1<\varepsilon,\ \gamma\in B,\ \mu(\alpha+\gamma)\in A,\ |w(1)|<\delta\}.$$

For the fixed value

$$\nu=(\alpha,\ w(\alpha),\ w_1(t_1),\ \gamma,\ w_1(\gamma),\ \mu(\alpha+\gamma),\ w(1))$$

it is not hard to note the following: first, the random variables I_1, I_3 and the vector (I_2,μ_1) are independent; second, for $\mu_1<\varepsilon$ the value I_2 does not change if the function f under the integral sign in I_2 is changed for $|x-w(\alpha)|>\varepsilon$ as to the function f_0 obtained have the derivative $f_0'(x,t)$ for all x (rather than only for $|x-w(\alpha)|\le\varepsilon$) and $t\in(\alpha,\alpha+\varepsilon^2)$ have a constant sign and satisfy the inequality $|f_0'(x,t)|\ge g(a_0,t-\alpha)$ (it is possible, for instance, to put $f_0'(x,t)=f'(w(\alpha),t)$ for $|x-w(\alpha)|>\varepsilon$); third, for $t\in[0,t_1]$ the process $w_1(t)$ is Gaussian with a correlation function $r(u,v)=\min\{u,v\}-uv/t_1$, hence by Lemma 2 the distribution of the random variable

$$I_0=\int\limits_{\alpha}^{\alpha+t_1} f_0(w(t),t)\,dt=\int\limits_0^{t_1} f_0(w(\alpha)+w_1(t_1),\alpha+t)\,dt$$

has density not exceeding $\rho(t_0)\equiv 4t_0^{-3/2}g^{-1}(a_0,t_0/2)$, since in this case

$$2\pi\sigma^2=2\pi\iint\limits_{0\leqslant u,v\leqslant t_1} r(u,v)\,g(a_0,u)\,g(a_0,v)\,du dv\geqslant$$
$$\geqslant 2\pi g^2(a_0,\ t/2)\iint\limits_{t_0/2\leqslant u,v\leqslant t_1} r(u,v)\,du dv\geqslant(1/16)\,t_0^3 g^2(a_0,\ t_0/2).$$

It follows from the above that

$$\mathbf{P}(\Omega\,|\,\nu,\ I_1,\ I_3)\leqslant\sup_y\mathbf{P}(I_2=I_0\in(y,y+h),\ \mu_1<\varepsilon\,|\,\nu)\leqslant$$
$$\leqslant\sup_y\mathbf{P}(I_0\in(y,y+h)\,|\,\nu)\leqslant\rho(t_0)\,h.$$

Therefore,

$$p(A,\ B)=h^{-1}\mathbf{MP}(\Omega|\nu,\ I_1,\ I_3)\leqslant$$

$$\leqslant\rho(t_0)\mathbf{P}(|w_1(t_1)|<\varepsilon,\ \gamma\in B,\ \mu(\alpha+\gamma)\in A,\ |w(1)|<\delta)=$$

$$=\rho(t_0)\mathbf{P}(|w(t_1)|<\varepsilon,\ \tau\in B,\ \mu(\tau)\in A,\ |w(1)|<\delta.$$

(The last equality is easily explained if one interchanges the intervals $(0,\alpha]$ and $(\alpha,\alpha+\gamma]$.) Using the Markov property of w, Lemma 3 and (8), we obtain next

$$p(A,B)\leqslant\rho(t_0)\int\limits_{t\in B}\int \mathbf{P}(|w(t_1)|<\varepsilon,\ \tau\in dt,\ w(\tau)\in dx)\times$$
$$\times\mathbf{P}(\mu(\tau)\in A,\ |w(1)|<\delta\,|\,\tau=t,\ w(\tau)=x)\leqslant\rho(t_0)\,Q(t_0,B)\,\Psi(A), \tag{9}$$

where

$$\Psi(A)=\min\left\{\int\limits_A\psi(a,\ T,\ \delta)\,da,\ 1,\ \delta\right\};\quad Q(t_0,B)=\mathbf{P}(|w(t_1)|<\varepsilon,\ \tau\in B). \tag{10}$$

Now we estimate $p(A,(0,\varepsilon^2))$. From the inequality $\rho(t_0\leq\rho(t)$ holding for $t\leq t_0$, as well as from Lemma 4, we obtain

$$\rho(t_0)\,Q(t_0,\ [(7/8)\,t_0,\ t_0))\leqslant\int\limits_{(7/8)t_0}^{t_0}\rho(t)\,\varphi_0(t)\,dt \qquad \text{for} \qquad t_0\leqslant\varepsilon^2,$$

which with (9) readily yield

$$p(A,\ (0,\ \varepsilon^2))=\sum_{j=0}^{\infty}p\left(A,\ \left[(7/8)^{j+1}\varepsilon^2,\ (7/8)^j\varepsilon^2\right)\right)\leqslant$$

$$\leqslant\Psi(A)\int\limits_0^{\varepsilon^2}\rho(z)\,\varphi_0(t)\,dt\equiv\Psi(A)\,Q_0(a_0).$$

From (9) and the relations $Q(t_0,B)\leq 1$ and

$$\int_0^{\varepsilon^2} \rho(t)\,\varphi_0(t)\,dt \geqslant \rho(\varepsilon^2)\int_0^{\varepsilon^2} \varphi_0(t)\,dt = c\rho(\varepsilon^2)$$

we immediately obtain

$$p(A,\ \{\varepsilon^2\}) \leqslant \rho(\varepsilon^2)\Psi(A) \leqslant cQ_0(a_0)\Psi(A);$$
$$p_0(A) \equiv p(A,\ (0,\ \varepsilon^2]) = p(A,\ (0,\ \varepsilon^2)) + p(A,\ \{\varepsilon^2\}) \leqslant c\Psi(A)Q_0(a_0). \quad (11)$$

In conclusion, we estimate $p_0((0,\infty))$. Since the continuous function

$$Q_0(a) = c\int_0^{\varepsilon^2(a)} t^{-3}g^{-1}(a,\ t/2)\exp(-\varepsilon^2(a)/2t)\,dt$$

is monotone decreasing in a, we can partition the interval $(0,\sqrt{T})$ into intervals $[a_{j+1},a_j)$, $j=1,2,\ldots$, where $a_1=\sqrt{T}$ and $a_j\downarrow 0$ as $j\to\infty$, in such a way that $Q_0(a_{j+1})\leq 2Q(a)$ for $a\in[a_{j+1},a_j)$. From this, (10) and (11) we have

$$p_0((0,\ \sqrt{T})) = \sum_{j=1}^{\infty} p_0([a_{j+1},\ a_j)) \leqslant \sum_{j=1}^{\infty} cQ_0(a_{j+1})\int_{a_{j+1}}^{a_j} \psi(a,\ T,\ \delta)\,da \leqslant$$
$$\leqslant \int_0^{\sqrt{T}} cQ_0(a)\,\psi(a,\ T,\ \delta)\,da \equiv \psi_0(\delta). \quad (12)$$

Note that

$$\psi_0(\delta) \geqslant cQ_0(\sqrt{T})\int_0^{\sqrt{T}} \psi(a,\ T,\ \delta)\,da \geqslant cQ_0(\sqrt{T})\min\{1,\ \delta\},$$

which together with (10) and (11) yield

$$p_0([\sqrt{T},\ \infty)) \leqslant c\Psi([\sqrt{T},\ \infty))Q_0(\sqrt{T}) \leqslant cQ_0(\sqrt{T})\min\{1,\ \delta\} \leqslant c\psi_0(\delta).$$

The last inequality plus (12) give us

$$p_0((0,\ \infty)) = p_0((0,\ \sqrt{T})) + p_0([\sqrt{T},\ \infty)) \leqslant c\psi_0(\delta).$$

The preceding estimate can be rewritten as

$$q_0(y, h, \delta) \equiv h^{-1}\mathbf{P}(I \in (y, y+h), |w(1)| < \delta) \leqslant c\psi_0(\delta).$$

The assertion of the Theorem follows now from the obvious relations:

$$L(f, w) = \sup_{y,h>0} \lim_{\delta\to\infty} q_0(y, h, \delta) \leqslant c \lim_{\delta\to\infty} \psi_0(\delta);$$

$$L(f, w_0) = \sup_{y,h>0} \overline{\lim_{\delta\to 0}}\, h^{-1}\mathbf{P}(I \in (y, y+h) \,|\, |w(1)| < \delta) =$$

$$= \sup_{y,h>0} \overline{\lim_{\delta\to 0}}\, q_0(y, h, \delta)/[\Phi(\delta, 1) - \Phi(-\delta, 1)] \leqslant c \overline{\lim_{\delta\to 0}}\, \psi_0(\delta)/\delta .$$

The author takes an opportunity to thank I.S. Borisov and the participants of the Seminar at the Probability Theory Department at the Mathematics Institute of the Siberian Branch of the USSR Academy of Sciences for the useful discussions.

REFERENCES

[1] S. Sawyer. "Rates of Convergence for some Functionals in Probability." *Ann. Math. Stat.*, 43, 1 (1972): 273-284.

[2] I.S. Borisov. "On the Rate of Convergence of Distributions of Functionals of Integral Type." *Theory Prob. Applications*, 21, 2 (1976): 283-299.

[3] A.I. Sakhanenko. "Odno uslovie suschestvovaniya ogranichennoj plotnosti u raspredelenij funktsionalov integral'nogo tipa" (On a Condition for Existence of Bounded Density for Distributions of Functionals of Integral Type). *Tezisy dokladov II Vilniusskoj konferentsii po teorii veroyatnosti i matematicheskoj statistike*. Vol. 2: 149-150. Vilnius, 1977.

[4] I.S. Borisov. "On Conditions of Existence of Bounded Density for Additive Functionals on Some Random Processes." *Theory Prob. Applications*, 25, 3 (1980): 454-465.

[5] W. Feller. *An Introduction to Probability Theory and Its Application.* New York: Wiley & Sons, 1968.

QUEUES WITH CUSTOMERS OF SEVERAL TYPES

S.G. Foss

For a single-server queue, at which n types of customers arrive, a service algorithm is developed which minimizes the average cost of idle period duration. The problem of constructing such an algorithm has been considered in [1,2,3] for queues with Poisson arrivals.

1. THE DESCRIPTION OF THE QUEUEING SYSTEM AND THE STATEMENT OF THE MAIN RESULT

We assume that all the random variables to be considered in this paper are given on a common probability space $\langle \Omega, F, P \rangle$. We use the following notations: $Z_+ = \{0,1,2,\dots\}$, $Z_+^n = Z_+ \times Z_+ \times \dots \times Z_+$ (n-multiple direct product), $E^n = Z_+^n \setminus \{(0,0,\dots,0)\}$.

We consider n independent sequences as

$$\{\bar{m}_{1,j}\}_{j=1}^{\infty}, \quad \{\bar{m}_{2,j}\}_{j=1}^{\infty}, \quad \dots, \quad \{\bar{m}_{n,j}\}_{j=1}^{\infty} , \tag{1}$$

(n+1)-dimensional random vectors, independent and identically distributed in each sequence; each vector $\bar{m}_{i,j}$ is of the form

$$\bar{m}_{i,j} = (\tau_{i,j}, m_{i,j}^{(1)}, m_{i,j}^{(2)}, \ldots, m_{i,j}^{(n)}) , \qquad (2)$$

where the first coordinate $\tau_{i,j}$ assumes nonnegative real values and the remaining n coordinates $m_{i,j}^{(1)}, m_{i,j}^{(2)}, \ldots, m_{i,j}^{(n)}$ are nonnegative integers. Also, we specify the n-dimensional random vector $\bar{m}_{0,0}=(m_{0,0}^{(1)}, \ldots, m_{0,0}^{(n)})$ not depending on the set $\{\bar{m}_{i,j}, 1\leq i\leq n, j\geq 1\}$, all the coordinates of which assume nonnegative integer values, $P\{\bar{m}_{0,0}=(0,0,\ldots,0)\}=0$.

For $i,\ell=1,2,\ldots,n$ we assume that

$$M\{m_{0,0}^{(i)}\} < \infty \ ; \qquad a_i = M\{\tau_{i,1}\} < \infty \ ; \qquad g_i^{(\ell)} = M\{m_{i,1}^{(\ell)}\} < \infty . \qquad (3)$$

In the system, customers of n types are being served. The service system consists of a server and a buffer which can be thought of as the room in which customers are waiting to be served. At time $t=0$ the server is free and the buffer contains $m_{0,0}^{(1)}, \ldots, m_{0,0}^{(n)}$ customers of the 1st, 2nd, ..., n^{th} types, respectively. At time $t=0$ the service of one of these customers begins. A new customer goes into the buffer through one of n channels and only at those instants of time when the server ends serving the subsequent customer from the queue; each time the number of the channel coincides with the type of the customer the service of which has been completed at any given time. More precisely, let at some time $t\geq 0$ the server get empty and customers be waiting in the queue.

Then, one of these customers enters immediately from the queue to the server in accordance with the service algorithm. Suppose it is the customer of type-s, assuming also that $\ell \geq 0$ customers of type-s have already been served. Then the service time of the $(\ell+1)^{th}$ customer of type-s is $\tau_{s,\ell+1}$; and at time $(t+\tau_{s,\ell+1})$ at which the service ends, $m^{(1)}_{s,\ell+1},\ldots,m^{(n)}_{s,\ell+1}$ customers of the 1st, $2^{nd},\ldots,n^{th}$ types arrive through the sth channel into the buffer. If, however, at time t there are no new customers in the queue, from then on no customers arrive (we say that this time t is a stopping time).

Each customer is being served sequentially, without break, at most one customer can be served at the same time. The order of service of the customers is determined by a "switching" function T which is defined as the set of functions $T=\{T^{(\ell)}\}^{\infty}_{\ell=1}$, where for $\ell=1,2,\ldots$

$$T^{(\ell)}: E^n \to \{1,2,\ldots,n\} . \qquad (4)$$

Suppose that at some time the ℓth customer has been served, and $(r_1,r_2,\ldots,r_n)$ customers are waiting in the queue to be served. Then the customer of the $T^{(\ell+1)}(r_1,\ldots,r_n)^{th}$ type enters for service. We denote the class of such switching functions by K_0.

For $T\in K_0$ let $\nu(T)$ be the total number of customers being served before the stopping time and let $\eta^{(T)}_i(t)$ $(i=1,2,\ldots,n;\ t\geq 0)$ be the number of customers of type-i waiting in the queue at time t. We define the functional $L:K_0\to[0,\infty)$ as follows:

$$L(T) = \left\{ \int_0^\infty \sum_{i=1}^n c_i \eta_i^{(T)}(t)\, dt \right\}, \tag{5}$$

where c_i $(1 \leqslant i \leqslant n)$ denote some positive integers and $T \in K_0$. By c_i we mean the cost of waiting in the queue for a customer of type-i per unit time.

For brevity we say that the jth customer of type-i which is accepted for service is the (i,j)-customer. Let $(d_1, d_2, \ldots, d_n)$ denote the permutation of the set $(1,2,\ldots,n)$ and $1 \leq s \leq n$. The switching function $T \in K_0$ is said to be a $(d_1, d_2, \ldots, d_s)$-function if for all $\ell = 1,2,\ldots$ the following equalities are satisfied: $T^{(\ell)}(r_1, \ldots, r_n) = d_1$ for $r_{d_1} \leqslant 1$ and $T^{(\ell)}(r_1, \ldots, r_n) = d_i$ for $i = 2,3,\ldots,s$ for $r_{d_1} = r_{d_2} = \cdots = r_{d_{i-1}} = 0$, $r_{d_1} \geq 1$.

Let $T \in K_0$. If there exists a permutation $(d_1, d_2, \ldots, d_n)$ of the set $(1,2,\ldots,n)$, such that T is a $(d_1, d_2, \ldots, d_n)$-function, we say that T has a priority service.

<u>THEOREM</u>. Let $m\{\nu(T)\} < \infty$ for some $T \in K_0$. Then there exists a priority switching function T_0 such that

$$L(T_0) \leqslant L(T) \tag{6}$$

for any switching function $T \in K_0$.

We give the form of the switching function T_0. Let an arbitrary permutation $(d_1, d_2, \ldots, d_n)$ of the set $(1,2,\ldots,n)$ be given. For $i = 1,2,\ldots,n-1$; $j,k \neq d_1, d_2, \ldots, d_i$ we define $a_j^{(i)}$ and $g_j^{(i,k)}$ by the formulas (22) and (23) (see Section 3) and put

$$b_k^{(i)} = \sum_{j \neq d_1, \ldots, d_i} c_j g_k^{(i,j)} \; ; \qquad b_k = \sum_{j=1}^n c_j g_k^{(j)} \; .$$

Then T_0 is a $(d_1,d_2,\ldots,d_n)$-function in which $d_1,d_2,\ldots,d_n$ are chosen according to the following rule: d_1 is such that

$$(b_{d_1}-c_{d_1})/a_{d_1} = \min_{1\le k\le n} \{(b_k-c_k)/a_k\}$$

and for $i=1,2,\ldots,n-1$ the quantity d_{i+1} is such that

$$(b^{(i)}_{d_{i+1}}-c_{d_{i+1}})/a^{(i)}_{d_{i+1}} = \min_{k\ne d_1,\ldots,d_i} \{(b^{(i)}_k-c_k)a^{(i)}_k\} .$$

2. AUXILIARY RESULTS

In proving the Theorem we shall need to use switching functions of a more general type, assuming that for any $\ell=1,2,\ldots,;$ $(r_1,\ldots,r_n)\in E^n$ the value $T^{(\ell)}(r_1,\ldots,r_n)$ can be random but independent of the future. The precise meaning of the notion of an independence from the future will be explained below. We denote by $K(K\supseteq K_0)$ the class of these switching functions. For a more convenient presentation, we prove the Theorem in a more general formulation -- for all switching functions of the class K.

First we introduce the class K' of switching functions. By a switching function $T\in K'$ we mean an arbitrary sequence of random functions $T=\{T^{(\ell)}\}_{\ell=1}^{\infty}$, in which $T^{(\ell)}$: $\Omega\times E^n\to\{1,2,\ldots,n\}$, satisfying the property: let $(r_1,\ldots,r_n)\in E^n$; then for $\ell=1,2,\ldots;$ $k\in\{1,2,\ldots,n\}$, if $P\{T^{(\ell)}(r_1,\ldots,r_n)=k\}>0$, then $r_k\ge 1$.

Let N^n_+ denote the family of sets $(\ell_1,\ell_2,\ldots,\ell_n)$ such that each of the ℓ_i $(i=1,2,\ldots,n)$ is either a nonnegative integer or $+\infty$. For $(\ell_1,\ell_2,\ldots,\ell_n)\in N^n_+$ let

$B(\ell_1,\ldots,\ell_n)=\sigma\{\bar{m}_{i,j},\ 1\le i\le n,\ 1\le j\le \ell_i\}$ denote the σ-algebra generated by the set of random vectors $\{\bar{m}_{1,j}\}_{j=1}^{\ell_1}$, $\{\bar{m}_{2,j}\}_{j=1}^{\ell_2}$, $\ldots$, $\{\bar{m}_{n,j}\}_{j=1}^{\ell_n}$ and let

$$D(\ell_1,\ldots,\ell_n) = \sigma\{\bar{m}_{i,j},\ 1\le i\le n,\ j>\ell_i\} .$$

We introduce the family of σ-algebras $\{G(\ell_1,\ldots,\ell_n)$, $(\ell_1,\ldots,\ell_n)\in N_+^n\}$ such that for any $(\ell_1,\ldots,\ell_n)\in N_+^n$ and $(\ell_1',\ldots,\ell_n')\in N_+^n$ the following properties are satisfied:

1) $F\supseteq G(\ell_1,\ldots,\ell_n)\supseteq B(\ell_1,\ldots,\ell_n)$;

2) independent σ-algebras are

$$G(\ell_1,\ldots,\ell_n) \quad \text{and} \quad D(\ell_1,\ldots,\ell_n) ; \tag{7}$$

3) if $\ell_1\le\ell_1',\ldots,\ell_n\le\ell_n'$, then $G(\ell_1,\ldots,\ell_n)\subseteq G(\ell_1',\ldots,\ell_n')$.

Let for $j=1,2,\ldots,$; $\bar{k}_j=(k_1,k_2,\ldots,k_j)$ be the set of integers, such that for $\ell=1,2,\ldots,j$

$$k_\ell \in \{1,2,\ldots,n\} .$$

By $q_i=q_i(\bar{k}_j)$ $(1\le i\le n)$ we denote the number of k_ℓ $(1\le \ell\le j-1)$ equal to i.

Now we introduce the class of switching functions $K\subset K'$: the switching function T of K' belongs to the class K, if for any $j=1,2,\ldots,$ for any $\bar{k}_j$ satisfying (8), the event $\{T^{(1)}=k_1,\ \ldots,\ T^{(j)}=k_j\}$ belongs to the σ-algebra $G(q_1,\ldots,q_n)$.

We note that the family of σ-algebras $\{G(\ell_1,\ \ldots,\ \ell_n)\}$ can be given, for example, by putting $\{G(\ell_1,\ldots,\ell_n)\}=\{B(\ell_1,\ldots,\ell_n)\}$. Properties (7) in this case are satisfied, and the class K is nonempty, since $K\supseteq K_0$.

We give now another example to illustrate how to specify $\{G(\ell_1,\dots,\ell_n)\}$. For any $\ell=1,2,\dots$; $(r_1,\dots,r_n)\in E^n$, let $T^{(\ell)}(r_1,\dots,r_n)$ be a random variable choosing (at random) one of the nonzero coordinates of the vector $(r_1,\dots,r_n)$, the family of random variables $\{T^{(\ell)}(r_1,\dots,r_n),\ \ell=1,2,\dots,$ $(r_1,\dots,r_n\in E^n\}$ being mutually independent as well as being independent of the set (1). Then if for $(\ell_1,\dots,\ell_n)\in N_+^n$ we specify $G(\ell_1,\dots,\ell_n)$ as the σ-algebra generated by the family of random variables and vectors $\{T^{(\ell)}(r_1,\dots,r_n),\ \bar{m}_{i,j},$ $1\le i\le n,\ i\le j\le \ell_j,\ 1\le \ell\le \ell_1+\dots+\ell_n,\ (r_1,\dots,r_n)\in E^n\}$, then $\{G(\ell_1,\dots,\ell_n)\}$ satisfies the property (7) and $T\in K$.

The functional L is defined on the class K by the formula (5).

We cite here a property characterizing the class of switching functions K'. Let $T\in K'$. We denote by $\lambda(T)$ the stopping time and by $\nu_i(T)$ $(1\le i\le n)$ the total number of customers of type-i served before the time $\lambda(T)$ for the switching function T (in particular, $\lambda(T)$ and $\nu_i(T)$ can be equal to $+\infty$).

<u>LEMMA 1</u>. Let for some $T_1\in K'$, $P\{\nu(T_1)<\infty\}=1$. Then for any $T_2\in K'$ the random vectors $(\lambda(T_1),\ \nu_1(T_1),\dots,\ \nu_n(T_1))$ and $(\lambda(T_2),\ \nu_1(T_2),\dots,\ \nu_n(T_2))$ coincide a.s.

<u>Proof</u>. For any $T\in K'$,

$$\lambda(T) = \sum_{i=1}^{n} \sum_{k=1}^{\nu_i(T)} \tau_{i,k} .$$

Therefore, it suffices to show that for any $T_2\in K'$ the random vectors $(\nu_1(T_1),\dots,\nu_n(T_1))$ and $(\nu_1(T_2),\dots,\nu_n(T_2))$

coincide almost surely (a.s.).

Next we take an arbitrary elementary state $\omega\in\Omega$ of a switching function, $\nu(T_1)<\infty$. Let $\nu_1(T_1)=\ell_1,\ldots,\ \nu_n(T_1)=\ell_n$. We show that for all $i=1,2,\ldots,n$ for this state ω the inequality $\nu_i(T_2)=\ell_i$ is satisfied. Suppose the contrary. For $i=1,2,\ldots,n$, $t\geq 0$, let $\nu_i^{(t)}(T_2)$ be the number of customers of type-i served before the time t for the switching function T_2. Then there exists a time $t\geq 0$ and $i\in\{1,2,\ldots,n\}$ such that after the time t the $(\ell_i+1)^{th}$ service of customer of type-i begins and $\nu_k^{(t)}(T_2)\leq\ell_k$ for $k=1,2,\ldots,n$. Then the number of customers of type-i waiting to be served at the time t for the switching function T_2, is

$$m_{0,0}^{(i)}+\sum_{k=1}^{n}\sum_{j=1}^{\nu_k^{(t)}(T_2)} m_{k,j}^{(i)}-l_i\leqslant m_{0,0}^{(i)}+\sum_{k=1}^{n}\sum_{j=1}^{l_k} m_{k,j}^{(i)}-l_j=0$$

because $\nu_1(T_1)=\ell_1,\ldots,\ \nu_n(T_1)=\ell_n$. Hence for the switching function T_2 at the time t the $(\ell_i+1)^{th}$ service of customers of type-i cannot begin because there are no such customers in the queue. Therefore, we arrive at a contradiction: thereby

$$P\{\nu_1(T_1)\geq\nu_1(T_2),\ldots,\ \nu_n(T_1)\geq\nu_n(T_2)\}=1$$

and $P\{\nu(T_2)<\infty\}=1$. If we use the same arguments as we did in replacing T_1 by T_2 and T_2 by T_1, we obtain

$$P\{\nu_1(T_1)\leq\nu_1(T_2),\ldots,\ \nu_n(T_1)\leq\nu_n(T_2)\}=1\ .$$

//

REMARK 1. We shall assume in the sequel that at least for one (thereby for all) $T \in K'$ the inequality $M\{\nu(T)\}<\infty$ (and therefore $M\{\lambda(T)\}<\infty$) is satisfied. This can be restated in terms of (3). Let $Q=\{g_i^{(\ell)}\}$ be a matrix of the order n.

LEMMA 2. In order that for all $T \in K'$ the inequality $M\{\nu(T)\}<\infty$ to be satisfied, it is necessary and sufficient that all eigennumbers of the matrix Q be less than 1 in magnitude.

Proof. Let $b_i=M\{\nu(T)\}$ be the mean total number of services in the system, where $m_{0,0}^{(k)}=0$ for $k \neq i$, $1 \leq k \leq n$ and $m_{0,0}^{(i)}=1$ a.s. By the inequality $M\{m_{0,0}^{(\ell)}\}<\infty$ for $\ell=1,2,\ldots,n$, in order that the condition of the Lemma be satisfied, it is necessary and sufficient that $\max_{1 \leq i \leq n} b_i<\infty$. We note that

$$b_i = 1 + \sum_{j=1}^{n} g_i^{(j)} + \sum_{j=1}^{n} g_i^{(j)}(2) + \ldots + \sum_{j=1}^{n} g_i^{(j)}(l) + \ldots, \tag{9}$$

where $g_i^{(j)}(\ell)$ are elements of the matrix $Q^{\ell}=Q \times Q \times \cdots \times Q$ (ℓ times). Then Eq. (9) can be written in the vector form: for $\bar{b}=(b_1,\ldots,b_n)$, $\bar{a}=(1,1,\ldots,1)$;

$$\bar{b} = \left(\sum_{l=0}^{\infty} Q^l \right) \bar{a} = (I-Q)^{-1} \bar{a},$$

where $Q^0=I$ is the identity matrix of the order n. A necessary and sufficient condition for the series $\sum_{\ell=0}^{\infty} Q^{\ell}$ to converge is that all the eigennumbers of the matrix Q be less than 1 in magnitude [4, p. 173]. //

3. THE PROOF OF THE THEOREM

For any $T \in K'$ we introduce the family of random variables $\{f_{i,j}^{(T)};\ 1\leq i\leq n,\ j\geq 1\}$ such that $f_{i,j}^{(T)}$ is the number of customers served up to the moment at which the (i,j)-customer has been served, if this is the case; and $f_{i,j}^{(T)}=\infty$ otherwise.

For any $T',T''\in K'$ we introduce the family of random variables

$$\{\gamma_{k,\ell,i,j}^{(T',T'')};\ 1\leq k,\ i\leq n;\ j,\ell\geq 1\}$$

such that

$$\gamma_{k,\ell,i,j}^{(T',T'')} = \begin{cases} 1 & \text{if } f_{k,\ell}^{(T')} < f_{i,j}^{(T')}, \quad f_{k,\ell}^{(T'')} > f_{i,j}^{(T'')}, \\ 0 & \text{otherwise}. \end{cases}$$

We note that $\gamma_{i,\ell,i,j}^{(T',T'')}=0$ a.s. for $1\leq i\leq n,\ j,\ell\geq 1$.

Let $T',\ T''\in K''$. By $A'_{i,j}$ $(A''_{i,j})$ we denote the random interval of time during which the (i,j)-customer is being served, and by $\eta'_k(i,j)$ $(\eta''_k(i,j))$ the number of customers of type-k waiting to be served during the interval of time $A'_{i,j}$ $(A''_{i,j})$, for the switching functions T' (T''); if the (i,j)-customer is not served, we let

$$\eta'_k(i,j) = 0 \qquad (\eta''_k(i,j) = 0)$$

and

$$A'_{i,j}(0,0) \quad (A''_{i,j} = (0,0)) \qquad (1\leq i,\ k\leq n;\ j\geq 1).$$

<u>LEMMA 3</u>. Let $T',\ T''\in K$. Then:

a) the random variable $\gamma_{i,j,k,\ell}^{(T',T'')}$ does not depend on the vectors $\bar{m}_{i,j}$ and $\bar{m}_{k,\ell}$;

b) the random variable $\eta'_k(i,j)$ does not depend on the vector $\bar{m}_{i,j}$ for $1\leq i,\ k\leq n;\ j,\ell\geq 1$.

<u>Proof</u>. We note that for any $T\in K$, for any $1\leq i,\ k\leq n$;

$j,\ell,r\geq 1$ the event $\{f_{i,j}^{(T)}=r;\ f_{k,\ell}^{(T)}>r\}$ belongs to the σ-algebra $G(\ell_1,\dots,\ell_n)$, where $\ell_s=\infty$ for $s\neq i$, $s\neq k$ and $\ell_i=j-1$, $\ell_k=-1$. Since

$$\{\gamma_{i,j,k,l}^{(T',T'')}=1\}=\bigcup_{r_1=1}^{\infty}\bigcup_{r_2=1}^{\infty}\{f_{i,j}^{(T')}=r_1;\ f_{k,l}^{(T')}>r_1\}\{f_{k,l}^{(T'')}=r_2;\ f_{i,j}^{(T'')}>r_2\},$$

then $\{\gamma_{i,j,k,\ell}^{(T',T'')}=1\}\in G(\ell_1,\dots,\ell_n)$.

The proof of condition "b" is similar. For $1\leq k$, $i\leq n$ and $T',T''\in K'$ let

$$N_{k,i}^{(T',T'')}=M\left\{\sum_{r=1}^{\infty}\sum_{j=1}^{\infty}\gamma_{k,r,i,j}^{(T',T'')}\right\}<\infty \quad \text{(by Remark 1);}$$

$$F_{k,i}^{(T',T'')}=N_{k,i}^{(T',T'')}-N_{i,k}^{(T',T'')}.$$

We note that

$$F_{k,k}^{(T',T'')}=0\ , \qquad F_{k,i}^{(T',T'')}=-F_{i,k}^{(T',T'')}\ .$$

Let $b_k=\sum_{\ell=1}^{n}c_\ell g_k^\ell$ for $1\leq k\leq n$.

<u>LEMMA 4</u>. Let T', $T''\in K$. Then

$$L(T')-L(T'')=\sum_{i=1}^{n}\sum_{k=1}^{n}F_{k,i}^{(T',T'')}\cdot a_i\cdot(b_k-c_k). \tag{10}$$

<u>Proof</u>. Let

$$\delta_{\ell,k}=\begin{cases}1 & \text{if } \ell=k\ ,\\ 0 & \text{if } \ell\neq k\ .\end{cases}$$

Then we have the equalities

$$\eta_k'(i,j)-\eta_k''(i,j)=\sum_{l=1}^{n}\sum_{r=1}^{\infty}\gamma_{l,r,i,j}^{(T',T'')}\cdot(m_{l,r}^{(k)}-\delta_{l,k})-$$
$$-\sum_{l=1}^{n}\sum_{r=1}^{\infty}\gamma_{i,j,l,r}^{(T',T'')}\cdot(m_{l,r}^{(k)}-\delta_{l,k}).$$

We wish to explain why these equalities arise. Let

$\gamma_{i,j,\ell,r}^{(T',T'')}=1$. Then before the (i,j)-customer is served for T'' the $\delta_{\ell,k}$ customer of type-k with the tag r is being served, that is, the number of customers of type-k in the queue decreases by $\delta_{\ell,k}$ for T'' with respect to T', but as a result of such "extra" service $m_{\ell,r}^{(k)}$ customers of type-k can be served. The remaining terms appear in the similar manner.

By Lemma 3

$$M\left\{\sum_{j=1}^{\infty}\left(\eta_k'(i,j)-\eta_k''(i,j)\right)\right\}=\sum_{l=1}^{n}F_{l,i}^{(T',T'')}\cdot g_l^{(k)}-F_{k,i}^{(T',T'')}.$$

Using the independence of $\eta_k'(i,j)$ and $\eta_k''(i,j)$ from $\tau_{i,j}$, we obtain

$$\begin{aligned}L(T')-L(T'') &= M\left\{\int_0^{\infty}\sum_{k=1}^{n}c_k\eta_k'(t)\,dt-\int_0^{\infty}\sum_{k=1}^{n}c_k\eta_k''(t)\,dt\right\}=\\ &=\sum_{i=1}^{n}\sum_{j=1}^{\infty}M\left\{\int_{A'_{i,j}}\sum_{k=1}^{n}c_k\eta_k'(t)\,dt-\int_{A''_{i,j}}\sum_{k=1}^{n}c_k\eta_k''(t)\,dt\right\}=\\ &=\sum_{i=1}^{n}\sum_{k=1}^{n}\sum_{j=1}^{\infty}c_k\cdot M\left\{\tau_{i,j}\cdot\left(\eta_k'(i,j)-\eta_k''(i,j)\right)\right\}=\\ &=\sum_{i=1}^{n}\sum_{k=1}^{n}a_i\cdot c_k\cdot M\left\{\sum_{j=1}^{\infty}\left(\eta_k'(i,j)-\eta_k''(i,j)\right)\right\}=\\ &=\sum_{i=1}^{n}\sum_{k=1}^{n}a_i\cdot c_k\left(\sum_{l=1}^{n}F_{l,i}^{(T',T'')}\cdot g_l^{(k)}-F_{k,i}^{(T',T'')}\right)=\\ &=\sum_{i=1}^{n}\sum_{k=1}^{n}F_{k,i}^{(T',T'')}\cdot a_i\,(b_k-c_k).\end{aligned}$$

<u>LEMMA 5</u>. For any switching function $T\in K'$, the permutations $(d_1,d_2,\ldots,d_n)$ of the set $(1,2,\ldots,n)$ and of the number $s\in(1,2,\ldots,n)$, there exists a switching function $T_s\in K'$ such that

a) T_s is a $(d_1,d_2,\ldots,d_s)$-function;

b) for $1\leq k\leq n$, $i\in\{d_{s+1},\ldots,d_n\}$, $j,\ell\geq 1$, $\gamma_{k,\ell,i,j}^{(T',T'')}=0$ a.s.

Proof. Let $\Omega=\{\omega\}$ be a probability space. For each elementary state $\omega\in\Omega$ of the switching function T, the order of service, the service times and sets of customers entering at the end of the service of the preceding customer are well defined. Noting Remark 1, we can assume that a finite number of customers

$$\nu_1(T) = \ell_1,\ldots, \quad \nu_n(T) = \ell_n , \quad (\ell_1,\ldots,\ell_n) \in Z_+^n$$

is served for a specified $\omega\in\Omega$.

By enumerations we mean the families of (random) variables $\{f_{i,j}, 1\leq i\leq n, j\geq 1\}$, $\{q_{i,j}, 1\leq i\leq n, j\geq 1\}$, to be used below.

For $1\leq i$, $k\leq n$; $j,\ell\geq 1$, let $f_{i,j}=f_{i,j}^{(T)}$;

$$q_{i,j} = \begin{cases} f_{k,\ell} & \text{if the } (i,j)\text{-customer appeared in the queue after the } (k,\ell)\text{-customer has been served;} \\ 0 & \text{if the } (i,j)\text{-customer was waiting in the queue to be served at an initial time;} \\ \infty & \text{if } f_{i,j}=\infty. \end{cases} \tag{11}$$

We note that $q_{i,j}<f_{i,j}$. Let for $1\leq i\leq n$

$$f_{i,0} = q_{i,1} . \tag{12}$$

The algorithm for the construction of T_s consists of a finite number of shift operations applied sequentially. The operation of the (d_k,ℓ)-shift $(1\leq k\leq n, \ \ell\geq 1)$ is applied for $\ell_{d_k}\geq\ell$; therefore $f_{d_k,\ell}<\infty$ and $q_{d_k,\ell}<\infty$.

Let

$$r = \min\{n \in Z_+ / n \geq \max\{q_{d_k,l};\ f_{d_k,l-1}\} + 1;$$
$$n = f_{i,j} \quad \text{for} \quad i \neq d_1;\ i \neq d_2,\ \ldots,\ i \neq d_{k-1}\}.$$

We note that since $q_{d_k,\ell} < f_{d_k,\ell}$; $f_{d_k,\ell-1} < f_{d_k,\ell}$, then $r \leq f_{d_k,\ell}$.

Now we enumerate the $\{f_{i,j}\}$ and $\{q_{i,j}\}$ as follows: for $1 \leq i \leq n$, $j \geq 1$

$$f'_{i,j} = \begin{cases} r & \text{if } (i,j) = (d_k,\ell), \\ f_{i,j}+1 & \text{if } r \leq f_{i,j} < f_{d_k,\ell}, \\ f_{i,j} & \text{otherwise}; \end{cases} \tag{13}$$

we enumerate $\{q_{i,j}\}$ in accordance with (11) and $\{f_{i,0}\}$ in accordance with (12).

For $\ell_{d_1} > 0$, by the (d_1)-operation we mean the sequential application of the shifts $(d_1,1)$, $(d_1,2)$, ..., (d_1,ℓ_{d_1}). For $\ell_{d_1} = 0$ the (d_1)-operation is given by the identity operation which leaves the enumeration of $\{f_{i,j}\}$ and $\{q_{i,j}\}$ unchanged.

We define the operations $(d_1,\ldots,d_k)$ $(2 \leq k \leq n)$ by induction. Let the (d_1)-, (d_1,d_2)-, ..., $(d_1,\ldots,d_{k-1})$-operations be defined. Then for $\ell_{d_k} = 0$ the $(d_1,\ldots,d_k)$-operation coincides with the $(d_1,\ldots,d_{k-1})$-operation. For $\ell_{d_k} > 0$ we make the $(d_k,1)$-shift and next the $(d_1,\ldots,d_{k-1})$-operation; then we proceed similarly with $(d_k,2)$-, ..., (d_k,ℓ_{d_k})-customers.

Thus we obtain the $(d_1,\ldots,d_k)$-operation.

The result of the application of the (d_1)-operation is the switching function $T_1(T)$ and the respective enumeration of $\{f_{k,j}^{(T_1)}\}$; the result of the sequential application of the (d_1)-, (d_1,d_2)-, ..., $(d_1,\ldots,d_k)$-operations $(2 \leq k \leq n)$ is the switching function $T_k(T)$ and the respective enumeration $\{f_{i,j}^{(T_k)}\}$. It is seen from the construction of this algorithm that the conditions of the Lemma are satisfied, since condition "b" is satisfied for each shift operation.

LEMMA 6. Let $T \in K$ and $T_s = T_s(T)$ $(1 \leq s \leq n)$ be constructed as in Lemma 5. Then $T_s \in K$.

Proof. We need to show that for any $j \geq 1$, $1 \leq s \leq n$ the event $\{T_s^{(1)} = k_1, \ldots, T_s^{(j)} = k_j\}$ belongs to the σ-algebra $G(q_1, \ldots, q_n)$. Since the order of serving the first j customers during the transition from T to T_s changes no more than j^s times during the shift operations, the inclusion $T_s \in K$ follows from the following fact: if $T \in K$, $1 \leq d \leq n$, $\ell \geq 1$, T_1 is the result of the application of the (d, ℓ)-shift operation to T, then $T_1 \in K$. Let $j \geq 1$; let the set $\bar{k}_j$ satisfying (8) be given; and let r be defined as in Lemma 5; also let $A = \{T_1^{(1)} = k_1, \ldots, T_1^{(j)} = k_j\}$.

We note that the events

$$\{T^{(1)} = k_1, \ldots, T^{(j)} = k_j, r > j\};$$

$$\{T^{(1)} = k_1, \ldots, T^{(j)} = k_j, r = i\};$$

$$\{T^{(1)} = k_1, \ldots, T^{(j)} = k_j, r = i, f_{dl}^{(T)} = s\};$$

$$\{T^{(1)} = k_1, \ldots, T^{(j)} = k_j, r = i, f_{d,l}^{(T)} > j\}$$

belong to the σ-algebra $G(q_1, \ldots, q_n)$ for $1 \leq i \leq s \leq j$.

Let us express A in terms

$$A = (A \cap \{r > j\}) \cup (A \cap \{r = j\}) \cup (A \cap \{r < j\}) .$$

We note that

$$A \cap \{r > j\} = \{T^{(j)} = k_1, \ldots, T^{(j)} = k_j, r > j\} \in G(q_1, \ldots, q_n)$$

and $A\cap\{r=j\}=\emptyset$ for $k_j\neq d$; otherwise

$$A\cap\{r=j\}=$$
$$=\bigcup_{m=1}^{n}\{T^{(1)}=k_1,\ldots,T^{(j-1)}=k_{j-1},T^{(j)}=m,\ r=j\}\in G(q_1,\ldots,q_n).$$

Next, for $1\le i\le s\le j-1$

$$A_{i,s}=A\cap\{r=i,\ f_{d,l}^{(T)}=s\}=\{T^{(1)}=k_1,\ldots,T^{(i-1)}=k_{i-1},$$
$$T^{(i)}=k_{i+1},\ldots,T^{(s-1)}=k_s,\ T^{(s)}=d,\ T^{(s+1)}=k_{s+1},\ldots,T^{(j)}=$$
$$=k_j,\ r=i,\ f_{d,l}^{(T)}=s\}\in G(q_1,\ldots,q_n)$$

for $k_i=d$ and $A_{i,s}=\emptyset$ for $k_i\neq d$,

$$A_{i,j}=A\cap\{r=i,\ f_{d,l}^{(T)}\geq j\}=\{T^{(1)}=k_1,\ldots,T^{(i-1)}=k_{i-1},T^{(i)}=$$
$$=k_{i+1},\ldots,T^{(j-1)}=k_j,\ r=i,\ f_{d,l}^{(T)}>j-1\}\in G(q_1',\ldots,q_n')$$

for $k_i=d$ and $A_{i,j}=\emptyset$ for $k_i\neq d$, where $q_m'=q_m$ for $m\neq d$ and $q_d'=q_d-1$.

Therefore $A_{i,j}\in G(q_1,\ldots,q_n)$ and

$$A\cap\{r<j\}=\left(\bigcup_{i=1}^{j-1}\bigcup_{s=i}^{j-1}A_{i,s}\right)\cup\left(\bigcup_{i=1}^{j-1}A_{ij}\right)\in G(q_1,\ldots,q_n).$$

//

We construct now the required permutation $(d_1,d_2,\ldots,d_n)$ of the set $(1,2,\ldots,n)$ by induction. First we define the number d_1. For $k=1,2,\ldots,n$ let $b_k=\sum_{j=1}^{n}c_jg_k^{(j)}$. Let $(d_1,d_2,\ldots,d_n)$ denote some arbitrary random permutation. We take at random $T\in K$ and construct the switching function $T_1=T_1(T)$ in accordance with Lemma 5. By Lemma 5 we have the following equalities:

a) $\gamma_{d_1,\ell,i,j}^{(T,T_1)}=0$ a.s. for $\ell,j\geq 1$, $1\le i\le n$, and therefore $N_{d_1,i}^{(T,T_1)}=0$;

b) $\gamma^{(T,T_1)}_{k,\ell,i,j}=0$ a.s. for $1\le i$, $k\le n$; $i\ne d_1$; $k\ne d_1$; $\ell,j\ge 1$, and therefore $F^{(T,T_1)}_{k,i}=0$.

Using Lemma 4, we obtain

$$L(T)-L(T_1)=\sum_{k=1}^{n} F^{(T,T_1)}_{k,d_1}\left(a_{d_1}(b_k-c_k)-a_k(b_{d_1}-c_{d_1})\right).$$

Since all the numbers $F^{(T,T_1)}_{k,d_1}$ are nonnegative for $1\le k\le n$, to have the inequality $L(T)-L(T_1)\ge 0$ satisfied, it suffices to choose $d_1\in\{1,2,\ldots,n\}$ such that

$$(b_{d_1}-c_{d_1})/a_{d_1} = \min_{1\le k\le n}\{(b_k-c_k)/a_k\} \quad . \tag{13}$$

Let for some $s\in\{1,2,\ldots,n-1\}$ the numbers $d_1,d_2,\ldots,d_s$ ($d_i\ne d_j$ for $i\ne j$) be such that for any switching function $T\in K$ and for the switching functions $T_i(T)$ constructed from T ($i=1,2,\ldots,s$; T_i are $(d_1,d_2,\ldots,d_i)$-functions; the construction follows along the lines of the construction of Lemma 5) the inequalities $L(T)\ge L(T_1)\ge\cdots\ge L(T_s)$ are satisfied. We find the number d_{s+1} such that the $(d_1,d_2,\ldots,d_s,d_{s+1})$-function $T_{s+1}=T_{s+1}(T)$ satisfies the inequality $L(T_s)\ge L(T_{s+1})$.

The further considerations we lead, for the sake of simplicity, assuming that $(d_1,d_2,\ldots,d_s)=(1,2,\ldots,s)$. This assumption does not cause the loss of generality of our arguments, since, if it is not satisfied, we can reenumerate the types of customers so that the type-i in the new enumeration coincides with the type-d_i in the preceding enumeration.

Let $K_s\subset K$ denote the class of $(1,2,\ldots,s)$-functions. We separate all the customers into two classes: 1) customers of the 1st, 2nd, ..., s^{th} types; 2) customers of the $(s+1)^{th}$, $(s+2)^{th}$, ..., n^{th} types.

We fix $T \in K_s$ and also suppose that $x=(i,j)$ is a customer, $\mathcal{B}_{x,1}=\mathcal{B}_{x,1}(\omega)$ is the (random) set of customers arriving at the queue after the customer x has been served. For $\ell \geq 2$ let

$$\mathcal{B}_{x,\ell} = \mathcal{B}_{x,\ell}(\omega) = \bigcup_{\substack{y \in \mathcal{B}_{x,\ell-1};\\ y \text{ is of class } 1}} \mathcal{B}_{y,1} \quad .$$

We assume that $\mathcal{B}_x = \bigcup_{\ell=1}^{\infty} \mathcal{B}_{x,\ell} = \mathcal{B}'_x \cup \mathcal{B}''_x$, where $\mathcal{B}'_x$ are customers of class 1 and $\mathcal{B}''_x$ are customers of class 2.

For $s+1 \leq i \leq n$, $j \geq 1$ we denote by $\tau^{(s)}_{i,j}$ the time required for serving the (i,j)-customer and all the customers of class 1 of the set $\mathcal{B}'_{(i,j)}$. Let $A^{(s,T)}_{i,j}$ be a random interval of time during which all these customers are being served (we assume $A^{(s,T)}_{i,j}=(0,0)$, if the (i,j)-customer is not served for the switching function T). Let $m^{(s,s+1)}_{i,j}, \ldots, m^{(s,n)}_{i,j}$ be the number of customers of the $(s+1)^{th}$, $(s+2)^{th}, \ldots, n^{th}$ types of the set $\mathcal{B}''_{(i,j)}$.

For $s+1 \leq i$, $k \leq n$; $j \geq 1$ let

$$M\{\tau^{(s)}_{i,j}\} = a^{(s)}_i \; ; \qquad M\{m^{(s,k)}_{i,j}\} = g^{(s,k)}_i \; . \tag{14}$$

<u>REMARK 2</u>. For any fixed $1 \leq s \leq n-1$, $j \geq 1$, $s+1 \leq i \leq n$ the total cost $C^{(i,j)}$ of waiting in line for the customers of the set $\mathcal{B}_{(i,j)}$ during the interval of time $A^{(s,T)}_{i,j}$ has identical distributions, and hence, in particular, identical means for various switching functions $T \in K_s$.

Because before the initial moment of serving the first

customer of class 2 other customers are being served identically for $T \in K_s$, if we consider only the class K_s of switching functions, we can assume without loss of generality that at the initial moment $t=0$ the customers of class 2 only were waiting to be served.

Let $T \in K_s$. We consider the set of random vectors $\{\bar M^{(T)}_{i,j},\ 1 \le i \le n-s,\ j \ge 1\}$, where

$$\bar M^{(T)}_{k,j} = (\tau^{(s)}_{i+s,j},\ m^{(s,s+1)}_{i+s,j},\ \dots,\ m^{(s,n)}_{i+s,j}) . \tag{15}$$

Since $T \in K_s$, the vectors $\{\bar M^{(T)}_{i,j}\}$ are mutually independent and for any $i=1,2,\dots,n-s$ they are identically distributed in each sequence $\{\bar M^{(T)}_{i,j}\}^{\infty}_{j=1}$. We note that for various switching functions $T_1, T_2 \in K_s$ the vectors $\bar M^{(T_1)}_{i,j}$ and $\bar M^{(T_2)}_{i,j}$ are identically distributed for any $i=1,2,\dots,n-s$, $j=1,2,\dots$.

For $T \in K_s$ we define the σ-algebras $\{\tilde G_T(\bar\ell),\ \bar\ell \in N^{n-s}_+\}$ and $\{\tilde D_T(\bar\ell),\ \bar\ell \in N^{n-s}_+\}$.

Let $\bar\ell=(\ell_1,\dots,\ell_{n-s}) \in N^{n-s}_+$,

$$S(\bar l) = \{(i,j),\ s+1 \le i \le n,\ 1 \le j \le l_{i-s}\};$$

$$F(\bar l) = \bigcup_{(i,j)\in S(\bar l)} \mathcal{B}_{(i,j)};$$

$\bar u(\bar\ell)=(u_1(\bar\ell),\ \dots,\ u_s(\bar\ell))$ being the number of customers of class 1 (respectively, of the 1st, 2nd, ..., s^{th} types) of the set $F(\bar\ell)$.

We assume that $A \in \tilde G_T(\bar\ell)$ $(A \in \tilde D_T(\bar\ell))$, if

$$A \cap \{\bar u(\bar\ell)=\bar r\} \in G(r_1,\ \dots,\ r_s,\ \ell_1,\ \dots,\ \ell_{n-s}) ;$$

$$(A \cap \{\bar u(\bar\ell)=\bar r\} \in D(r_1,\ \dots,\ r_s,\ \ell_1,\ \dots,\ \ell_{n-s}))$$

for any vector $\bar r=(r_1,\dots,r_s) \in N^s_+$. We note that the σ-algebras

$\tilde{G}_T(\bar{\ell})$ and $\tilde{D}_T(\bar{\ell})$ are independent for any $\bar{\ell}\in N_+^{n-s}$, and if $\bar{\ell},\bar{\ell}'\in N_+^{n-s}$, $\bar{\ell}\leq\bar{\ell}'$, then $\tilde{G}_T(\bar{\ell})\subseteq\tilde{G}_T(\bar{\ell}')$.

For $T\in K_s$ we introduce the class of switching functions $\tilde{K}_s=\tilde{K}_s(T)$: the switching function $T'=\{T'^{(\ell)}\}_{\ell=1}^{\infty}$, where for $\ell=1,2,\ldots$, $T'^{(\ell)}:\Omega\times E^{n-s}\to\{1,2,\ldots,n-s\}$, belongs to the class $\tilde{K}_s$, if for any set $\bar{k}_j$ $(j=1,2,\ldots)$ satisfying (8) for $n-s$ instead of n, and for $q_i=q_i(\bar{k}_j)$ $(1\leq i\leq n-s)$, $\bar{q}=(q_1,\ldots,q_{n-s})$ we have the inclusion $\{T'^{(1)}=k_1,\ \ldots,\ T'^{(j)}=k_j\}\in\tilde{G}_T(\bar{q})$.

Next we consider for $T\in K_s$ the queue in which customers of $(n-s)$-types arrive, given by the set of vectors $\{\bar{M}_{i,j}^{(T)}\}$, with switching functions of the class $\tilde{K}_s(T)$. We set the switching functions T into correspondence with the switching function $T^*=\{T^{*(\ell)}\}_{\ell=1}^{\infty}$ in the following manner: for any set $(\ell_1,\ldots,\ell_{n-s})\in E^{n-s}$ for any $k\geq1$ we put

$$T^{*(k)}(\ell_1,\ldots,\ell_{n-s}) = T^{(k+t_T)}(0,\ldots,0,\ \ell_1,\ldots,\ell_{n-s}), \quad (16)$$

where t_T is the random variable equal to the number of services of customers of class 1 before the kth service of customers of class 2 for the switching function T. It can be verified immediately that $T^*\in\tilde{K}_s(T)$.

For $1\leq i\leq n-s$, let $T\in K_s$, $T'\in\tilde{K}_s(T)$, $\eta_i^{(T')}(t,T)$ be the number of customers of the ith type, which have arrived before the instant of time t at the system with $(n-s)$-types of customers which have not been yet served. Next, let $\{c_i'=c_{i+s};\ 1\leq i\leq n-s\}$;

$$L(T',T)=M\left\{\int_0^{\infty}\sum_{i=1}^{n-s}c_i'\eta_i^{(T')}(t,T)\,dt\right\}. \quad (17)$$

By the definition of T^* we have the equality

$$L(T) = L(T^*, T) + \sum_{i=s+1}^{n} \sum_{j=1}^{\infty} M\{C^{(i,j)}\}. \tag{18}$$

For the class $\tilde{K}_s(T)$, one can formulate, using different notation, the results corresponding to Lemmas 3-6 for the class K, which can be proved entirely similarly. We state, for example, Lemma 4 in terms of the class $\tilde{K}_s(T)$. For $T', T'' \in \tilde{K}_s(T)$ for $1 \le i$, $k \le n-s$, $j, \ell \ge 1$ we introduce the random variables $\gamma_{i,j,k,\ell}^{(T',T'')}$, the integers $N_{k,\ell}^{(T',T'')}$ and $F_{k,\ell}^{(T',T'')}$.
For $s+1 \le k \le n$ let

$$b_k^{(s)} = \sum_{\ell=s+1}^{n} c_\ell g_k^{(s,\ell)} .$$

<u>LEMMA 4'</u>. For any $T', T'' \in \tilde{K}_s(T)$

$$L(T', T) - L(T'', T) = \sum_{i=1}^{n-s} \sum_{k=1}^{n-s} F_{k,i}^{(T',T'')} a_{s+i}^{(s)} \left(b_{s+k}^{(s)} - c_{s+k}\right).$$

Let $(f_1, \ldots, f_{n-s})$ denote the permutation of the set $(1,2,\ldots,n-s)$, $1 \le i \le n-s$. For $T' \in \tilde{K}_s(T)$ one can introduce in accord with the construction of Lemma 5, $(f_1, \ldots, f_i)$-functions $T_i(T')$ such that (by Lemma 6) $T_i(T') \in \tilde{K}_s(T)$ and for $1 \le k \le n-s$, $r \in \{f_{i+1}, \ldots, f_{n-s}\}$, $\ell, j \ge 1$, $\gamma_{k,\ell,r,j}^{T',T_i(T'))} = 0$ a.s.

For $T \in K$ let $T_s = T_s(T)$ be a $(1,2,\ldots,s)$-function, and let $T_{s+1} = T_{s+1}(T)$ be a $(1,2,\ldots,s,d_{s+1})$-function (where $s+1 \le d_{s+1} \le n$). Then $T_s \in K_s$, $T_{s+1} \in K_s$.

We express the functions defined by the equality (16) in terms of $T_s^* \in \tilde{K}_s(T_s)$, $T_{s+1}^* \in \tilde{K}_s(T_{s+1})$, respectively for $T=T_s$ and $T=T_{s+1}$.

Let $T_{s+1}^{**} = T_1(T_s^*) \in \tilde{K}_s(T_s)$. Then by the definition of $L(\cdot,\cdot)$

$$L(T^*_{s+1}, T_s) = L(T^*_{s+1}, T_{s+1}) \ . \tag{19}$$

Noting Lemma 4' and using the considerations used in finding (13), we obtain that if the integer d_{s+1} is chosen such that the equality

$$\left(b^{(s)}_{d_{s+1}} - c_{d_{s+1}}\right) \Big/ a^{(s)}_{d_{s+1}} = \min_{s+1 \leqslant k \leqslant n} \left\{\left(b^{(s)}_k - c_k\right)/a^{(s)}_k\right\} \tag{20}$$

is satisfied, then $L(T^{**}_{s+1}, T_s) \leq L(T^*_s, T_s)$. It follows from the equalities (18) and (19) that for this choice of the integer d_{s+1} the inequality $L(T_{s+1}) \leq L(T_s)$ is satisfied. Note that in the case where $(d_1, d_2, \ldots, d_s) \neq (1,2,\ldots,s)$, the integer d_{s+1} is chosen in accord with the equality

$$\left(b^{(s)}_{d_{s+1}} - c_{d_{s+1}}\right) \Big/ a^{(s)}_{d_{s+1}} = \min_{k \neq d_1,\ldots,d_s} \left\{\left(b^{(s)}_k - c_k\right)/a^{(s)}_k\right\}, \tag{21}$$

where $b^{(s)}_k$ and $a^{(s)}_k$ are determined by the formulas (22) and (23) below.

Let $(f_1, \ldots, f_n)$ denote the arbitrary permutation of the set $(1,2,\ldots,n)$ such that for some $1 \leq s \leq n-1$ the equality $(f_1, \ldots, f_s) = (d_1, \ldots, d_s)$ is satisfied. Let $s < i,\ j \leq n$. Then

$$a^{(s)}_{f_i} = a_{f_i} + \sum_{l=1}^{s} a_{f_l} g^{(f_l)}_{f_i} + \sum_{l=1}^{s} a_{f_l} \sum_{k=1}^{s} g^{(f_k)}_{f_i} g^{(f_l)}_{f_k} + \\ + \ldots + \sum_{l=1}^{s} a_{f_l} \sum_{k_1=1}^{s} \ldots \sum_{k_r=1}^{s} g^{(f_{k_1})}_{f_i} \ldots g^{(f_l)}_{f_{k_r}} + \ldots; \tag{22}$$

$$g^{(s,f_j)}_{f_i} = g^{(f_j)}_{f_i} + \sum_{k=1}^{s} g^{(f_k)}_{f_i} g^{(f_j)}_{f_k} + \ldots + \sum_{k_1=1}^{s} \ldots \sum_{k_l=1}^{s} g^{(f_{k_1})}_{f_i} \ldots g^{(f_j)}_{f_{k_l}} + \ldots \tag{23}$$

We consider the matrices

$$X_s = \begin{pmatrix} g_{f_{s+1}}^{(f_1)} & \cdots & g_{f_{s+1}}^{(f_s)} \\ \vdots & & \vdots \\ g_{f_n}^{(f_1)} & \cdots & g_{f_n}^{(f_s)} \end{pmatrix} = \left(g_{f_k}^{(f_l)}\right)_{\substack{s+1\leqslant k\leqslant n;\\ 1\leqslant l\leqslant s}}$$

$$Q_s = \left(g_{f_k}^{(f_l)}\right)_{\substack{1\leqslant k\leqslant s;\\ 1\leqslant l\leqslant s}} \quad Y_s = \left(g_{f_k}^{(f_l)}\right)_{\substack{s+1\leqslant k\leqslant n;\\ s+1\leqslant l\leqslant n}}$$

$$Z_s = \left(g_{f_k}^{(f_l)}\right)_{\substack{1\leqslant k\leqslant s\\ s+1\leqslant l\leqslant n}}; \quad Y^{(s)} = \left(g_{f_k}^{(s,f_l)}\right)_{\substack{s+1\leqslant k\leqslant n\\ s+1\leqslant l\leqslant n}}$$

and the vectors

$$\bar{a}^{(s)} = \left(a_{f_{s+1}}^{(s)}; \ldots; a_{f_n}^{(s)}\right); \bar{a}_s = \left(a_{f_{s+1}}; \ldots; a_{f_n}\right); \bar{b}_s = \left(a_{f_1}; \ldots; a_{f_s}\right).$$

Equations (22) and (23) written in the vector and matrix forms respectively, are:

$$\bar{a}^{(s)} = \bar{a}_s = X_s \left(\sum_{l=0}^{\infty} Q_s^l\right) \bar{b}_s = \bar{a}_s + X_s (I_s - Q_s)^{-1} \bar{b}_s; \tag{24}$$

$$Y^{(s)} = Y_s + X_s \left(\sum_{l=0}^{\infty} Q_s^l\right) Z_s = Y_s + X_s (I_s - Q_s)^{-1} Z_s, \tag{25}$$

where $Q_s^0 = I_s$ and for $\ell = 1,2,\ldots,$ $Q_s^\ell = Q_s \times Q_s \times \cdots \times Q_s$ (ℓ times).

We note that if one reenumerates arbitrarily inside the group of types of customers tagged $f_1, f_2, \ldots, f_s$, the coefficients $g_{f_i}^{(s,f_j)}$ and $a_{f_i}^{(s)}$ do not change.

We prove now the following assertion: let

$$\left(b_{f_{s+1}}^{(s)} - c_{f_{s+1}}\right) \Big/ a_{f_{s+1}}^{(s)} = \left(b_{f_{s+2}}^{(s)} - c_{f_{s+2}}\right) \Big/ a_{f_{s+2}}^{(s)} =$$

$$= \min_{s+1\leqslant j\leqslant n} \left\{\left(b_{f_j}^{(s)} - c_{f_j}\right) \Big/ a_{f_j}^{(s)}\right\} < \left(b_{f_{s+3}}^{(s)} - c_{f_{s+3}}\right) \Big/ a_{f_{s+3}}^{(s)}$$

and let the $(f_{s+i})^{th}$ type be chosen as the type of customers, which has the $(s+1)^{th}$ priority (that is, we put $d_{s+1} = f_{s+1}$).

Then

$$\left(b_{f_{s+2}}^{(s+1)} - c_{f_{s+2}}\right) \Big/ a_{f_{s+2}}^{(s+1)} < \left(b_{f_{s+3}}^{(s+1)} - c_{f_{s+3}}\right) \Big/ a_{f_{s+3}}^{(s+1)}.$$

We can prove without loss of generality for $s=0$ and for $f_i=1$. We note that for $2 \le k,\ \ell \le n$,

$$g_k^{(1,l)} = g_k^{(l)} + g_k^{(1)} g_1^{(l)} 1/1 - g_1^{(1)};\ a_k^{(1)} = a_k + g_k^{(1)} a_1 1/1 - g_1^{(1)};$$

$$b_k^{(1)} = \sum_{l=2}^{n} c_l g_k^{(1,l)} = \sum_{l=2}^{n} c_l g_k^{(l)} + g_k^{(1)}/1 - g_1^{(1)} \sum_{l=2}^{n} c_l g_1^{(l)}.$$

Then

$$h_k = \left(b_k^{(1)} - c_k\right)/a_k^{(1)} = \frac{\left(b_k - c_1 g_k^{(1)}\right) + g_k^{(1)}/\left(1 - g_1^{(1)}\right)\left(b_1 - c_1 g_1^{(1)}\right) - c_k}{a_k^{(1)}} =$$

$$= \frac{(b_k - c_k) + g_k^{(1)}/\left(1 - g_1^{(1)}\right)(b_1 - c_1)}{a_k + g_k^{(1)}/\left(1 - g_1^{(1)}\right) a_1}\ .$$

We make use of the following relations. Let $x,y,z,t>0$. Then if $x/y=z/t$, we have $(x+z)/(y+t)=x/y$, and if $x/y<z/t$, then $x/y<(x+z)/(y+t)<z/t$. Therefore,

$$h_{f_2} = (b_1-c_1)/a_1\ ; \qquad h_{f_3} > (b_1-c_1)/a_1\ ,$$

which was to be proved.

Next, we classify the types of customers into groups as follows: the kth type enters the 1st group, if

$$(b_k-c_k)/a_k = \min_{1 \le j \le n} \{(b_j-c_j/a_i\}\ .$$

We construct the remaining groups by induction. Let the 1st, 2nd, ..., $(\ell-1)^{th}$ groups be constructed, $s_1, s_2, \ldots, s_{\ell-1}$ being the number of types of customers in each group, respectively We put $s=s_1+s_2+\cdots s_{\ell-1}$. If $s=n$, the construction has been completed. Let $s<n$ and let the types of customers of the first $(\ell-1)$ groups have numbers $d_1, \ldots, d_s$. We form the ℓth group as follows: the kth type enters the ℓth group

$(k \neq d_1, \ldots, d_s)$, if

$$(b_k^{(s)} - c_k)/a_k^{(s)} = \min_{j \neq d_1, \ldots, d_s} \{(b_j^{(s)} - c_j)/a_j^{(s)}\} .$$

To summarize, we have shown that <u>all the types of customers fall into r groups $(1 \le r \le n)$, each type of customers of the ℓth group has priority over the types of customers of the $(\ell+1)^{th}$ group $(1 \le \ell \le r)$; inside each of the groups the priorities can be assigned arbitrarily; each priority switching function thus constructed satisfies the assertion of the Theorem</u>. This completes the proof. //

4. SOME COROLLARIES

In Klimov [1,2] and in Rykov [3] the following queue has been considered. A single-server queue which performs n types of operations has a Poisson input with parameter $\delta > 0$. Each customer independently of others with probability $p_i \left(1 \le i \le n, \sum_{i=1}^{n} p_i = 1\right)$ is a customer of type-i. At the initial time the queue consists of $\bar{m}_{0,0} = (m_{0,0}^{(1)}, \ldots, m_{0,0}^{(n)})$ customers of distinct types. No break is allowed. In addition to the customers, at the moment at which the jth service of customers of type-i (service duration is $\tau_{i,j}$) ended, $\bar{\psi}_{i,j} = (\psi_{i,j}^{(1)}, \ldots, \psi_{i,j}^{(n)})$ new customers of distinct types arrive at the system: It is assumed that the random variables $\{\tau_{i,j}, 1 \le i \le n, j \ge 1\}$ and the random vectors $\{\bar{\psi}_{i,j}, 1 \le i \le n, j \ge 1\}$ are mutually independent; for any fixed $i = 1, 2, \ldots, n$ they are identically distributed and do not depend on the input. We consider the class K_0 of switching functions. The customers are being served in the same way as in the systems described in Section 1.

For $i=1,2,\ldots,n$ let there be some nonnegative integers and for $T\in K_0$, $t\geq 0$ let $\zeta_i^{(T)}(t)$ be the number of customers of the ith type, which arrived before the time t, which have not been yet served before t for the switching function T. In [1-3] the problem of minimizing the functional was stated:

$$N(T) = M\left\{\int_0^{\lambda^0(T)} \sum_{i=1}^{n} c_i \zeta_i^{(T)}(t)\,dt\right\},$$

where $\lambda^0(T)$ denotes the moment of termination of the first period of occupancy (in the terminology of this paper, a stopping time). The priority switching function minimizing the the functional $N(T)$ has been constructed in the class K_0 of switching functions.

We show that the problem of minimizing the functional $N(T)$ in the system with a Poisson input described above, reduces to the problem of minimizing the functional $L(T)$ in the system given by the set of random vectors of the form (1); and, therefore, the results of [1-3] follow from the results of this paper.

For $T\in K_0$ and $t>0$ let $r^{(T)}(t)$ be the last moment at which the service of the customers terminates, not exceeding t; let $\phi_i^{(T)}(t)$ $(1\leq i\leq n)$ be the number of customers of the ith type, which have arrived at the system before $r^{(T)}(t)$, the service of which has not been yet started before t; let

$\nu_i^0(T)$ $(1\le i\le n)$ be the number of customers of the ith type, served before the moment $\lambda^0(T)$, in the system with a Poisson input. We show that

$$C(T) = M\left\{\int_0^{\lambda^0(T)} \sum_{i=1}^{n} c_i\left(\zeta_i^{(T)}(t) - \varphi_i^{(T)}(t)\right) dt\right\}$$

does not depend on the switching function $T\in K_0$, i.e., $C(T)=C\equiv$constant.

We note that (see, for example, [3]) the random vectors $(\lambda^0(T), \nu_1^0(\ ), \ldots, \nu_n^0(T))$ are identically distributed for distinct $T\in K_0$.

For $i=1,2,\ldots,n$ $j=1,2,\ldots,$ $T\in K_0$ in the case when the event $\nu_i^0(T)\ge j$ occurs (i.e., the jth service of customers of the ith type terminates before the moment of time $\lambda^0(T)$), let $A_{i,j}(T)$ be a random interval of time during which the jth service of customers of the ith type occurs; let $U_{i,j}(T)$ be the (random) set of customers which have arrived during the interval of time $A_{i,j}(T)$; let $Y_{i,j}(T)$ be the total cost of idle performance during the interval of time $A_{i,j}(T)$ of the customers of the set $U_{i,j}(T)$; otherwise, let $Y_{i,j}(T)=0$. Since the input is Poisson and for any $T_1,T_2\in K_0$ the random variables $\nu_i^0(T_1)$ and $\nu_i^0(T_2)$ are identically distributed, the random variables $Y_{i,j}(T_1)$ and $Y_{i,j}(T_2)$ are also identically distributed (for any fixed $i=1,2,\ldots,n$, $j=1,2,\ldots$). Therefore $V_{i,j}=M\{Y_{i,j}(T)/\nu_i^0(T)\ge j\}$ does not depend on the switching function $T\in K_0$. Next, since the random variables in the sequence $\{\tau_{i,j}\}_{j=1}^{\infty}$ are identically distributed for any $i=1,2,\dots,n$, we obtain that $V_{i,j}$ does not depend on j. Let $V_i=V_{i,j}$. Then

$$C(T)=M\left\{\int_0^{\lambda_0(T)}\sum_{i=1}^{n}c_i\left(\zeta_i^{(T)}(t)-\varphi_i^{(T)}(t)\right)dt\right\}=M\left\{\sum_{i=1}^{n}\sum_{j=1}^{\infty}Y_{i,j}(T)\right\}=$$

$$=\sum_{i=1}^{n}\sum_{j=1}^{\infty}M\{Y_{i,j}(T)/\nu_i^0(T)\geqslant j\}P\{\nu_i^0(T)\geqslant j\}=$$

$$=\sum_{i=1}^{n}V_i\sum_{j=1}^{\infty}P\{\nu_i^0(T)\geqslant j\}=\sum_{i=1}^{n}V_iM\{\nu_i^0(T)\}.$$

Since $M\{\nu_i^0(T)\}$ does not depend on T, then $C(T)=C\equiv\text{const}$. Thus the problem of minimizing the functional $N(T)$ reduces to the problem of minimizing the functional

$$M\left\{\int_0^{\lambda_0(T)}\sum_{i=1}^{n}c_i\varphi_i^{(T)}(t)\,dt\right\}.$$

REMARK 3. Let $I=\{\alpha\}$ be some index set and for each $\alpha\in I$ let n sequences of random vectors $\{\bar{m}_{1,j}^{(\alpha)}\}_{j=1}^{\infty},\dots,\{\bar{m}_{n,j}^{(\alpha)}\}_{j=1}^{\infty}$ satisfying (1) be given. For any $\alpha,\beta\in I$, $i=1,2,\dots,n$, $j=1,2,\dots$ let the random vectors $\bar{m}_{i,j}^{(\alpha)}$ and $\bar{m}_{i,j}^{(\beta)}$ be identically distributed. We define for $\alpha\in I$ and $T\in K_0$ the functional

$$L(T/\alpha)=M\left\{\int_0^{\lambda(T)}\sum_{i=1}^{n}c_i\eta_i^{(T)}(t)\,dt\right\}$$

in the system given by the set $\{\bar{m}_{i,j}^{(\alpha)};\ 1\le i\le n;\ j\ge 1\}$. Then $I(T/\alpha)$ does not depend on $\alpha\in I$.

We define now a set of random vectors of the form (1) in the system with a Poisson input. For $T\in K_0$ let $q_{i,j}^{(1)}(T)$, $\dots$, $q_{i,j}^{(n)}(T)$ be respectively the number of customers of the

1st, 2nd, ..., n^{th} types, which have arrived from the input during the time of the jth service of customers of the ith type (provided the service occurs) (in the system with a Poisson input). Let

$$\overline{m}_{i,j}^{(T)} = \left(\tau_{i,j},\ \psi_{i,j}^{(1)} + q_{i,j}^{(1)}(T),\ \ldots,\ \psi_{i,j}^{(n)} + q_{i,j}^{(n)}(T)\right).$$

Then, by the Poisson property of the input, the set of random vectors $X(T)=\left\{\overline{m}_{i,j}^{(T)},\ 1\le i\le n;\ j\ge 1\right\}$ satisfies the condition (1) and for any $T_1, T_2 \in K_0$; $i=1,2,\ldots,n$; $j=1,2,\ldots$ the random vectors $\overline{m}_{i,j}^{(T_1)}$ and $\overline{m}_{i,j}^{(T_2)}$ are identically distributed.

We note that if for $T\in K_0$, $i=1,2,\ldots,n$, $j=1,2,\ldots$ the quantity $\eta_i^{(T)}(t)$ denotes the number of customers of the ith type in the server at the moment of time t for the switching function T in the system given by the set of vectors $X(T)$, then $\eta_i^{(T)}(t)=\phi_i^{(T)}(t)$ a.s. and, therefore

$$L(T/T) = M\left\{\int_0^{\lambda^0(T)} \sum_{i=1}^{n} c_i\varphi_i^{(T)}(t)\,dt\right\}.$$

We fix arbitrarily $T'\in K_0$. By the theorem stated in this paper, it is possible to construct the priority switching function $T_0\in K_0$ such that

$$L(T_0/T') = \min_{T\in K_0} L(T/T') .$$

But, since by Remark 3 $L(T_0/T')=L(T_0/T_0)$, $L(T_0/T_0)=N(T_0)-C$ and for any switching function $T\in K_0$ $L(T/T')=L(T/T)$, $L(T/T)-N(T)-C$, then, therefore, $N(T_0)=\min\limits_{T\in K_0} N(T)$. Thus, the priority switching function T_0 constructed in accord with (13) and (21) minimizes the functional $N(T)$ in the class K_0.

The author expresses his deep gratitude to B.A. Rogozin for his valuable comments and counsel.

REFERENCES

[1] G.P. Klimov. "Time-sharing Service Systems. I." *Theory Prob. Applications*, 19, 3 (1974): 532-551.

[2] G.P. Klimov. "Time-sharing Service Systems. II." *Theory Prob. Applications*, 23, 2 (1978): 314-321.

[3] V.V. Rykov. "Primenenie regeneriruyushchikh protsessov pri issledovanii upravlyaemykh sistem" (On the Use of Regenerating Processes in Investigating Controllable Systems). *Trudy III Vsesoyuznogo shkoly-soveshchaniya po teorii massovogo obsluzhivaniya*, 135-148. Pushchino-na-Oke: Izdatel'stvo MGU, 1976.

[4] P. Lancaster. *Theory of Matrices*. New York: Academic Press, 1969.

TRANSLATION SERIES IN MATHEMATICS AND ENGINEERING

V.F. Dem'yanov, and L.V. Vasil'ev
NONDIFFERENTIABLE OPTIMIZATION

1984, approx. 300 pp.
ISBN 0-911575-09-X Optimization Software, Inc.
ISBN 0-387-90951-6 Springer-Verlag New York Berlin Heidelberg Tokyo
ISBN 3-540-90951-6 Springer-Verlag Berlin Heidelberg New York Tokyo

V.A. Dubovitskij
ULAM PROBLEM OF OPTIMAL SUPERPOSITION OF LINE SEGMENTS

1984, approx. 100 pp.
ISBN 0-911575-04-9 Optimization Software, Inc.
ISBN 0-387-90946-X Springer-Verlag New York Berlin Heidelberg Tokyo
ISBN 3-540-90946-X Springer-Verlag Berlin Heidelberg New York Tokyo

V.V. Ivanishchev, and A.D. Krasnoshchekov
CONTROL OF VARIABLE STRUCTURE NETWORKS

1984, approx. 200 pp.
ISBN 0-911575-05-7 Optimization Software, Inc.
ISBN 0-387-90947-8 Springer-Verlag New York Berlin Heidelberg Tokyo
ISBN 3-540-90947-8 Springer-Verlag Berlin Heidelberg New York Tokyo

N.I. Nisevich, G.I. Marchuk, I.I. Zubikova, and I.B. Pogozhev
MATHEMATICAL MODELING OF VIRAL DISEASES

1984, approx. 400 pp.
ISBN 0-911575-06-5 Optimization Software, Inc.
ISBN 0-387-90948-6 Springer-Verlag New York Berlin Heidelberg Tokyo
ISBN 3-540-90948-6 Springer-Verlag Berlin Heidelberg New York Tokyo

V.G. Lazarev, Ed.
PROCESSES AND SYSTEMS IN COMMUNICATION NETWORKS

1984, approx. 220 pp.
ISBN 0-911575-08-1 Optimization Software, Inc.
ISBN 0-387-90950-8 Springer-Verlag New York Berlin Heidelberg Tokyo
ISBN 3-540-90950-8 Springer-Verlag Berlin Heidelberg New York Tokyo

B.A. Berezovskij
BINARY RELATIONS IN MULTICRITERIAL OPTIMIZATION

1984, approx. 180 pp.
ISBN 0-911575-11-1 Optimization Software, Inc.
ISBN 0-387-90953-2 Springer-Verlag New York Berlin Heidelberg Tokyo
ISBN 3-540-90953-2 Springer-Verlag Berlin Heidelberg New York Tokyo